Novel Structural and Functional Material Properties Enabled by Nanocomposite Design

Novel Structural and Functional Material Properties Enabled by Nanocomposite Design

Editors

Jürgen Eckert
Daniel Kiener

MDPI • Basel • Beijing • Wuhan • Barcelona • Belgrade • Manchester • Tokyo • Cluj • Tianjin

Editors

Jürgen Eckert
Austrian Academy of Sciences (ÖAW) and
Department of Materials Science
Leoben
Austria

Daniel Kiener
Department Materials Science
Leoben
Austria

Editorial Office
MDPI
St. Alban-Anlage 66
4052 Basel, Switzerland

This is a reprint of articles from the Special Issue published online in the open access journal *Nanomaterials* (ISSN 2079-4991) (available at: https://www.mdpi.com/journal/nanomaterials/special_issues/nano_nanocomposite).

For citation purposes, cite each article independently as indicated on the article page online and as indicated below:

LastName, A.A.; LastName, B.B.; LastName, C.C. Article Title. *Journal Name* **Year**, *Volume Number*, Page Range.

ISBN 978-3-0365-6724-2 (Hbk)
ISBN 978-3-0365-6725-9 (PDF)

Cover image courtesy of Jana Wilmers & Michael Wurmshuber

© 2023 by the authors. Articles in this book are Open Access and distributed under the Creative Commons Attribution (CC BY) license, which allows users to download, copy and build upon published articles, as long as the author and publisher are properly credited, which ensures maximum dissemination and a wider impact of our publications.

The book as a whole is distributed by MDPI under the terms and conditions of the Creative Commons license CC BY-NC-ND.

Contents

About the Editors . . . . **vii**

Jürgen Eckert and Daniel Kiener
Special Issue "Novel Structural and Functional Material Properties Enabled by
Nanocomposite Design"
Reprinted from: *Nanomaterials* **2023**, *13*, 586, doi:10.3390/nano13030586 1

**Marzhan M. Kubenova, Kairat A. Kuterbekov, Malik K. Balapanov, Rais K. Ishembetov,
Asset M. Kabyshev and Kenzhebatyr Z. Bekmyrza**
Some Thermoelectric Phenomena in Copper Chalcogenides Replaced by Lithium and Sodium
Alkaline Metals
Reprinted from: *Nanomaterials* **2021**, *11*, 2238, doi:10.3390/nano11092238 5

**Irena Petrinic, Janja Stergar, Hermina Bukšek, Miha Drofenik, Sašo Gyergyek,
Claus Hélix-Nielsen and Irena Ban**
Superparamagnetic Fe_3O_4@CA Nanoparticles and Their Potential as Draw Solution Agents in
Forward Osmosis
Reprinted from: *Nanomaterials* **2021**, *11*, 2965, doi:10.3390/nano11112965 53

**Michael Burtscher, Mingyue Zhao, Johann Kappacher, Alexander Leitner,
Michael Wurmshuber, Manuel Pfeifenberger, et al.**
High-Temperature Nanoindentation of an Advanced Nano-Crystalline W/Cu Composite
Reprinted from: *Nanomaterials* **2021**, *11*, 2951, doi:10.3390/nano11112951 71

Junga Moon, Huaide Jiang and Eun-Cheol Lee
Physical Surface Modification of Carbon-Nanotube/Polydimethylsiloxane Composite
Electrodes for High-Sensitivity DNA Detection
Reprinted from: *Nanomaterials* **2021**, *11*, 2661, doi:10.3390/nano11102661 85

**Zeng Gao, Congxin Yin, Dongfeng Cheng, Jianguang Feng, Peng He, Jitai Niu
and Josip Brnic**
Sintering Bonding of SiC Particulate Reinforced Aluminum Metal Matrix Composites by Using
Cu Nanoparticles and Liquid Ga in Air
Reprinted from: *Nanomaterials* **2021**, *11*, 1800, doi:10.3390/nano11071800 97

Abhilash Gunti, Parijat Pallab Jana, Min-Ha Lee and Jayanta Das
Effect of Cold Rolling on the Evolution of Shear Bands and Nanoindentation Hardness in
$Zr_{41.2}Ti_{13.8}Cu_{12.5}Ni_{10}Be_{22.5}$ Bulk Metallic Glass
Reprinted from: *Nanomaterials* **2021**, *11*, 1670, doi:10.3390/nano11071670 111

**Giovanni Spinelli, Rosella Guarini, Rumiana Kotsilkova, Evgeni Ivanov
and Vittorio Romano**
Experimental, Theoretical and Simulation Studies on the Thermal Behavior of PLA-Based
Nanocomposites Reinforced with Different Carbonaceous Fillers
Reprinted from: *Nanomaterials* **2021**, *11*, 1511, doi:10.3390/nano11061511 129

Jana Wilmers, Miranda Waldron and Swantje Bargmann
Hierarchical Microstructure of Tooth Enameloid in Two Lamniform Shark Species, *Carcharias
taurus* and *Isurus oxyrinchus*
Reprinted from: *Nanomaterials* **2021**, *11*, 969, doi:10.3390/nano11040969 159

Kunlin Wu, Bing-Chiuan Shiu, Ding Zhang, Zhenhao Shen, Minghua Liu and Qi Lin
Preparation of Nanoscale Urushiol/PAN Films to Evaluate Their Acid Resistance and Protection of Functional PVP Films
Reprinted from: *Nanomaterials* **2021**, *11*, 957, doi:10.3390/nano11040957 . . . . **175**

Dmitrii Shuleiko, Mikhail Martyshov, Dmitrii Amasev, Denis Presnov, Stanislav Zabotnov, Leonid Golovan, et al.
Fabricating Femtosecond Laser-Induced Periodic Surface Structures with Electrophysical Anisotropy on Amorphous Silicon
Reprinted from: *Nanomaterials* **2021**, *11*, 42, doi:10.3390/nano11010042 . . . . **191**

About the Editors

Jürgen Eckert

Jürgen Eckert (Univ.-Prof. Dr.-Ing. habil. Dr. h.c.) is Professor of Material Physics at the Montanuniversität Leoben and Director of the Erich Schmid Institute of Materials Science of the Austrian Academy of Sciences. His research interests span the full breath of metastable materials, composites, and metallic glasses from synthesis to application.

Daniel Kiener

Daniel Kiener (Univ.-Prof. Dipl.-Ing. Dr. mont.) is Professor of Micro- and Nanomechanics of Materials at the Montanuniversität Leoben. His research interests focus on mechanical and functional structure property relations in nanomaterials, nanocomposites, and nanoporous structures, and their tailoring by defect engineering. Therefore, he also develops and applies miniaturized in situ testing methods to directly link fundamental material processes to global characteristics.

Editorial

Special Issue "Novel Structural and Functional Material Properties Enabled by Nanocomposite Design" †

Jürgen Eckert [1,2,*] and Daniel Kiener [1,*]

1. Department of Materials Science, Montanuniversität Leoben, 8700 Leoben, Austria
2. Erich Schmid Institute of Materials Science, Austrian Academy of Sciences, 8700 Leoben, Austria
* Correspondence: juergen.eckert@unileoben.ac.at or juergen.eckert@oeaw.ac.at (J.E.); daniel.kiener@unileoben.ac.at (D.K.)
† This article belongs to the Special Issue Novel Structural and Functional Material Properties Enabled by Nanocomposite Design.

Nanocomposites bear the potential to enable novel material properties that considerably exceed the capabilities of their individual constituent phases, thereby enabling the exploration of white areas on material property charts. In this inaugural Special Issue for the newly released subsection of Nanocomposites in the journal *Nanomaterials*, we aim to provide an overview of the current state of the art in enabling novel structural and functional material properties by presenting a better understanding and implementation of nanocomposite design. The covered properties of interest encompass the whole material usage span. This starts with structural modifications of nanocomposites by employing different synthesis routes, assesses their microstructure-dependent mechanical properties such as strength, ductility, and high-temperature stability. Furthermore, we address the functional characteristics of nanocomposites, such as soft magnetic properties or thermoelectricity, as well as the tailored property adjustment by design strategies (bioinspired design, chemical sensitivity, and bio sensing). Thus, the included contributions detail methods for the synthesis, characterization, modeling, and in-depth understanding of the mechanisms governing the outstanding properties of this fascinating material class.

In total, ten manuscripts including one review article, are published in this Special Issue, addressing the abovementioned characteristics and many more interesting aspects of nanocomposite materials. In the remainder, we present a brief overview of the some of the exciting insights of this Special Issue.

Petrinic et al. [1] synthesized citric-acid-covered superparamagnetic magnetite nanoparticles by co-precipitation for use as a draw solution agent on forward osmosis. The coated nanoparticles performed almost four-fold more strongly as an osmotic agent than the pure citric acid, making them a potential candidate for a broad range of concentration applications where current technologies are still limited.

In their study, Burtscher et al. [2] utilized severe plastic deformation to create a W–Cu nanocomposite. Through a combination of electron microscopy and elevated temperature nanoindentation experiments, they were able to assess the temperature-dependent elastic and plastic properties, including strain rate sensitivity and activation volumes, which are fundamental pre-requisites for future employment in extreme environments, such as fusion reactors.

Moon et al. [3] employed physical rather than chemical modifications of carbon-nanotube/polydimethylsiloxane composite electrode surfaces to enhance the detection of biomolecules such as DNA. By dip-coating the electrodes with functionalized multi-wall carbon nanotubes, the detection limit could be increased by a factor of more than 1000 compared with the original composite, paving the way towards lower detection limits of biosensor systems.

Citation: Eckert, J.; Kiener, D. Special Issue "Novel Structural and Functional Material Properties Enabled by Nanocomposite Design". *Nanomaterials* **2023**, *13*, 586. https://doi.org/10.3390/nano13030586

Received: 18 January 2023
Accepted: 26 January 2023
Published: 1 February 2023

Copyright: © 2023 by the authors. Licensee MDPI, Basel, Switzerland. This article is an open access article distributed under the terms and conditions of the Creative Commons Attribution (CC BY) license (https://creativecommons.org/licenses/by/4.0/).

The report by Gao et al. [4] addresses the sintering bonding of SiC-particle-reinforced Al matrix composites using Cu nanoparticles and liquid Ga as self-fluxing fillers. Based on microstructural and mechanical analysis, they report that the metal matrix composites can be tuned with respect to joint strength and gas tightness by adjusting the Cu nanoparticles and the sintering temperature.

Gunti et al. [5] assessed the effect of different rolling conditions on the hardness and shear band revolution of the Virtreloy 1 bulk metallic glass using nanoindentation to different maximum load levels. The increase in pop-in events during nanoindentation decreased with the amount of cold rolling, indicating a more homogenous deformation of the metallic glass during rolling, which will benefit future forming operations of such materials.

In their study, Spinelli et al. [6] targeted the enhanced thermal conductivity of polylactic acid. They used graphene nanoplatelets as filler material in the polymer. By adding 9 wt% and ensuring favorable alignment of the filler platelets, the thermal conductivity increased by more than 250%, opening novel views towards heat transfer applications.

Wilmers et al. [7] studied the hierarchical morphology of the enameloid of two different shark teeth: biological hard and tough nanocomposite consisting almost entirely out of brittle phases. Analyzing the structural patterns in comparison to amniote enamel enabled the identification of microstructural design principles for ensuring certain biomechanical functions, thereby deriving strategies for the design of bioinspired composite materials with superior properties.

In the study by Wu et al. [8], different amounts of urushiol, extracted from raw lacquer and known for its acid-resisting properties, were added to a fixed amount of polyacrylonitrile, subsequently electrospun into nanoscale fibers, and deposited as thin films. The addition of urushiol improved the mechanical strength and chemical stability, rendering the fibers unique materials with strong acid resistance and weak acid dissolution properties for future use.

Shuleiko et al. [9] used femtosecond laser machining to introduce nanoscale one-dimensional laser-induced periodic surface structures on amorphous hydrogenated silicon. Furthermore, the laser treatment caused partial crystallization of the amorphous substrate into nanocrystalline Si. Taken together, this material surface modification strongly enables enhanced conductivity and introduces an orientation dependence.

The review by Kubenova et al. [10] focused on the thermoelectric properties of copper chalcogenides, with an emphasis on property modifications by substitution with sodium and lithium alkali metals. The cationic sub-lattice is responsible for the very low thermal conductivity, whereas nonstoichiometric defects provide high electronic conductivity; the combination of these aspects contributes to the high thermoelectric figure of merit, approaching a value of three.

Conflicts of Interest: The authors declare no conflict of interest.

References

1. Petrinic, I.; Stergar, J.; Bukšek, H.; Drofenik, M.; Gyergyek, S.; Hélix-Nielsen, C.; Ban, I. Superparamagnetic Fe_3O_4@CA Nanoparticles and Their Potential as Draw Solution Agents in Forward Osmosis. *Nanomaterials* **2021**, *11*, 2965. [CrossRef] [PubMed]
2. Burtscher, M.; Zhao, M.; Kappacher, J.; Leitner, A.; Wurmshuber, M.; Pfeifenberger, M.; Maier-Kiener, V.; Kiener, D. High-Temperature Nanoindentation of an Advanced Nano-Crystalline W/Cu Composite. *Nanomaterials* **2021**, *11*, 2951. [CrossRef] [PubMed]
3. Moon, J.; Jiang, H.; Lee, E.-C. Physical Surface Modification of Carbon-Nanotube/Polydimethylsiloxane Composite Electrodes for High-Sensitivity DNA Detection. *Nanomaterials* **2021**, *11*, 2661. [CrossRef] [PubMed]
4. Gao, Z.; Yin, C.; Cheng, D.; Feng, J.; He, P.; Niu, J.; Brnic, J. Sintering Bonding of SiC Particulate Reinforced Aluminum Metal Matrix Composites by Using Cu Nanoparticles and Liquid Ga in Air. *Nanomaterials* **2021**, *11*, 1800. [CrossRef] [PubMed]
5. Gunti, A.; Jana, P.; Lee, M.-H.; Das, J. Effect of Cold Rolling on the Evolution of Shear Bands and Nanoindentation Hardness in $Zr_{41.2}Ti_{13.8}Cu_{12.5}Ni_{10}Be_{22.5}$ Bulk Metallic Glass. *Nanomaterials* **2021**, *11*, 1670. [CrossRef] [PubMed]

6. Spinelli, G.; Guarini, R.; Kotsilkova, R.; Ivanov, E.; Romano, V. Experimental, Theoretical and Simulation Studies on the Thermal Behavior of PLA-Based Nanocomposites Reinforced with Different Carbonaceous Fillers. *Nanomaterials* **2021**, *11*, 1511. [CrossRef] [PubMed]
7. Wilmers, J.; Waldron, M.; Bargmann, S. Hierarchical Microstructure of Tooth Enameloid in Two Lamniform Shark Species, *Carcharias taurus* and *Isurus oxyrinchus*. *Nanomaterials* **2021**, *11*, 969. [CrossRef] [PubMed]
8. Wu, K.; Shiu, B.-C.; Zhang, D.; Shen, Z.; Liu, M.; Lin, Q. Preparation of Nanoscale Urushiol/PAN Films to Evaluate Their Acid Resistance and Protection of Functional PVP Films. *Nanomaterials* **2021**, *11*, 957. [CrossRef]
9. Shuleiko, D.; Martyshov, M.; Amasev, D.; Presnov, D.; Zabotnov, S.; Golovan, L.; Kazanskii, A.; Kashkarov, P. Fabricating Femtosecond Laser-Induced Periodic Surface Structures with Electrophysical Anisotropy on Amorphous Silicon. *Nanomaterials* **2021**, *11*, 42. [CrossRef]
10. Kubenova, M.M.; Kuterbekov, K.A.; Balapanov, M.K.; Ishembetov, R.K.; Kabyshev, A.M.; Bekmyrza, K.Z. Some Thermoelectric Phenomena in Copper Chalcogenides Replaced by Lithium and Sodium Alkaline Metals. *Nanomaterials* **2021**, *11*, 2238. [CrossRef]

Disclaimer/Publisher's Note: The statements, opinions and data contained in all publications are solely those of the individual author(s) and contributor(s) and not of MDPI and/or the editor(s). MDPI and/or the editor(s) disclaim responsibility for any injury to people or property resulting from any ideas, methods, instructions or products referred to in the content.

Review

Some Thermoelectric Phenomena in Copper Chalcogenides Replaced by Lithium and Sodium Alkaline Metals

Marzhan M. Kubenova [1,*], Kairat A. Kuterbekov [1], Malik K. Balapanov [2], Rais K. Ishembetov [2], Asset M. Kabyshev [1] and Kenzhebatyr Z. Bekmyrza [1]

[1] Faculty of Physics and Technical Sciences, L.N. Gumilyov Eurasian National University, Nur-Sultan 010008, Kazakhstan; kkuterbekov@gmail.com (K.A.K.); assetenu@gmail.com (A.M.K.); kbekmyrza@yandex.kz (K.Z.B.)
[2] Physical and Technical Institute, Bashkir State University, 450076 Ufa, Russia; balapanovmk@mail.ru (M.K.B.); ishembetovrkh@rambler.ru (R.K.I.)
* Correspondence: kubenova.m@yandex.kz

Abstract: This review presents thermoelectric phenomena in copper chalcogenides substituted with sodium and lithium alkali metals. The results for other modern thermoelectric materials are presented for comparison. The results of the study of the crystal structure and phase transitions in the ternary systems Na-Cu-S and Li-Cu-S are presented. The main synthesis methods of nanocrystalline copper chalcogenides and its alloys are presented, as well as electrical, thermodynamic, thermal, and thermoelectric properties and practical application. The features of mixed electron–ionic conductors are discussed. In particular, in semiconductor superionic copper chalcogenides, the presence of a "liquid-like phase" inside a "solid" lattice interferes with the normal propagation of phonons; therefore, superionic copper chalcogenides have low lattice thermal conductivity, and this is a favorable factor for the formation of high thermoelectric efficiency in them.

Keywords: thermoelectric materials; copper sulfide; crystal structure; conductivity; diffusion; thermal conductivity; Seebeck coefficient; superionic conductors

1. Introduction

In recent years, mixed electron–ion conductors, and in particular, semiconductor superionic copper selenide and sulfide, have become the objects of intensive research by scientists involved in the development and study of thermoelectric materials (TE) due to the "discovery" of their "liquid-like state" of ions, which reduces the lattice thermal conductivity of the crystal to record low values [1,2]. The development of nanotechnology has made the possibilities of modifying materials to improve their useful properties virtually limitless. For example, the addition of even a small fraction of nanosized particles makes the bulk material nanocomposite and noticeably improves its thermoelectric characteristics. This does not complicate the technology of its production [3,4].

The heat-to-electricity energy conversion efficiency of an ideal thermoelectric generator is determined by the Carnot efficiency and materials' performance as [4]:

$$\eta = \left(\frac{T_{hot} - T_{cold}}{T_{hot}}\right)\left[\frac{\sqrt{1 + ZT_{avg}} - 1}{\sqrt{1 + ZT_{avg}} + \left(\frac{T_{cold}}{T_{hot}}\right)}\right], \quad (1)$$

where T_{hot} is the hot-side temperature and T_{cold} is the cold-side temperature; ZT_{avg} is the average dimensionless thermoelectric figure-of-merit, which is a critical measure for materials' performance. ZT value of material is calculated from the formula

$$ZT = \alpha^2 \sigma T / k, \quad (2)$$

where α, σ and k are the Seebeck coefficient, electrical conductivity and thermal conductivity of a material, correspondingly.

Achieving the optimal combination of all three material properties at the same time is a challenge. In addition, for the practical application of thermoelectric material, the manufacturability of production, the availability and cheapness of raw materials, stability of properties, mechanical resistance, environmental friendliness of production, and other factors are important. Currently, among industrially produced materials, the most widespread is doped bismuth telluride $(Bi_{1-x}Sb_x)_2(Se_{1-y}Te_y)_3$, which has a figure of merit $ZT \approx 1$ at room temperature. Despite the fact that many materials have been obtained under laboratory conditions that are superior in thermoelectric characteristics to bismuth telluride, for the reasons stated above, it has remained the most demanded commercial thermoelectric material for several decades [3,4].

The excellent thermoelectric properties of copper and silver chalcogenides have long been known [5–7], but their practical application is hindered by the high rate of copper diffusion. At elevated temperatures, rapid degradation of thermoelements occurs due to copper release. For this reason, in the 1980s, the developments of American physicists on the use of silver-doped copper selenide in thermoelectric elements [8] were curtailed; similar problems are described in [9,10]. The boom of interest of specialists in this class of materials arose recently after the publication of the article [1] by a group of Chinese researchers with the group of G.J. Snyder (USA). It focused on the superionic, "liquid-like" crystalline nature of copper selenide, which helps to reduce the lattice thermal conductivity [11]. This and subsequent publications actually created a promising new direction—the design of effective thermoelectric materials through a decrease in thermal conductivity, by creating conditions that suppress the propagation of phonons, but do not impede electronic transport [2]. The classification of thermoelectric materials includes the concept of "superionic thermoelectric materials". In addition, to increase ZT, a targeted change in the synthesis conditions for the modification of known materials and nanostructuring are often used today [2–4,12–15].

Over the past 3–4 years of intensive research, the figure of merit ZT has been significantly increased: for copper selenide to $ZT = 2.1$ at 700 °C [16], for copper sulfide $Cu_{1.97}S$ to $ZT = 1.9$ at 700 °C [17]. However, the problem of these materials remains the risk of rapid degradation of the material [18], which remarkably reduces the practical significance of the above-mentioned works.

Active studies of the anomalously fast diffusion of ions in solids began with the development of ideas about the defect structure (ideas of Schottky, Frenkel, and K. Wagner [19,20]). It was found that fast diffusion in superionics is due to the peculiarities of their crystal structure. Moreover, high ionic conductivity is observed in crystals with strong structural disorder. For example, in silver iodide, iodine ions form a rigid lattice skeleton, and silver ions constantly move through the numerous voids of the lattice. The degree of disorder in the silver sublattice is higher than in liquid, and when the crystal melts, the rate of silver diffusion decreases. The fact of correlation between the degree of lattice disordering and the value of ionic conductivity is described, for example, in [20,21].

Studying copper sulfides and copper selenides substituted with lithium, it was found that the ionic conductivity and diffusion rate of copper in materials decrease by almost an order of magnitude compared to the initial binary compositions, and the thermoelectric coefficient remains high and even increases [22–29]. This significantly improves the prospects of copper chalcogenides for practical use. In 2017, the work of a large international group of researchers (USA, Canada, China) [30] was published, in which, in continuation of works [24,25], they studied copper selenide doped with lithium. The lithium was reacted with a fine powder of $Cu_{1.9}Se$ by mechanical alloying using a ball miller for 90 min under Ar. To consolidate the powder into a pellet, the powder was hot-pressed at 45 MPa and 700 °C for 1 h under an Ar atmosphere using a high-density graphite die in an induction furnace. The work of Kang, S.D. et al. [30] confirmed that doping with lithium improves the stability of copper selenide while maintaining high thermoelectric characteristics. For the composition $Li_{0.09}Cu_{1.9}Se$, they obtained the maximum $ZT \approx 1.4$ at 727 °C. In a recent work by Ge, Z.H. et al. [31], the thermoelectric properties of copper sulfide $Cu_{1.8}S$ (digenite) doped with sodium are described. Doping with sodium and nanostructuring allowed them to increase the figure of merit ZT of copper sulfide from 0.6 to 1.1 at 500 °C.

Superionic thermoelectric materials have some peculiarities; for example, in [32–35], a sharp abrupt increase in the value of ZT was observed in the region of the superionic phase transition. This fact deserves special attention and further detailed research, since it gives hope to use it as another way to improve the thermoelectric characteristics. In the chalcogenides, the region of the superionic phase transition can occupy several tens of degrees, and the transition onset temperature depends on nonstoichiometry and the presence of an impurity that stabilizes the superionic state at a lower temperature [19]. Lowering the phase transition temperature can be used to transfer the working area of the thermoelectric to the region of lower temperatures, in which copper diffusion is significantly reduced while maintaining a high ZT value, which should greatly reduce material degradation.

In this respect, lithium-substituted copper sulfides and selenides look more suitable, since they can be single-phase even at room temperature [25]. Sodium-substituted sulfides, with a noticeable sodium content, are mixtures of various sulfide phases at room temperature [36–40], or individual chemical substances are formed with properties far from properties of initial thermoelectric material; for example, $NaCu_4S_4$ [41] represents the behavior of an ideal metal. However, multiphase sodium-containing copper sulfides obtained in the nanocrystalline state turn out to be nanocomposites, which favorably affects their thermoelectric and thermal properties.

In the past 10–15 years, the main trend in improving thermoelectric efficiency is an inhibition of a thermal conductivity of a material [2–4,42]. Most of the ways to reduce thermal conductivity while maintaining high electrical conductivity of the material are somehow related to the nanostructuring of the material. According to a review article on nanostructured thermoelectrics by P. Pichanusakorn and P. Bandaru [42], while total nano-object materials such as superlattices and nanowires do promise a significant decrease in lattice thermal conductivity k_L—mainly due to a decrease in mean free path λ, problems in large-scale and reliable synthesis in fact preclude their widespread use, so bulk materials with embedded nanoscale elements now look more suitable for practical applications. In our opinion, this conclusion is still relevant, and nanocomposites and compacted mixtures of nanocrystalline materials require more attention from researchers in order to better understand their properties and learn how to control them to improve their thermoelectric properties.

The general goal of this review is to consider copper chalcogenides and their alloys as thermoelectric materials from the point of view of the researcher of the superionic state, since it is this state that provides these materials with beautiful thermoelectric properties and excellently low thermal conductivity. In addition, we wanted to show convenient electrochemical methods for working with copper and silver chalcogenides, which are not in demand by the current generation of researchers. These ideas developed in the works of C. Wagner, I. Yokota, and S. Miyatani; for example, the coulometric titration method and methods for measuring transport characteristics in direct dependence on Fermi level are very suitable for tuning thermoelectric power by changing the copper content. However, it is the superionic state that complicates the practical application of copper and silver chalcogenides, since the mobility of cations similar to the mobility of atoms in a liquid leads to the release of the metal from the sample during a long operation at high temperatures. Therefore, we also considered here the possibilities of overcoming this important problem: associated with a decreasing in ionic conductivity of the material and with the design of thermoelectric modules, developed taking into account the threshold of metal release from chalcogenide.

In addition, we consider here some other TE materials, such as half-Heusler phases, skutterudites, etc., so that the reader has the opportunity to compare and can more fully evaluate the superionic copper chalcogenides as thermoelectric materials.

2. Materials

The host thermoelectric materials that are the objects of our review, namely, copper chalcogenides $Cu_{2-x}X$ (X = S, Se), have both high electronic conductivity and high ionic conductivity. They are called mixed conductors and also superionic semiconductors. Their high electronic conductivity is caused by the presence of a large number of vacancies in the cation sublattice, which determines the p-type conductivity. They have a wide region of homogeneity over the copper sublattice; that is, they are phases of variable composition, and retain the type of crystal lattice when the stoichiometric index of copper changes from 2 to 1.75. This makes it possible to control the electrical, thermoelectric, diffusion, and other properties of these compounds by changing the degree of nonstoichiometric composition, which tunes concentration of electron holes and carrier's effective mass, too.

The high cationic conductivity of copper chalcogenides is possible due to the action of several factors, and the main ones are the strong disorder of the cation sublattice in the superionic phase and the presence of a connected network of free interstitial sites, which are shallow potential wells in the path of mobile ions. The value of the ionic conductivity in the superionic phase of Cu_2Se reaches 2.2 S/cm at 400 °C, and the self-diffusion coefficient equals 2.5×10^{-5} cm^2/s at 350 °C [22]. Due to the high concentration of free interstitial sites in the cation sublattice, the activation energy for copper diffusion does not include the energy of defect formation and consists only of the activation energy for migration of cations. For copper selenide, the activation energy of ionic conductivity is 0.15 ± 0.01 eV [22], and for hexagonal superionic phase of copper sulfide, it is 0.19 ± 0.02 eV [24].

2.1. Crystal Structure and Phase Transitions

2.1.1. Cu-S System

Copper and sulfur form a wide variety of compounds, ranging from chalcocite (Cu_2S) to villamaninite (CuS_2) with other intermediate phases: djurleite ($Cu_{1.96}S$), roxbyite ($Cu_{1.8125}S$), digenite ($Cu_{1.8}S$), anilite ($Cu_{1.75}S$), geerite ($Cu_{1.6}S$), spionkopite ($Cu_{1.39}S$), yarrowite ($Cu_{1.12}S$), and covellite (CuS) [43,44]. There are still controversial points on the crystal structure of low-temperature sulfide phases, but the discussion of these disagreements is not included in the subject of our review; therefore, we will only briefly present information from the latest structural works on these materials.

Chalcocite γ-Cu_2S below 104 °C is described by the monoclinic space group P21/c with a unit cell containing 48 Cu_2S formula units [45]. Between 104 °C and 435 °C, β-Cu_2S chalcocite is hexagonal with the space group $P6_3/mmc$. Above 435 °C, high-temperature chalcocite α-Cu_2S transforms into a cubic close-packed structure of digenite $Fm\overline{3}m$ [19,43]. At room temperature, chalcocite usually exists as a mixture with djurleite, since the two phases readily transform into each other. Djurleite ($Cu_{1.965}S \div Cu_{1.934}S$) has a monoclinic lattice (space group $P21/n$) and is stable up to 93 ± 2 °C [45], then reversibly decomposes into hexagonal chalcocite $Cu_{1.988}S$ and hexagonal digenite $Cu_{1.84}S$ [46].

Digenite ($Cu_{1.8}S$) exists in two forms: a low-temperature phase (below 91 °C) and a high-temperature cubic form with the space group $Fm\overline{3}m$ (above 91 °C) [47]. According to Roseboom [46], the copper content in digenite increases with temperature and reaches the composition Cu_2S at 435 °C. Below 72 °C, digenite forms a metastable phase, which transforms into orthorhombic anilite [43]. The high-temperature cubic modification contains four copper atoms (ions) with uncertain coordination in the unit cell. The equilibrium distribution of copper atoms over the voids of the "framework" of the lattice can be calculated based on the principle of the maximum configuration entropy. In [48], the indexing of X-ray reflections and the distribution of cations in $Cu_{1.8}S$ at room temperature was carried out according to this principle. It was found that copper ions preferably occupy tetrahedral positions (1/4, 1/4, 1/4) (within the zinc blende structure) and trigonal positions (1/3, 1/3,

1/3). Octahedral positions remain vacant. The influence of temperature and the degree of deviation from stoichiometry on the coordination of mobile cations in various polymorphic modifications is currently insufficiently reflected in the available literature. At the same time, the system of mobile cations certainly affects the thermodynamics and kinetics of polymorphic transformations and the semiconducting properties of copper sulfide [20].

Anilite ($Cu_{1.75}S$) is relatively stable and forms at a temperature of 75 ± 3 °C [43]. Anilite occurs naturally as a mineral. Its crystal structure is rhombic with cell parameters a = 7.89 Å, b = 7.84 Å, c = 11.01 Å. Sulfur atoms form a rigid skeleton of the lattice, while copper atoms are ordered in interstices. The ordered distribution of copper atoms leads to a slight displacement of sulfur atoms from the nearest cubic positions. With a sulfur content of more than 36.36% at room temperature, anilite coexists with covellite (CuS) [43]. Covellite (CuS) is hexagonal with $P6_3/mmc$ space group [18,43], and it does not exist as a single phase at room temperature [43]. Potter [49] showed that covellite is stoichiometric within 0.0005 of the $n(Cu)/n(S)$ mole ratio. When anilite is heated above 75 ± 2 °C, a mixture of cubic digenite and covellite is formed [43]. The results of Dennler et al. [18] showed that CuS is not stable at temperatures above 180 °C either in air or in an N_2 atmosphere, and the material was observed to decompose to Cu_2S and S. It is in accordance with the earlier paper of D. Shah [50], which concluded decomposition of CuS to Cu_2S and S in air.

Roxbyite ($Cu_{1.8125}S$), according to Mumme, W.G. et al. [50], has a triclinic lattice, with space group $P\bar{1}$, with unit cell dimensions a = 13.4051 (9) Å, b = 13.4090 (8) Å, c = 15.4852 (3) Å, α = 90.022 (2), β = 90.021 (2), γ = 90.020 (3) [51]. The structure of roxbyite is based on a hexagonal close-packed framework of sulfur atoms with the copper atoms occupying these layers, all having triangular coordination. Other layers sandwiched between the close-packed sulfur layers consist purely of double or split layers of Cu atoms. Some of these Cu atoms have twofold linear coordination, but mostly they have three- and fourfold coordination to the sulfur atoms in the close-packed layers that lie above and below them. The crystal structure of roxbyite bears a strong kinship to those of low chalcocite and djurleite [51].

The crystal structures of minerals spionkopite ($Cu_{1.39}S$) and yarrowite ($Cu_{1.12}S$) were first described by Goble [52] in 1980. Yarrowite has a hexagonal lattice with a = 3.800 (1) Å, c = 67.26 (4) Å, Z = 3. Spionkopite also is hexagonal with a = 22.962 (3) Å, c = 41.429 (1) Å, Z = 18. Space groups for both minerals are $P3m1$, $P\bar{3}m1$, or $P321$. Yarrowite and spionkopite have well-developed subcells that strongly resemble the unit cell of covelline.

2.1.2. Cu-Na-S System

Solid Solutions on Base of Copper Sulfide

The wide range of homogeneity of superionic copper sulfide along the metal sublattice (from Cu_2S to $Cu_{1.75}S$) allows doping with other metals while maintaining the type of crystal structure, allowing one to obtain homogeneous samples with the desired useful properties.

The effect of sodium doping on transport phenomena in copper sulfide was investigated by Z. H. Ge et al. [31]. The thermoelectric properties of bulk samples of copper sulfide $Na_xCu_{1.8}S$ (x = 0, 0.005, 0.01, 0.03, 0.05), consolidated using the technology of spark plasma sintering from a nanopowder with an average nanoparticle size of 3 nm, synthesized by mechanical fusion, are described. The limit of sodium solubility in the crystal structure of the sulfide is revealed as x = 0.01. The purpose of the doping was to reduce the conductivity and increase the Seebeck coefficient. According to measurements of the Hall effect, in the samples doped with sodium, the concentration of carriers decreases, compared to pure $Cu_{1.8}S$. In addition, the presence of many nanosized pores and grains was found, which led to a decrease in thermal conductivity by a factor of 2–3. As a result, a high value of ZT = 1.1 at 500 °C for $Na_{0.01}Cu_{1.8}S$ was achieved in this work, mainly due to a decrease in thermal conductivity, which is almost twice as high as for pure $Cu_{1.8}S$, and is comparable in magnitude with modified PbS materials (ZT ≈ 1.2 at 650 °C). The sodium solubility in the interstices of the $Cu_{1.8}S$ lattice is 0.28%; at a higher sodium concentration (for the $Na_{0.05}Cu_{1.8}S$ alloy), inclusions of the Na_2S and $Cu_{1.96}S$ phases were observed.

Note that light sodium doping was recently studied by Z. Zhu et al. [53] for copper selenide. Doping with Na by mechanical fusion introduces multiple micropores, which can optimize heat transfer through strong phonon scattering at the interfaces between micropores and grains. The introduction of micropores is an effective way to improve thermoelectric characteristics, which is similar to another effective method of introducing a secondary nanophase [3,16,42]. According to the results of our dissertation work, it is the presence of additional nanoscale phases that explains the high thermoelectric characteristics of the sulfide alloys studied in the work.

Binary Compounds of Na-S System

The properties of Na_2S are well studied. In general, it has a crystalline structure of fluorite and a cubic modification, ionic conductivity (~0.1 S cm^{-1} at 723 °C) is high even for superionic conductors, and weakly depends on cationic substitution, which indicates a high disorder ("melting") of the cationic sublattice. The energy of motion was found by B. Bertheville et al. [54] to be 0.61 ± 0.05 eV for a cation vacancy. As is known, the lattice parameter is 6.5373 Å, the degree of filling of the cation sublattice is 0.988 [55], the strong disordering of the structure provides a low lattice component of thermal conductivity. In these systems, the properties of the transition of an electron occurs without a change in the momentum of an electron; the difference in the energies of electrons between the bottom of the conduction band and the top of the valence band is 2.23–3.05 eV [55,56]. In [57], by M. Kizilyalli et al., it is reported that new Na_2S structures of cubic and rhombic symmetry were obtained at high temperatures. The approximate unit cell parameters were found to be a = 11.29 Å for the cubic form and a = 15.94 Å, b = 16.00 Å, and c = 16.18 Å for the orthorhombic form.

$Na_2Cu_4S_3$

$Na_2Cu_4S_3$ phase was found by Savelsberg G. and Schäfer H. [58,59]. $Na_2Cu_4S_3$ has a monoclinic lattice type, with the space group $C2/m$. The lattice parameters are a = 1563 (3), b = 386 (2), c = 1033 (2) pm, β = 107.6°. These atoms together form sulfur atoms in layers, which are stacked in the direction of the c-axis through separated octahedrally coordinated Na atoms. After that, all copper atoms are triple coordinated with sulfur. In addition to the two sulfur atoms, there are six copper neighbors, as sulfur atoms have layers surrounded by three copper atoms. Burschka C. and Naturforsch Z. [60] described the crystal structure of $Na_3Cu_4S_4$, which also belongs to the thiocuprate class.

$NaCu_5S_3$

The crystal structure of $NaCu_5S_3$ was studied by H. Effenberger et al. [61]. The ternary system was obtained by the hydrothermal synthesis method. $NaCu_5S_3$ is hexagonal with space group $P\,6322$-D 6, Z = 2 [61]. The structural parameters are a = 6.978(5) Å, c = 7.209(6) Å. The formation energy of the system is 0.382 eV. $NaCu_5S_3$ decomposes into $Cu_{1.75}S + Na_3(CuS)_4 + Cu$. The S atom has an irregular coordination figure created by two neighboring Na atoms and four copper atoms. W. Yong et al. [62] studied the optical properties of the $NaCu_5S_3$ thiocuprate. The Cu-S structure doped with Na alkali metal ions was synthesized by the hydrothermal method. The diffuse reflectance spectrum shows that the band gap of Cu_2S nanocrystals is 1.21 eV; after doping with Na$^+$, the color of the product strongly changed. The optical band gap measured at the edge of the absorption band of $NaCu_5S_3$ was 0.49 eV, which indicates a decrease in the photoelectric properties in the visible region of the spectrum.

$NaCu_4S_4$

Zhang X. et al. [63] studied the compound $NaCu_4S_4$. The structure of $NaCu_4S_4$ is reported to consist of a two-dimensional Cu/S framework of trigonal symmetry. The compounds are carried out through Cu-S bonds with the participation of metal atoms from a layer of the GaS type of sulfur atoms. If all monosulfides and disulfides have a charge of 2^-, then the charges on the metal decrease to $Na(Cu^+)_2(Cu^{2+})(S_2)S_2$. However, the chalcogen present in ternary systems, in addition to copper, is also in a mixed valence state. This situation is similar to the situation in CuS, where the formal charge of Cu is 1^+ and the average charge of S is 1^-. Thus, $NaCu_4S_4$ represents the behavior of an ideal metal. In $NaCu_4S_4$, the [CuS] framework has a 0.25^- total charge, and the average charge S decreases even more to 1.25^-, still short 2^- for the filled S^{2-} p-band. The addition of extra electrons to the sulfur [CuS] bands results in a less localized state with a significant degree of delocalization.

In the works of Klepp K. et al. [64,65], the crystal structure of $Na_4Cu_2S_3$ [64] and $Na_7Cu_{12}S_{10}$ [65] thiocuprates with discrete anions was investigated. $Na_4Cu_2S_3$ is tetragonal, with the space group $I4_1/a$ with a = 9.468(1) Å, c = 36.64(2) Å and Z = 16. An outstanding feature of $Na_4Cu_2S_8$ is the formation of discrete V-shaped thiocuprate anions $[S-Cu-S-Cu-S]^{4-}$ with copper in an almost linear coordination by sulfur. The bond length (d_{Cu-S} = 2.15 Å) and angles are in good agreement with the infinite anionic chains of KCuS [66]. The existence of a compound with NaCuS composition was not confirmed.

2.1.3. Cu-Li-S System

The ternary system Cu-Li-S has been poorly studied. Only Cu_2S, Li_2S, and LiCuS compounds were extensively investigated. It was shown in the works of Balapanov et al. [24,25] that in the $Li_xCu_{2-x}S$ system, solid solutions are formed based on the superionic f.c.c. phase of copper sulfide Cu_2S up to x = 0.25. At a lithium content above $x \approx 0.15$, a solid solution is formed already at room temperature; in the range 0 < x <0.15 at room temperature, a mixture of phases is observed, which, upon heating, gradually transforms into a solid solution based on the cubic phase of copper sulfide with the space group $Fm\bar{3}m$. The crystal structure and phase relationships at a lithium content above x = 0.25 have not been studied. In the Cu-Li-Se system, there is information about the crystal structure and phase transitions for the composition $Li_{0.25}Cu_{1.75}Se$ [23]. In contrast to ternary sulfide, ternary selenide exhibits a more complex pattern of phase transitions, and phase transition to f.c.c. structure shifts above 500 °C, while $Li_{0.25}Cu_{1.75}S$ structure is cubic at room temperature [25]. In the paper of Balapanov et al. [25], by neutron diffraction studies, it is shown that gradual disordering in the cation sublattice with the temperature increasing leads to changes in the symmetry of the crystal lattice of the superionic conductor $Li_{0.25}Cu_{1.75}Se$ at 127 and 227 °C. At temperatures close to these values, anomalies in the temperature dependence of both ionic and electron conductivity are observed. Let us consider in more detail the available information on the crystal structure and phase transitions in the Li-Cu-S system.

Li_2S

Lithium sulfide has the antifluorite structure at ambient conditions, space group $Fm\bar{3}m$, Z = 4 [67] with cell parameter a = 5.7158(1) Å. It undergoes a diffuse ("Faraday") phase transition to a fast ion conduction region at about 527 °C and is referred to as a superionic conductor [68,69]. Its high ionic conductivity ($\approx$0.15 S/cm at 727 °C) as a consequence of a Frenkel defect formation without any significant distortion of the f.c.c. sulfur sublattice. The Li diffusion process is carried out by hopping between regular tetrahedral and interstitial octahedral sites. The mean residence times on the regular Li sites were estimated to be 17.3 ps at 900 °C, 6.7 ps at 1000 °C, and 4.3 ps at 1090 °C [69]. The phase transition at 527 °C is also confirmed by studies of Brillouin scattering [70].

Elastic neutron diffraction of Li_2S, measured as a function of temperature in [68], shows the onset of a diffuse phase transition near 627 °C to a superionic state. Inelastic neutron scattering has been used to investigate the harmonic lattice dynamics of Li_2S at 15 K. The authors of [68] conclude that the present data set suggests that in Li_2S, the simple defect structure, the occupation of $(\frac{1}{2}, \frac{1}{2}, \frac{1}{2})$ sites, is created. A shell model has been successfully fitted to the data. The results of the band structure and electron density in Li_2S

are presented in the paper of Tsuji J. [71]. From the results of calculations of the electronic structure, it was found that Li_2S is a semiconductor with an indirect band gap, while a similar Na_2S compound is a semiconductor with a direct band gap.

Solid Solutions on Base of Copper Sulfide

Judging by the similarity of the crystal structure and the closeness of the lattice parameters of the binary compounds Li_2S and Cu_2S, as well as the closeness of the ionic radii of Cu^+ and Li^+ ions, it can be assumed that they can form solid solutions, at least at high temperatures. In similar systems with heavier alkaline cations (Cu_2S-K_2S, Cu_2S-Rb_2S, Cu_2S-Tl_2S), incommensurate quasi-one-dimensional structures of the ACu_7S_4 type (A = Tl, K, Rb) [72,73] are formed, which are chemically equivalent to the $Li_{0.25}Cu_{1.75}S$ compound (which, however, has a cubic structure). In the range from $-243\ °C$ to $127\ °C$, six to seven phase transitions with superstructures are observed in these compounds. Their properties are explained using the theory of charge density waves. Such quasi-one-dimensional structures are not synthesized with sodium, but other crystal types are realized as thiocuprates. Apparently, such structures are energetically unfavorable with light cations of alkali metals. A similar situation is observed for structures of the type [74] with a monoclinic $C12/m1$ lattice—there are homologous compounds, but futile attempts to obtain an isostructural compound with sodium [75,76]. It can be stated that at a large value of the ratio of the radii of the chalcogen/metal ions, other crystal structures are more favorable.

The authors of [24–29] experimentally investigated the crystal structure and transport properties of $Li_xCu_{2-x}S$ solid solutions, which exhibit superionic conductivity. X-ray diffraction studies [25] revealed that $Li_xCu_{2-x}S$ ($x \leq 0.25$) compounds are solid solutions on the base of α-Cu_2S at temperature higher than certain temperature, which depends on the chemical composition of the phase. According to the work of Balapanov et al. [25], the presence of lithium in the lattice in a sufficiently high concentration ($0.10 < x < 0.25$) significantly reduces the temperature of the phase transition in $Li_xCu_{2-x}S$ to the cubic $Fm\bar{3}m$ phase. Lithium ions in the lattice preferably occupy octahedral $32(f)_{II}$ positions, while copper ions are mostly distributed in tetrahedral $32(f)_I$ positions. The authors of [25] conclude that since the "easy diffusion paths" of copper ions pass through the octahedral voids, this reduces the ionic conductivity of $Li_xCu_{2-x}S$ ($x \leq 0.25$) solid solutions compared to binary copper sulfide.

In the range of 20–500 °C, four phase transitions are observed in $Li_{0.25}Cu_{1.75}Se$ by Balapanov et al. [23]. The phase transformation (PT) in $Li_{0.25}Cu_{1.75}Se$ occurs at temperature (130 ÷ 140) °C from triclinic to monoclinic syngony. At 230 ÷ 242 °C, the monoclinic phase is followed by the rhombohedral modification. Both of these PTs are accompanied by drops on the calorimetric curve. At about 380 °C, observed anomalies in temperature dependencies of the ionic conductivity of the chemical diffusion coefficient and jump of ionic Seebeck coefficient have been induced by the PT from the rhombohedral to hexagonal phase of $Li_{0.25}Cu_{1.75}Se$. Neutron diffraction studies revealed the cubic structure of the $Li_{0.25}Cu_{1.75}Se$ compound (with space group $Fm\bar{3}m$) at 500 °C. Copper ions are statistically distributed over tetrahedral and trigonal voids of rigid Se sublattice, and lithium ions randomly occupy $32(f)$ positions.

A detailed study of the thermal and electrochemical behavior of the $Cu_{4-x}Li_xS_2$ phase (x = 1, 2, 3) was carried out by Chen E.M. and Poudeu P. [77]. In this work, it was shown that Cu_3LiS_2 (x = 1) and $LiCuS$ (x = 2) crystallize with unique crystal structures of low symmetry at room temperature. While the XRD pattern at room temperature of the sample with x = 3 is comparable to that of the cubic structure of Li_2S, additional peaks observed on the XRD patterns of samples with x = 1 and x = 2 suggest lower symmetry structures for both phases near room temperature (Figure 1). Taking into account differential scanning calorimetry (DSC) measurements, the authors determined structural phase transition at 140 °C for samples with x = 1 and x = 2 from a low-symmetry rhombohedral modification to the high-temperature cubic modification of the binary compound. In addition, for this compound, a tendency of a decrease in thermal conductivity with an increase in temperature and an increase in the Cu:Li ratio was revealed, which is an essential factor for improving these systems for thermoelectric purposes.

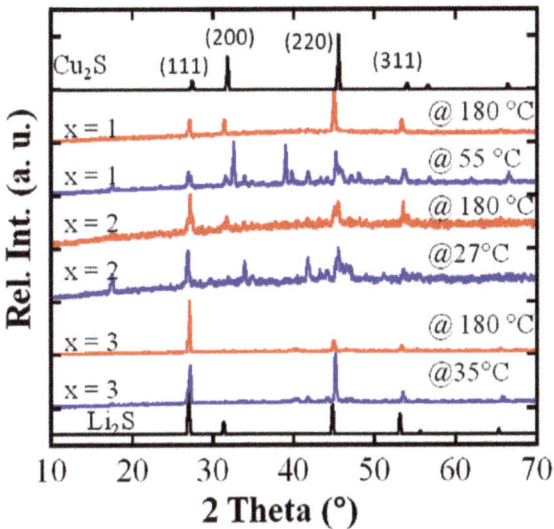

Figure 1. Temperature dependences of the X-ray patterns of the $Cu_{4-x}Li_xS_2$ series powders in comparison with the calculated diagrams of the Cu_2S and Li_2S powders. Reprinted with permission from ref. [77]. Copyright 2015 Elsevier.

In a paper by S.D. Kang et al. [30], the phase transition in $Cu_{2-x}Se$ doped lithium is discussed. The authors note that significant changes are happening with Li doping. In the parent compound $Cu_{2-x}Se$, the superionic transition happens across a temperature range 103–137 °C for $x = 0.01$, in which the low temperature phase (complete structure remains unknown by the authors) and the high temperature phase (cubic, $Fm\bar{3}m$) are mixed. The transition shows strong hysteresis even at the slowest practical ramping rates, showing behavior characteristic of a first-order transition. The biggest change upon substituting Cu with Li is the splitting of the phase transition into two transitions, involving an additional intermediate phase. The high-temperature superionic phase is reached at a higher temperature of around 227 °C upon heating. The total transformation enthalpy (i.e., integrated area of the peaks) remains large, indicating that the cation sublattice melting nature of the transition still persists with Li doping. The distribution of the transformation enthalpy over a wide range of temperatures in both peaks likely indicates that both transitions happen gradually through a phase mixture region in the phase diagram. Powder X-ray diffraction also shows the existence of the intermediate phase. At 152–177 °C, when the peak characteristic of the low temperature phase is almost diminished, some peaks that are not present in the high temperature phase persist. These peaks can be indexed together with the other strong peaks using a monoclinic unit cell (a = 6.379 Å; b = 5.815 Å; c = 6.155 Å; β = 97.91°, which is similar to what has been suggested from an earlier study on $Li_{0.25}Cu_{1.75}Se$ by Bickulova et al. [78] and Balapanov et al. [23]. The compound LiCuS has been studied quite well.

LiCuS

Kieven D. et al. [79] in 2011 studied sputtered LiCuS films with thickness ~200 nm on quartz glass for use in solar cells. In the work, both optical transmission and reflection spectra were obtained. For an indirect and a forbidden direct type of transition, the band gap E_g was found as 2.0 ± 0.1 eV and 2.5 ± 0.1 eV, respectively. Later, the work of German scientists from the Max Planck Institute, A. Beleanu et al. [80], determined the crystal structure of LiCuS ternary sulfide. The crystal structure was determined using neutron and X-ray powder diffraction and solid-state NMR analysis on the ^{7}Li nucleus [80]. Polycrystalline $Li_{1.1}Cu_{0.9}S$ was obtained by the reaction of Li foil (99.999%) with CuS powder: $CuS + 1.1\ Li = Li_{1.1}Cu_{0.9}S + 0.1\ Cu$. The compound crystallizes in the orthorhombic structure of Na_3AgO_2 type with space group *Ibam*. The crystal structure of $Li_{1.1}Cu_{0.9}S$ can be obtained from the cubic structure of Li_2S by moving a part of Li along the *c* axis, and the Li atoms become linearly coordinated by S atoms. There are 24 atoms per unit cell, occupying four different crystallographic sites. S occupies *8j* (S), and the remaining metals, Li and Cu, occupy *8g* (M1), *4c* (M2), and *4b* (M3). All the metals sites are occupied by randomly mixed Li and Cu atoms. The lattice parameters decrease almost linearly with increasing lithium concentration, which is explained by the smaller ionic radius of lithium compared to the radius of copper. Performing in the work [80] the density functional theory calculations show that $Li_{1.1}Cu_{0.9}S$ is a direct band-gap semiconductor with an energy gap of 1.95 eV, which is consistent with experimental data.

In the work of the Egyptian author S. Soliman [81], theoretical calculations of the band gap (1.7 eV) for half-Heusler LiCuS compound by the Engel and Vosko method are presented. Calculations were performed for each structure distribution to determine the lowest energy structure. The electronic structure calculations were performed using the lowest energy distribution as follows. The calculations were based on a spatial symmetry arrangement according to space group 216 ($F\bar{4}3m$), where Li, Cu, and S occupied positions *4a*, *4b*, and *4d*, respectively. The lattice parameter $a_{calc} = 5.53$ Å. The calculations showed that the distribution of the atoms for LiCuS among the above-mentioned Wyckoff positions yielded the lowest energy per formula unit for the compound.

Thus, studying the current state of research on the crystal structure of superionic copper chalcogenides, it can be stated that ions are characterized by diffuse phase transitions occupying a wide temperature range as a rule. With increasing temperature, there is a continuous redistribution of mobile copper ions over different types of interstices in the anionic core of the crystal lattice. Mobile cations can be likened to a "cationic liquid" filling the voids of the structure. The presence of a "liquid-like phase" inside a "solid" lattice interferes with the normal propagation of phonons; therefore, superionic copper chalcogenides have low lattice thermal conductivity, which is a favorable factor for the formation of their high thermoelectric efficiency. As shown below, the insufficient stability of these compounds due to the high diffusion rate of copper can be increased by introducing impurity atoms or creating a composite structure by including nano-objects, such as carbon nanotubes, graphene fragments, etc., into the superionic matrix.

2.2. Methods for Synthesis of Perspective Thermoelectric Materials

At present, much attention is paid to the development and study of new highly efficient thermoelectric materials in the scientific world, which can be judged by the greatly increased number of publications in this field in the past decade. Modern strategies for the search and synthesis of promising thermoelectrics are based on the quantum theory of solids and the latest advances in nanotechnology [3,4,82,83]. Depending on the group of materials (film, bulk, superlattice, supramolecular, etc.) and on the temperature range, the achieved record ZT values lie in range from 1 to 4. Our review is mainly devoted to bulk thermoelectric materials, including composite materials. Their dimensionless thermoelectric efficiency ZT does not exceed 3, as can be seen in Figure 2, which shows some of the best achievements of recent years. We have presented here mainly chalcogenide materials.

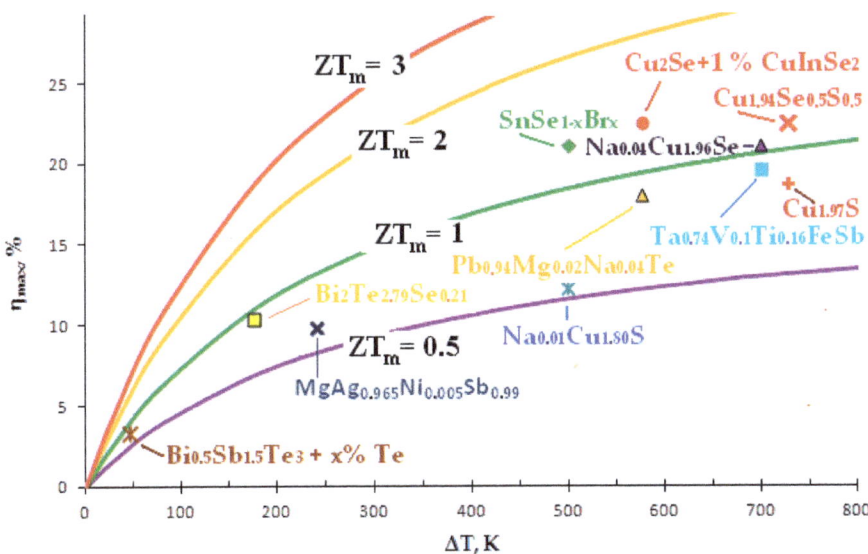

Figure 2. Thermoelectric efficiency of some thermoelectric materials: $Cu_{1.97}S$ [84], $Na_{0.04}Cu_{1.96}Se$ [53], $Na_{0.01}Cu_{1.80}S$ [31], $Cu_2Se + 1\%\ CuInSe_2$ [85], $Cu_{1.94}Se_{0.5}S_{0.5}$ [86], $Bi_{0.5}Sb_{1.5}Te_3 + x\%\ Te$ [87], $Ta_{0.74}V_{0.1}Ti_{0.16}FeSb$ [88], $Pb_{0.94}Mg_{0.02}Na_{0.04}Te$ [89], $MgAg_{0.965}Ni_{0.005}Sb_{0.99}$ [90], $SnSe_{1-x}Br_x$ [91], $Bi_2Te_{2.79}Se_{0.21}$ [92].

This section will briefly review synthesis methods and design techniques for bulk thermoelectric materials that have been frequently used over the past ten years.

The chemistry of the synthesis of semiconductor chalcogenides has developed very intensively in the past two decades. Priority was given to various methods of "cold" synthesis, which do not require large time and energy costs, and allow one to immediately obtain synthesis products in nanosized form.

M.R. Gao [93] describes more than 15 liquid-phase methods for the synthesis and modification of chalcogenide nanomaterials: liquid exfoliation method [94], hot-injection method [95], single-source precursor method [96,97], hydrothermal method [98], solvothermal method [99], mixed solvent method [100], microwave method [101], sonochemical method [102], electrodeposition method [103], electrospinning method [104], photochemical method [105], template-directed method [106], ion exchange reactions, and others.

The hydrothermal method has been widely used for the synthesis of a variety of functional nanomaterials with specific sizes and shapes [93]. The main advantages of hydrothermal and solvothermal processes are fast reaction kinetics, short processing times, phase purity, high crystallinity, low costs, and so on. According Gao et al. [93], the hydrothermal method has achieved great success in the preparation of nanocrystalline chalcogenides with various nanostructures, such as $β$-In_2S_3 nanoflowers, CuS micro-tubules, Sb_2S_3, Sb_2Se_3 and Sb_2Te_3 nanobelts, Ag_2Te nanotubes, Ag_2Se nanoparticles, and so on. In a hydrothermal process, the presence of a small quantity of organic ligands often plays a key role to determine the sizes, shapes, and structures of the nanocrystals. By carefully adjusting the *pH* value, monomer concentration, as well as the reaction temperature and reaction time, bismuth chalcogenides with various nanostructures such as nanostring-cluster hierarchical Bi_2Te_3 and Bi_2S_3 nanoribbons and Bi_2Te_3 nanoplates have been successfully synthesized using the hydrothermal methods. For instance, in Zhu et al.'s [53] work, the nanocrystalline $Na_{0.04}Cu_{1.96}Se$ bulk samples were synthesized with excellent ZT = 2.1 at 700 °C by combining hydrothermal synthesis and hot pressing. By the solvothermal method, lots of MC nanocrystals with an elegant control of the size and shape distributions and also the crystallinity have been synthesized, including wire-like Cu_2Te, Ag_2Te, and Bi_2S_3; belt-like Bi_2S_3; flower-like $γ$-In_2Se_3; dendrite-like $Cu_{2-x}Se$ and Cu_2S; and so on. The hot-injection method is very effective in synthesizing high-quality nanocrystals with good crystallinity and narrow size distributions [95].

Zhao Y. et al. [107] described three chemical synthesis methods for the synthesis of $Cu_{2-x}S$ (x = 1, 0.2, 0.03) nanocrystals (NC): sonoelectrochemical, hydrothermal, and dry thermolysis. The control of the chemical composition of NC was carried out by regulating the reduction potential in the sonoelectrochemical method, by regulating the PH value in the hydrothermal method, and by selecting the pre-treatment of precursors in the dry thermolysis method. The use of the electrochemical method of doping with lithium allowed the authors to obtain and study solid solutions $Li_xCu_{2-x}S$ ($0 < x < 0.25$) [24,26,28], $Li_xCu_{(2-x)-\delta}Se$ ($x \le 0.25$) [22,23], promising for thermoelectric applications. To obtain $Li_xCu_{(2-x)-\delta}Se$ nanopowders, Ishembetov R. Kh. et al. [108] used the method of electrohydrodynamic impact, which makes it possible to effectively grind even the hardest materials down to a few nanometers.

Despite the popularity of liquid-phase synthesis methods, high-temperature ampoule synthesis from elements followed by ball milling and hot pressing is also a popular method for producing high-performance thermoelectrics based on copper chalcogenides. D. Yang et al. [109] analyzed recent works on the synthesis of binary copper and silver chalcogenides. They note that for thermoelectric materials research, dense pelletized samples are required. The $Cu_{2-x}X$ (X = S, Se, Te) powders synthesized by the above-mentioned methods are compacted by hot pressing and spark plasma sintering (SPS). However, the temperature and electrical field involved may drive the mobile Cu ions [110], causing composition inhomogeneity and undesired microstructures in nonstoichiometric of $Cu_{2-x}X$ compounds. He et al. found that high packing density of Cu_2Te samples could be obtained by pressureless direct annealing without hot pressing or SPS, and the resulting density values were comparable to those prepared by SPS [110].

For synthesis of other types of thermoelectric materials, similar methods are used. The authors of the works of D. Kraemer et al. [90] synthesized a p-type MgAgSb alloy operating at temperatures from 20 °C to 245 °C, with very high conversion efficiency of 8.5%, using a simple one-step hot pressing technology. The thermoelectric material powder was prepared by ball milling procedure.

Among many thermoelectric materials, half-Heusler compounds (space group F43m, C1b)—principally with the composition XYZ, where X and Y are transition or rare earth elements and Z is a main group element are the most studied ones [111]. Localized $3d$ states of transition elements such as Fe, Co, Ni, etc., make the maximum of the valence band or the minimum of the conduction band flat and heavy. Thus, higher carrier concentrations, which require a higher dopant content, are required to optimize power factors. Their physical properties are largely determined by the valence electron count (VEC) [82]. High thermoelectric performance is generally achieved in the semiconducting half-Heusler phases with VEC = 18, while VEC > 18 usually leads to metallic conduction behavior [4]. Half-Heusler compounds possess several promising features, namely, a high Seebeck coefficient, moderate electrical resistivity, and good thermal stability. According to S.J. Poon [111], the RNiSn-type half-Heusler compounds, where R represents refractory metals Hf, Zr, and Ti, are the most studied to date. Since 2013, the verifiable ZT of half-Heusler compounds has risen from 1 to near 1.5 for both n- and p-type compounds in the temperature range of 500–900 °C [111]. TaFeSb-based half-Heusler phase with record high ZT of ~1.52 at 700 °C was prepared in 2019 by H. Zhu et al. [88] using two-step ball-milling and hot-pressing methods. In the work of H.T. Zhu et al. [112], the ZrCoBi alloy with a high ZT of ~1.42 at 700 °C and a high thermoelectric conversion efficiency of ~9% at the temperature difference of ~500 °C was obtained. In pair with ZrCoBi, the high-performance half-Heusler thermoelectric modules $CoSb_{0.8}Sn_{0.2}$ were used, which ensure self-propagating synthesis and optimization of the topological structure. Due to the nonequilibrium reaction process in n-type and p-type materials, dense dislocation matrices were introduced, which greatly reduced the lattice thermal conductivity. In the work of Austrian scientists G. Rogl et al. [113], synthesized half-Heusler compounds of the n- and p-type ($Ti_{0.5}Zr_{0.5}$ based on NiSn and NbFeSb) were obtained by the high-pressure torsion method to improve their thermoelectric characteristics due to a sharp decrease in the direction of ultralow thermal conductivity. This decrease is due to grain refinement and a high concentration of defects caused by deformation, that is, vacancies and dislocations, which are determined by severe plastic deformation.

One of the directions in the search for highly efficient thermoelectric materials remains skutterudites [83]. Skutterudites are a kind of cobalt arsenide minerals consisting of variable traces of iron or nickel substituting for cobalt to formulate $CoAs_3$. The chemical formula of skutterudites can be expressed as ReM_4X_{12}, where Re is a rare earth element, M is a transition metal element, and X is a non-metal element from Group V, such as phosphorus, antimony, or arsenic. These are lead- and tellurium-free thermoelectric materials that are highly efficient for intermediate temperature range applications. These materials have the figure of merits ZT values close to 1.3 for p-type and 1.8 for n-type, with good mechanical stabilities [114]. For CoSb skutterudites, obtained by spark plasma sintering (SPS) at 650 °C under a pressure of 50 MPa, the thermoelectric figure of merit ZT ~1.7 at 577 °C has been achieved [115]. Liu et al. [116] reviewed the recent progress made in $CoSb_3$-based materials and synergistic optimization of the thermal and electrical properties. Multi-filled skutterudites demonstrated inferior thermal conductivities, resulting in a considerable increase in ZT values [117].

Thus, by controlling the synthesis and forming nanocrystalline, nanostructured, and nanocomposite materials, it is possible to obtain a wide range of required properties of semiconducting chalcogenides for a variety of applications, as evidenced by numerous recent works [4,82,83,109].

3. Transport Phenomena in Mixed Electron–Ion Conductors

3.1. Electrical Properties of Copper Sulfide and Its Alloys

The reviewed materials on the base of copper chalcogenides as a rule have mixed electronic–ionic conductivity. However, the number of ion transfer usually does not exceed a few percent, and the determining mechanism of electrical conductivity is electron transfer. Copper chalcogenides have a fairly wide range of homogeneity over the cationic sublattice. The lack of cations in the lattice, according to the electroneutrality rule, leads to the formation of electron holes in the valence band and to p-type conductivity. Nonstoichiometric defects in $Cu_{2-\delta}X$ (X = S, Se, Te) compounds, the concentration of which is determined by the nonstoichiometric index δ, play the role of an alloying component (impurity). This impurity forms shallow levels in band gap, which are usually completely ionized at room temperature and above temperatures.

In $Cu_{2-\delta}X$ (X = S, Se, Te) compounds, usually the concentration of "impurity" holes in the valence band (n_p) is much higher than the concentrations of uncontrolled impurities and equilibrium point defects. The concentration n_i of intrinsic carriers is determined by the temperature and the band gap. For nonstoichiometric compositions, at temperatures far from fusion temperature, n_i is also significantly less than n_p. In this case, the temperature dependence of the electronic (hole) conductivity is determined by the temperature dependence of the mobility and has a metallic character. In compensated materials, for example, in $Li_xCu_{2-x}X$ (X = S, Se, Te), the temperature dependence of conductivity is semiconducting [19,20].

The peculiarities of the crystal structure and the presence of a mobile ionic subsystem create a number of problems in the interpretation of electronic kinetic effects in superionic semiconductors. Based on the work of V.M. Berezin [20], the following main aspects of these problems can be distinguished:

(a) The disordered crystal structure of the studied chalcogenides leads to the fact that besides the background of the periodic potential of the anionic sublattice, charge carriers are exposed to the fluctuation potential of the cation sublattice; thus, the description of the phenomena of the transfer of electrons and holes faces the same problems as in non-crystalline solids and liquids [118]. The specificity lies in the description of the total effect of two sublattices—periodic and disordered, on electrons and holes. In this case, (polaron) effects associated with the localization of electronic wave functions (Anderson transition) become possible in the electronic system [119]. The existence of similar localized states as applied to intercalate chalcogenides with a two-dimensional character of conductivity was studied, for example, in the works of A.N. Titov [120] and Yarmoshenko Y.M. [121].

(b) The ease of "overflow" of cations over the voids of the anion framework with a change in temperature or a change in the nonstoichiometry of the composition leads to a smearing of phase transitions and a continuous change in the parameters of the band structure during this redistribution (change in the effective mass of carriers, width of the gap), etc.

(c) The anharmonicity of vibrations of atoms of the crystal lattice and high coefficients of self-diffusion in a disordered sublattice call into question the applicability of the developed theory of scattering of current carriers in semiconductors in the harmonic approximation. The temperature dependences of the electron and hole mobilities must be refined experimentally and new approaches to their theoretical description must be sought.

To describe the electronic energy spectrum in superionic semiconductors, it is convenient to use the functions of the density of electronic states $g(\varepsilon)$, where ε is energy of electron. This is a universal function of the electronic system, the use of which is not related with the periodicity and defectiveness of the atomic crystal structure. For an arbitrary isotropic dispersion law $\varepsilon(p)$:

$$g(\varepsilon) = \frac{8\pi}{h^3} p^2 V \frac{dp}{d\varepsilon}, \tag{3}$$

where V is the electronic system volume [20]. For the case of electrons in the conduction band with a parabolic dispersion law, the density of electronic states is determined as:

$$g(\varepsilon) = \frac{8\sqrt{2}}{h^3} \pi V m^{*\frac{3}{2}} (\varepsilon - \varepsilon_c)^{1/2}, \tag{4}$$

where m^* is effective mass of electron.

Since the dispersion law $p(\varepsilon)$ in disordered systems cannot be correctly introduced, it is practically impossible to use Formula (4) to calculate $g(\varepsilon)$ in the materials under study. However, in [20], another method was indicated for introducing the density of states, which can be used for practical calculations in superionic chalcogenides. Let there be a degenerate electron system with the Fermi energy ε_F. A change in the number of electrons in this system by ΔN should correspond to a change in the position of the Fermi level by $\Delta \varepsilon_F$, which, taking into account two spin orientations, can be represented as:

$$\Delta \varepsilon_F = \frac{1}{2} \frac{\Delta N}{g(\varepsilon)}. \tag{5}$$

Expression (5) allows us to find the density of states near the Fermi level. As shown by Wagner [122], the change in the chemical potential μ_p (Fermi level) of electron holes in copper chalcogenides is related by the formula $\Delta \mu_e \equiv \Delta \varepsilon_F = e \Delta E$ to the electromotive force (e.m.f.) E of an electrochemical cell of type:

$$Cu/CuBr/Cu_{2-x}S/Pt, \tag{6}$$

where Cu is reversible metallic electrode, CuBr is electronic filter (material with unipolar Cu^+—ionic conductivity in range of 340–440 °C [123]), $Cu_{2-x}S$ is a sample, and Pt is inert metallic electrode. At temperatures above room temperature, all nonstoichiometric defects in nonstoichiometric copper chalcogenides are already ionized (which is often confirmed by the metallic type of temperature dependence of the conductivity); therefore, the change in the number of electron charge carriers can be expressed through the change in the nonstoichiometric index δ:

$$\Delta N = \Delta \delta N_A \frac{V}{V_m}, \tag{7}$$

where V is the sample volume, V_m is the molar volume, and N_A is the Avogadro number. The Formula (7) can be used for the experimental determination of change in the density of electronic states:

$$g(\varepsilon) = \frac{\Delta \delta \cdot V \cdot N_A}{2e \cdot \Delta E \cdot V_m}, \tag{8}$$

In the presence of two types of carriers in a semiconductor, its conductivity σ is described by the equation:

$$\sigma = e(n_e \mu_n + n_p \mu_p), \tag{9}$$

where n_e and μ_n are the concentration and mobility of electrons, respectively, n_p and μ_p are the same for holes, and e is the electron charge.

Taking into account the relationship between the hole concentration and the nonstoichiometry degree δ in such sample as $Cu_{2-\delta}S$, for example:

$$n_p = \delta N_A / V_m \tag{10}$$

where V_m is the molar volume of the phase and N_A is Avogadro's number. From Equation (7), one can obtain a formula to estimate the hole mobility from compositional dependence of the conductivity $\sigma(\delta)$:

$$\mu_p = V_m / [F(d\sigma/d\delta)], \tag{11}$$

where $F = eN_A = 96{,}480$ C/mol is the Faraday number.

3.1.1. Electronic Conductivity

The simplest way to study the properties of a semiconductor is to measure its electrical conductivity. As is known, the result of this measurement depends on the number of mobile charge carriers (electrons, holes), on the distribution of their thermal velocities, and on the deviation from the equilibrium distribution, which is caused by the applied electric field. The classical transport theory used is usually based on the approximation of the Boltzmann kinetic equation, which arose from the kinetic theory of gases, where the electrical conductivity is considered to be a free electron gas flow.

The Drude model, related to the drift mobility and conductivity of carriers, is discussed in connection with its applicability to explain electron transport in solids, some ideas that still remain an integral part of the classical theory of free electrons.

The usual expression for electrical conductivity is:

$$\sigma = neu, \tag{12}$$

where quantity $u = e\tau_m/m$ is the drift mobility of electrons.

Electrical conductivity is also often expressed in terms of the average mean free path of electrons $\lambda_m = v_T \tau_m$, defined as the distance traveled by an electron moving with thermal velocity v_T during the mean free path τ_m. Thus, in the Drude model, the electrical conductivity can be written as:

$$\sigma = \frac{ne^2 \tau_m}{m} = \frac{ne^2 \lambda_m}{m v_T} = \frac{ne^2 \lambda_m}{(3mk_B T)^{1/2}}, \tag{13}$$

where k_B is the Boltzmann constant.

The Lorentz model demonstrates the transfer of an electric charge associated with particles of an electron gas (electrical conductivity), as well as with the transfer of kinetic energy by the same electrons (electronic thermal conductivity) [124,125]. The resulting formula for electrical conductivity:

$$\sigma = \frac{4ne^2\lambda_m}{3(2\pi m k_B T)^{1/2}} \qquad (14)$$

From the point of view of the quantum mechanical concept, only those electrons that are near the Fermi level (Sommerfeld's model) contribute to the electrical conductivity. These states drift in space due to the external electric field at a high speed, approximately equal to the Fermi velocity (V_F), and only the movement of these electrons in the direction of the electric field can contribute to the conductivity. Thus, conductivity can be described:

$$\sigma = \frac{1}{3}e^2 V_F^2 \tau N(\varepsilon_F) \qquad (15)$$

This quantum mechanical equation shows that the conductivity depends on the Fermi velocity (V_F), the relaxation time (τ), and the degree of filling $N(\varepsilon_F)$ of the Fermi level, which is proportional to the density of states.

The Seebeck coefficient in the generate regime varies with the carrier concentration according to the formula (parabolic band, energy-independent scattering approximation):

$$\alpha = \frac{8\pi^2 k_B^2}{3eh^2} m^* T \left(\frac{\pi}{3n}\right)^{2/3}, \qquad (16)$$

where m^* is the effective mass of carrier [120].

One of the first studies of the electronic conductivity of copper sulfides was the work of J.B. Wagner and C. Wagner [122]. Using the relation of Fermi level in $Cu_{2-\delta}S$ with e.m.f. E of the cell (I), it was found in [122] that the ratio of the effective mass of electron holes to the mass of a free electron is 7 ± 2 at 435 °C in $Cu_{1.8}S$. Measurements of the conductivity indicate a significant increase in the mobility of electron holes with an increase in the copper deficit in the samples. In the work of I. Yokota [125] in 1953, for the example of $Cu_{2-\delta}S$, his diffusion theory of the transfer of ions and electrons in mixed conductors was presented. With the help of this theory, Yokota determined the threshold of the current density at which copper begins to separate from the $Cu_{2-\delta}S$ sample as:

$$a \equiv \frac{ejL}{2k_B T \sigma_p} > 1, \qquad (17)$$

where j is a current density, L is a sample length, and σ_p is the hole conductivity. Taking this knowledge into account is important for creating conditions for long-term operation of thermoelectric devices based on copper chalcogenides and similar materials.

Japanese scientists T. Ishikawa and Sh. Miyatani [126] in 1977 investigated the electronic and ionic conductivity and the Hall coefficient of binary copper chalcogenides depending on the electromotive force (e.m.f.) of the electrochemical cell (I) $Cu/CuBr/Cu_{2-x}S$ / Pt. It was established earlier by C. Wagner [122] that e.m.f. of cell of type (I) is determined by the expression:

$$E = \frac{\mu_{Cu} - \mu_{Cu}^0}{F}, \qquad (18)$$

where $\mu_{Cu} - \mu_{Cu}^0$ is the difference in chemical potentials of copper atom in the $Cu_{2-x}S$ sample and in metal copper electrode, correspondingly. In a steady state, the difference in chemical potentials is compensated by the resulting difference of electrical potentials, which is expressed by Equation (8). Equation (8) shows that cell e.m.f. E can be judged on the change in the chemical potential of copper atoms in the sample, and, consequently, on the copper content in the phase of variable composition. T. Ishikawa and Sh. Miyatani considered $E = \Delta(en)/\Delta\varepsilon\ hh$ as the relative height of the Fermi level in copper chalcogenides. There is $\varepsilon = eE/k_B T$, where E is electron energy.

A significant increase in the mobility of electron holes at 100 °C ~2 cm^2/V s and an effective mass of holes ~2.3 m_0 (where m_0 is the mass of a free electron) were found. The authors explain the degree of Cu excess as a result of the screening action of carriers (holes) on the vacancies of copper ions.

I. Yokota and Sh. Miyatani [127] in 1981 presented the phenomenological theory of ion–electronic conductivity of Cu$_2$S based on cross-conduction and ambipolar diffusion of Cu$_2$S. The cross-conductivities σ_{ie} and σ_{ei} for steady states at only electronic current passing and for only ionic current passing through specimen were measured by Sh. Miyatani for the β-phase of Cu$_2$S at 340 °C; it was found that $\sigma_{ei} \approx \sigma_{ie}$ and the inverse Onsager relation is held for Cu$_2$S, and the cross-conductivity values were at least 100 times lower than the ionic conductivity and the electronic conductivity.

Gafurov I.G. [128] studied electrical properties of Cu$_{2-x}$S doped with lithium. He established that substitution by lithium leads to decreasing electronic conductivity as well as to decreasing ionic conductivity. The Seebeck coefficient of Cu$_{(2-x)-\delta}$Li$_x$S ($0 < x < 0.25$) increases with a rise in nonstoichiometricity degree δ in the temperature interval 20 °C to 410 °C. Measured Hall mobilities of holes lie in the range of 6 ÷ 60 cm^2 V^{-1} s^{-1}. Debae temperatures θ_D of superionic cubic phase of Cu$_{(2-x)-\delta}$Li$_x$S are lower than θ_D = 145 K for Cu$_2$S; for example, θ_D = 101 K for Cu$_{1.75}$Li$_{0.25}$S.

Ishembetov et al. [108] in 2011 studied the effect of grain size on the electronic conductivity of copper selenide Cu$_{1.9}$Li$_{0.10}$Se. With a decrease in the grain size to 50 nm, a decrease in electrical conductivity by a factor of 2–3 was observed. Interestingly, at temperatures above 540 °C, the Seebeck coefficient of a nanostructured sample begins to decrease, which may be associated with the development of intrinsic conductivity. Kang et al. [30] in 2017 found that doping with lithium increases the conductivity and, as a result, improves the thermoelectric performance of copper selenide. They reported a maximum ZT > 1.4 for Li$_{0.09}$Cu$_{1.9}$Se composition, prepared by solid phase alloying, ball milling, and hot pressing. Copper sulfide and selenide are very similar in their electrical properties, so something similar should occur in the Cu$_2$S system. In the work [129] by M. Guan et al., the Cu$_{2-x}$Li$_x$S compounds (x = 0, 0.005, 0.010, 0.050, and 0.100) were studied. When $x < 0.05$, the Cu$_{2-x}$Li$_x$S samples are stable and pure phases, having the same monoclinic structure as the pristine Cu$_2$S at room temperature. The electrical and thermal conductivities were measured in the temperature range from 27 °C to 627 °C. The electrical conductivity in the Cu$_{2-x}$Li$_x$S is greatly improved with the Li doping content increasing due to the enhanced carrier concentrations. For Cu$_{1.95}$Li$_{0.05}$S, the σ increases to about 87 S cm^{-1} at 27 °C, about one order of magnitude higher than that of the Cu$_2$S matrix. Since doping with Li in Cu$_2$S increases the ion activation energy [24] and thereby lessens the influence of mobile ions on heat-carrying phonons, it leads to a significant increase in the thermal conductivity of Li-doped Cu$_2$S samples. The maximum value ZT = 0.84 was obtained at 627 °C for the composition Cu$_{1.99}$Li$_{0.01}$S, an improvement of about 133% compared to the Cu$_2$S matrix. The impurity of lithium significantly reduces the conductivity both by compensating for holes and by introducing additional scattering centers. Impurity atoms also reduce the thermal conductivity of the material. In our opinion, further improvement in the thermoelectric figure of merit of lithium-doped copper chalcogenides can be achieved with a higher degree of doping with lithium and a lower copper content than in the works of Kang et al. [30] and Guan et al. [129]. Then, the cation sublattice as a whole will contain a sufficient number of vacancies to maintain high conductivity. In other words, the material must be an uncompensated semiconductor so that its conductivity is not too low. Perhaps, nonstoichiometric chalcogenides rather than Cu$_{1.95}$S and Cu$_{1.90}$Se should be chosen for Li doping to receive homogeneous compositions, such as Cu$_{1.85}$S and Cu$_{1.85}$Se.

The transfer phenomena in solid solutions of Cu$_2$Se-Ag$_2$Se, Cu$_2$Se-Li$_2$Se, and Cu$_2$S-Li$_2$S systems, depending on temperature, chemical composition, and nonstoichiometry degree, were studied by R.Kh. Ishembetov [130]. In the work, the temperature dependences of electronic conductivity were obtained for solid solutions of the following compositions: Ag$_{0.23}$Cu$_{1.75}$Se, Ag$_{0.5}$Cu$_{1.5}$Se, Ag$_{1.2}$Cu$_{0.8}$Se, and Li$_{0.1}$Cu$_{1.75}$Se. It is shown that substitution with silver significantly changes the electrical properties of copper selenide. While the composition $Ag_{0.23}Cu_{1.75}Se$ exhibits a temperature dependence of conductivity of highly degenerate semiconductor, $Ag_{1.2}Cu_{0.8}Se$ has a semiconducting character of the conductivity. The presence of an isovalent impurity (silver or lithium) leads to the appearance of a metal–semiconductor transition with temperature variation; for example, about 340 °C for the $Li_{0.1}Cu_{1.75}Se$ composition, the monotonic decrease in conductivity with temperature increasing is changed to semiconductor behavior of conductivity, and for the $Ag_{0.5}Cu_{1.5}Se$ composition, at the same temperature, the semiconductor character of the dependence is changed to metallic. The dependence of the electronic conductivity σ_e on the silver content in the solid solutions $Ag_{1.2\pm\delta}Cu_{0.8}Se$, $Ag_{0.23\pm\delta}Cu_{1.75}Se$, and $Ag_{0.5-\delta}Cu_{1.5}Se$ and on the copper content in the solid solution $Li_{0.1}Cu_{1.75+\delta}Se$ has been studied. For the $Ag_{1.2}Cu_{0.8}Se$ composition, the dependence of the electronic conductivity on the degree of nonstoichiometry has a minimum. In the author's opinion, the observed minimum corresponds to the stoichiometric composition, which is explained by a decrease in the concentration of defects in the cation sublattice when the stoichiometric composition is approached. With the same composition, a change in the sign of the electronic Seebeck coefficient is observed, and the hole conductivity is replaced by electronic one. It proves that the $Ag_{1.2}Cu_{0.8}Se$ solid solution, in contrast to copper selenide, which exists only with a copper deficiency, can exist both with a deficiency and with an excess of metal relative to the stoichiometric composition. From the slope of the $\sigma_e(\delta)$ dependences, the mobilities of charge carriers in the α-$Ag_{1.2\pm\delta}Cu_{0.8}Se$ solid solutions at 400 °C are determined in paper [131]. For analysis, the $\sigma_e(\delta)$ dependence was divided into three regions, within which the slope of the dependence is approximately constant: region (1) of n-type conductivity ($\delta < 0$) and two regions of p-type conductivity $0 < \delta < 0.005$ (region 2) and $0.005 < \delta < 0.011$ (region 3). The obtained values of the hole mobility μ_p are 8.1 and 46 cm^2 V^{-1} s^{-1} in regions 1 and 2, respectively. Thus, an increase in the silver content within the homogeneity region decreases hole's mobility in the $Ag_{1.2\pm\delta}Cu_{0.8}Se$. Similarly, for the region of electronic conduction, under the assumption that one free electron participating in the conduction corresponds to one excess silver ion Ag^+ in the cation sublattice, the value of the electron mobility $\mu_n = 6.8$ cm^2 V^{-1} s^{-1} is obtained. For low temperature pseudo-tetragonal β-CuAgSe, in the paper of S. Ishiwata et al. [132], values of electron mobility were reported as ~20,000 cm^2 V^{-1} s^{-1} at -263 °C and ~2500 cm^2 V^{-1} s^{-1} at ambient temperature.

For hot-pressing consolidated CuAgSe nanoplatelets in work [133] by N.A. Moroz et al., it is observed that the thermopower of tetragonal β-CuAgSe gradually decreases with increasing temperatures from -50 µV K^{-1} at 27 °C to ~-20 µV K^{-1} at the vicinity of $T_{\beta\alpha}$ ~177 °C and jumps to a large positive value (+200 µV K^{-1}) at a temperature slightly above $T_{\beta\alpha}$ in the cubic α-CuAgSe. Earlier, Hong et al. [134] reported that high-temperature α-CuAgSe is a p-type semiconductor and exhibits low thermal conductivity, while β-CuAgSe shows metallic conduction with dominant n-type carriers and low electrical resistivity. The thermoelectric figure of merit ZT of the polycrystalline α-CuAgSe at 450 °C is ~0.95. The sign reversal from a negative value at low T to a positive value at high T was reported earlier [132], with the reversal temperature T_S ~127 °C. This feature is not observed in Cu$_2$Se and it reflects the existence of two types of carriers in the β-phase, as was claimed in work [132].

To determine the effective mass of electrons in the stoichiometric composition $Ag_{1.2}Cu_{0.8}Se$, the method developed by C. Wagner was applied in [130]. The value of the effective mass, at a temperature of 400 °C, found from processing the curves of coulometric titration according to the Wagner method [122], was $m^* = 0.08\ m_e$. This low effective mass is more characteristic for silver chalcogenides; much higher m^* values are observed for copper chalcogenides. Note that in Ag$_2$Se, the effective mass of electrons was found equal to 0.10 m_e in the state of equilibrium with selenium and 0.19 m_e for the state of equilibrium with silver [135]. Thus, strong substitution with silver in copper selenide leads to the appearance of lighter charge carriers. Calculations in this work using the Mott formula show that the reduced Fermi level in the $Li_{0.10}Cu_{1.75}Se$ alloy in the temperature range 20–60 °C has values $\eta = 20 \div 23$, and $\eta = 24 \div 50$ in the temperature range 130–230 °C.

Publications of experimental works on electron transport in compounds of the Na-Cu-S system are presented in small quantities, the first of which is the paper of Z. Peplinski et al. [136], published in 1982. It was obtained that from −258 °C to 27 °C, the mixed-valence compound $Na_3Cu_4S_4$ is metallic. The compound exhibits Pauli paramagnetism with a value of $\chi_m = 15 \times 10^{-5}$ emu/mol in the temperature range of −173–27 °C. For pressed samples, the conductivity was measured from 300 S cm^{-1} at 27 °C to 1500 S cm^{-1} at −258 °C. Measurements on single crystals revealed that the conductivity σ is highly anisotropic, with enhanced conductivity $\sigma_{||}$ parallel to the crystal needle axis, corresponding to pseudo-one-dimensional $[Cu_4S_4{}^{3-}]_\infty$ columns in the structure. For single crystals, $\sigma_{||}$ was measured from 15,000 S cm^{-1} at 27 °C to 300,000 S cm^{-1} at −258 °C. The authors of the work [136] assumed that the low values of conductivity observed for pressed tablets are the result of interparticle resistance. In the work of American scientists X. Zhang et al. [63] in 1995, the results of electrical conductivity for the composition $NaCu_4S_4$ are presented. The compound demonstrates metallic conductivity and temperature-independent Pauli paramagnetism with a value of $\chi_m = 6.2 \times 10^{-5}$ emu/mol. The conductivity decreases almost linearly with temperature, from ~3300 S/cm at −263 °C to ~240 S/cm at −48 °C, and the Seebeck coefficient is 3 µV/K, confirming the metallic nature of the p-type.

Interesting results of electrical and transport properties were obtained by Ge Z.H. [137] in 2016 for $Na_xCu_{1.8}S$ ($x = 0, 0.005, 0.01, 0.03, 0.05$). The obtained values of electrical conductivity (σ) of the samples are very high. The conductivity first increases and then decreases with temperature rise, where the turning moment is observed at ~100 °C. The σ values for $Na_xCu_{1.8}S$ samples continuously decrease with increasing Na content due to a decrease in the carrier concentration and mobility. The authors express the defect equation for Na entering the interstices of the $Cu_{1.8}S$ lattice as:

$$xNa \xrightarrow{Cu_{1.8}S} Cu_{Cu}^x + S_s^x + xNa_i + xe' \tag{19}$$

As shown in Equation (19), the Na$^+$ ion enters the $Cu_{1.8}S$ lattice, increasing the electron concentration. The authors note that electrons recombine with holes, decreasing the hole concentration. The concentration of Hall carriers decreases from 6.37×10^{22} to 4.87×10^{21} cm^{-3} with x increasing. The mobility of carriers increases from 15.3 to 31.9 cm^{-2} s^{-1}.

In the work of Zhu Z. et al. [53] in 2019, the electrophysical parameters are observed for $Na_xCu_{2-x}Se$ ($x = 0, 0.01, 0.02, 0.03,$ and 0.04) bulk samples sintered by hot pressing, for both low- and high-temperature phases. The resistivity of $Na_xCu_{(2-x)}Se$ increases monotonically with increasing temperature, except for the inflection point near the phase transition temperature, which indicates that the sample exhibits metallic conductivity. It was shown that Na doping decreases the conductivity from 188 S cm^{-1} for Cu_2Se to 130 S cm^{-1} for $Na_{0.04}Cu_{1.96}Se$ at 700 °C. The authors [53] explain this behavior of the conductivity by a decrease in the carrier mobility caused by additional scattering from the numerous micropores of the structure. Together with this factor, undoubtedly, the reason for the decrease in the conductivity must be additional scattering of carriers by impurity sodium atoms.

We also note the results of the work of Zhang Yi et al. [138] published in 2020, devoted to the study of influence of Na_2S doping ($x = 0, 0.5, 1, 2$ wt.%) on thermoelectric performance of digenite $Cu_{1.8}S$ in the temperature range 50–500 °C. It is well known that the pure composition of digenite demonstrates (Figure 3) extremely high electrical conductivity (σ) due to ionization of vacancies in the Cu sublattice. A transition from semiconducting to metallic conductivity is seen in Figure 4. The inflection point temperature is about 88 °C, which corresponds to the phase transition from the hexagonal phase to the cubic phase of digenite. The work was completed almost simultaneously with the work of Z. Zhu [53], i.e., independently, but what makes these works similar it is a high porosity of the material's structure due to sodium doping. However, the material synthesized by Zhu turned out to be much more efficient ($ZT \sim 1.3$ at 500 °C) due to the fact that the thermal conductivity of his material was lower, while the factor $\alpha^2 \sigma$ was almost close for both materials. It can be seen from the plots of the conductivity and the Seebeck effect in Ref. [138] that the concentration of charge carriers in the material is not optimized. It is possible that if not digenite, but jarleite (with the same 4 at.% Sodium) was chosen as the matrix for doping, the thermoelectric efficiency would be higher with the same low thermal conductivity due to the superionic state of the lattice and developed porosity.

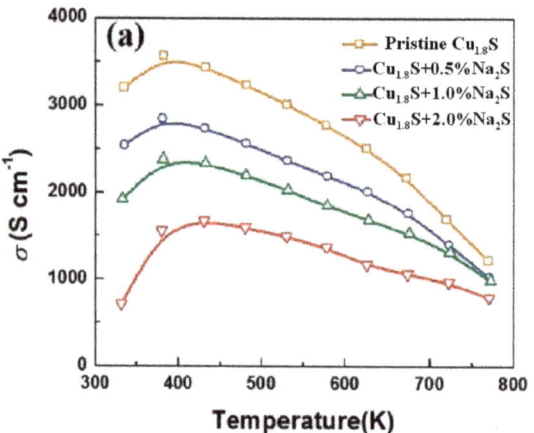

Figure 3. Temperature dependence of electrical conductivity of bulk $Cu_{1.8}S$ samples doped with Na_2S ($x = 0, 0.5, 1, 2$ wt.%). Reprinted with permission from ref. [138]. Copyright 2020 Elsevier.

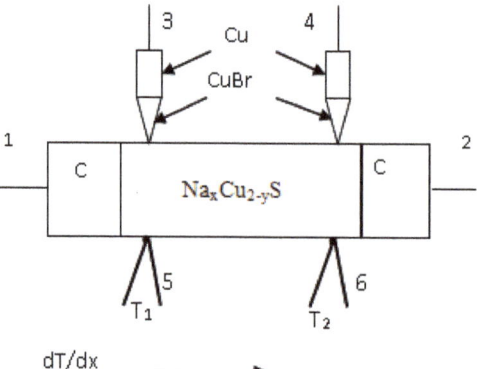

Figure 4. Schematic diagram of a cell for measuring the coefficients of electronic and ionic thermo-emf and electronic conductivity. In the picture, symbols C and c denote Cu probes and electrodes, CuBr is ionic conductor, $Na_xCu_{2-y}S$ is sample (example), T- thermocouple. The control of the constancy of the chemical composition and the equilibrium state of the samples during measurements was carried out by the EMF of the electrochemical cell Cu / CuBr / sample / Pt, measured between pairs of contacts 3-5 and 4-6, in two cross sections of the sample. Reliable clamping of potential probes 6 and thermocouples 4 to the sample was ensured using flexible wire 5, tightened by springs.

Thus, in almost all works considered in this section, the substitution of lithium, sodium, and silver for copper reduces the electronic conductivity of copper chalcogenides due to a decrease in the carrier mobility and the effect of compensation of the hole concentration upon ionization of the donor impurity. In the case of heavy doping, the band gap and the effective mass of the carriers change. The doping of copper chalcogenides with an impurity of sodium sulfide revealed the formation of micropores, which weakly decreases the electrical conductivity and greatly decreases the thermal conductivity, generally improving the thermoelectric figure of merit ZT. Optimization of the carrier concentration by controlling the composition nonstoichiometry and impurity concentration remain the main methods for achieving the maximum thermoelectric power of copper chalcogenides and their alloys.

3.1.2. Ionic Conductivity

Recently, solid-state ion-conducting superionic conductors have attracted increased attention of TE researchers. As is known, ionic transfer in ordinary solids does not exceed 10^{-10}–10^{-12} S cm^{-1}. However, an abnormally high ionic conductivity of superionic conductors ~1 S cm^{-1} is observed at temperatures significantly lower than their melting point, which is close to the conductivity of liquid electrolytes [19].

The crystal structure of the superionic phases of copper chalcogenides and similar compounds can be regarded as a rigid sublattice composed of chalcogen atoms (S, Se, Te) and disordered (melted) cationic sublattice, over which a liquid-like charged fluid of Cu ions diffuses [19,139]. In paper [1], the strategy was proposed to decrease lattice thermal conductivity below that of a glass by reducing not only the mean free path of lattice phonons but also eliminating some of the vibrational modes completely. According to authors [1], the idea of using the liquid-like behavior of superionic conductors may be considered an extension of the phonon glass electron crystal (PGEC) concept [140] and such materials could be considered phonon liquid electron crystal (PLEC) thermoelectrics. Local atomic jumps and rearrangement of the liquid inhibit the propagation of transverse waves and disrupt heat propagation by phonons [1]. Such liquid-like behavior results in ultralow lattice thermal conductivity and, for example, in the Cu_2Se α-phase, it diminishes to 0.4–0.6 W m^{-1} K^{-1} and, as a result, provides the high ZTs around 1.3–2.1 at 727 °C [139].

Just as Drude's idea of a gas of free electrons turned out to be fruitful for describing the current in metals, so the idea of "melting" of one of the crystal sublattices can become the foundation for the construction of a unified theory of diffusion in fast ionic conductors. The crowdion mechanism of diffusion has been known for a long time, but its implementation in ordinary crystals requires a very high activation energy of about 5–6 eV, which renders it negligible. However, something similar can occur in superionic conductors (SICs) under conditions of a "molten" sublattice of mobile ions. In a recent work by X. He et al. [141], it was shown by means of ab initio modelling for several lithium-ion conductors that "fast diffusion in superionic conductors does not occur through isolated ion hopping as is typical in solids, but instead proceeds through concerted migrations of multiple ions with low energy barriers". To characterize the extent of concerted migrations, X. He et al. [141] calculated the correlation factor related to the Haven ratio. Whereas a correlation factor of 1.0 corresponds to isolated single-ion diffusion, the correlation factor is calculated as 3.0, 3.0, and 2.1 for $Li_{10}GeP_2S_{12}$, $Li_7La_3Zr_2O_{12}$, and $Li_{1.3}Al_{0.3}Ti_{1.7}(PO_4)_3$, respectively, in the AIMD simulations at 627 °C, corresponding to correlated hopping of approximately two to three ions on average in these SICs. Therefore, X. He et al. summarize that the concerted migration is the dominant mechanism for fast diffusion in SICs, as it is in liquids [142,143].

Since the crystal structure forms both the electronic and ionic properties of solids, they cannot be unrelated. Among other factors, lattice vibrations play an important role in the interaction of the electronic and ionic subsystems. Among the works on clarifying the relationship between the electronic, phonon, and ionic subsystems, in our opinion, the works of K. Wakamura [144] and H. Kikuchi et al. [145] should be noted. Wakamura [144] discusses the significant role of a narrow band gap in the formation of high ionic conductivity with low activation energy and low superionic transition temperature. He provides an analysis of experimental results for a wide class of materials, which shows the presence of a strong correlation between the above parameters. The high dielectric constant (for example, ε_∞ = 9.7 for Cu_2S at 23 °C) also correlates with high ionic conductivity, which

Wakamura explains by the screening of the Coulomb interaction between ions by electrons. Kikuchi et al. [145] investigated the relationship between the ionic conductivity and the electronic structure in some copper and silver chalcogenides. It is known that Ag_2Te has an antifluorite structure, and silver ions move along tetrahedral positions through adjacent octahedral positions. H. Kikuchi assumed that the antifluorite structure $Fm\bar{3}m$ is the ground state, and the local crystal structure $F\bar{4}3m$ can be considered as a transition state in the process of cation diffusion jumping. The diffusion jump changes the local crystal structure from $Fm3m$ to $F\bar{4}3m$. Therefore, the activation energy of diffusion can be estimated as the difference between the binding energies of cations in the two aforementioned crystal lattices. It is noted that Ag_2Te and Cu_2Te have different degrees of p-d hybridization, and the d states of silver are much weaker associated with the p states of tellurium atoms and, therefore, are less localized. The random distribution of silver in the cation sublattice has no noticeable effect on the electronic structure of the ground state in Ag_2Te. On the contrary, the band structure of Cu_2Te is sensitive to the arrangement of copper atoms in the crystal lattice, which means a strong degree of p-d hybridization in Cu_2Te. Thus, the ionic conductivity of Ag_2Te higher than that of Cu_2Te.

In the works of A.A. Lavrent'ev et al. [146] and Domashevskaya E.P. et al. [147], it was shown that, although the d-states of copper in chalcogenides are split in energy, and although they hybridize significantly with the p-states of anions, most of them fall into an energy region where there is a dip in the density of states of the anion and almost no hybridization. An increase in the symmetry of the environment of the anion leads to an increase in the fraction of copper d states not participating in hybridization. Copper selenide, which has a cubic anionic sublattice, has the lowest degree of hybridization. In our opinion, this means that copper is weakly bound to the anionic core and therefore has a low diffusion activation energy, as a result of which the high-temperature Cu_2Se phase is an excellent superionic conductor for copper ions.

For investigations of superionic TE materials, electrochemical methods are very convenient. To measure partial electronic and ionic conductivities, to control and change the chemical composition of copper chalcogenides within their homogeneity region, the electrochemical cell contains electronic probes and electrodes (Pt) and ionic probes and electrodes (Cu/CuBr) [125,126] (Figure 4). This method is commonly referred to as the Hebb–Wagner method. By passing a direct current through circuit $Cu/Cu_{2-\delta}X/CuBr/Pt$, the copper deficit δ in $Cu_{2-\delta}X$ (X = S, Se, Te) sample can be changed continuously with high precision (see above Equation (6)) and the e.m.f. E of the cell provides us with an information about the change in the chemical potential of copper atoms in the sample [122].

The voltage applied between the current electrodes at the conductivity measurements by Hebb–Wagner method must be lower than the decomposition potential of the phase under study (see Equation (16) above). An essential assumption underlying the applicability of the Hebb–Wagner method is the assumption that the chemical potential of copper ions in the sample becomes constant when a direct current is passed through the sample for a long time. To measure the electronic component of the conductivity of a mixed electronic–ionic conductor, it is necessary to measure the equilibrium value of the voltage drop across the electronic probes (Pt) when a direct current is passed through the electronic electrodes (Pt) [126]. When the current is switched on through the sample, the process of concentration polarization occurs in it, which is controlled by the chemical diffusion coefficient. For a 2 cm sample length, the equilibration time can range from a few seconds to several hours. If the ionic conductivity is comparable in magnitude to the electronic conductivity, then the error in the conductivity value using the nonequilibrium value of the potential difference between the potential probes can be large. This situation can be observed, for example, for the stoichiometric composition of copper sulfide in the superionic phase.

In order to search for ways to reduce the ionic conductivity, leading to the degradation of thermoelements, the substitution of copper in copper chalcogenides with lithium was investigated by Balapanov M. et al. [24], as well as the effect of grain sizes on the ionic conductivity [27,148]. It was found that in copper sulfide, the substitution of copper cations by lithium cations leads to a noticeable degradation in the properties of ion transport. In Table 1, parameters of ion transport in $Li_xCu_{2-x}S$ ($0 \leq x \leq 0.25$) superionic phases near a temperature of 355 °C are presented from [24].

Table 1. Parameters of ion transport in solid solutions $Li_xCu_{2-x}S$ near a temperature of 355 °C. Reprinted from reference [24]. Copyright 2003 John Wiley and Sons.

Structure	σ_i (S cm^{-1})	E_a (eV)	$\widehat{E_a}$ (eV)
Cu_2S	2.0	0.19 ± 0.02	0.23 ± 0.02
$Cu_{1.95}Li_{0.05}S$	0.57	0.50 ± 0.10	0.54 ± 0.03
$Cu_{1.90}Li_{0.10}S$	0.71	0.37 ± 0.06	0.30 ± 0.04
$Cu_{1.85}Li_{0.15}$	0.26	0.45 ± 0.04	0.51 ± 0.09
$Cu_{1.80}Li_{0.20}S$	0.20	0.59 ± 0.10	0.49 ± 0.02
$Cu_{1.75}Li_{0.25}S$	0.58	0.33 ± 0.01	0.28 ± 0.04

Figure 5 shows the temperature dependences of the ionic conductivity for a few $Li_xCu_{2-x}S$ ($x \leq 0.25$) solid solutions, from which it can be seen that the compositions $Li_{0.15}Cu_{1.85}S$ and $Li_{0.2}Cu_{1.8}S$ have the lowest ionic conductivity. If the ionic conductivity of Cu_2S at 350 °C is 2.4 S cm^{-1} at an activation energy of 0.19 eV, then for $Li_{0.15}Cu_{1.85}S$, it is 0.26 S cm^{-1} at an activation energy of 0.45 eV. A significant decrease in ionic conductivity is a positive moment for the thermoelectric application of $Li_{0.15}Cu_{1.85}S$ [149]. The authors [24] explain the concentration dependence of the ionic conductivity in the binary system Cu_2S-Li_2S with the well-known "mixed-mobile ionic effect", which mainly is observed in superionic glasses, but it takes place in solids, too [150]. The chemical diffusion coefficients also tend to diminish with increasing lithium content due to a decrease in ionic conductivity. Ionic conductivity and chemical diffusion in the studied compounds have almost close values of the activation energy (Table 1). Substitution of lithium for copper in $Cu_{2-x}Li_xS$ alloys led to a decrease in the temperature of the superionic phase transition and to an increase in the activation energy of ionic conductivity. The authors [24] suggest the absence of d-electrons in lithium as one of the reasons for the growth in the activation energy of the ionic conductivity. From an atomic crystal point of view, the reason for the deterioration of the diffusion properties of copper sulfide and copper selenide upon substitution with lithium is that lithium ions sadly occupy intermediate $32f_{II}$ positions connecting tetrahedral and octahedral interstices in the anionic framework of the crystal lattice, through which channels of fast diffusion of cations pass. Overlapping of fast diffusion channels leads to an increase in the diffusion activation energy and a decrease in the cation mobility [26].

In works [151,152], it was found that copper and silver ions make comparable contributions to the total ionic conductivity of Cu_2X-Ag_2X solid solutions (where X = S, Se, Te). It was found that substitution by silver results in increasing ionic conductivity [151–153] of copper chalcogenides, while substitution by lithium leads to a strong decreasing in the ionic conductivity [22,24,26]. Balapanov M.Kh. et al. [154] studied the thermal diffusion of silver atoms and the Soret effect in $Ag_{(2-x)+\delta}Cu_xSe$ alloys (x = 0.1, 0.2, 0.4). The Soret effect (the gradient of metal concentration along the sample in a thermal field) was measured as $d\delta/dT = (0.2 \div 0.6) \times 10^{-5}$ K^{-1}. It was found that the heat of ionic transfer is much higher than the heat of electron transfer and is close to the activation energy of ionic conductivity.

In the work [27], the ionic conduction and the chemical diffusion in dependence on average grain sizes of $Cu_{1.75}Se$, $Li_{0.25}Cu_{1.75}Se$, and $Li_{0.25}Cu_{1.75}S$ samples at 140–240 °C were studied. The ionic conduction of $Cu_{1.75}Se$ is shown to increase with the average grain size. With the coarsening of grains, the activation energy for the ionic conduction decreases in $Cu_{1.75}Se$ and increases in $Li_{0.25}Cu_{1.75}S$. Usually for solids, grain boundary diffusion exhibits lower activation energy than bulk diffusion. The increase in the ionic conduction with the grain size in $Cu_{1.75}Se$ is explained by Balapanov M. [27] as the activation energy of cation diffusion through the grain bulk being lower than that for diffusion along the intergrain layers. Whereas in superionics with a high activation energy of the diffusion, grain boundaries provide accelerated diffusion, in the superionic conductor of the channel type with a low activation energy of diffusion throughout the grain bulk, the opposite occurs, since the fast diffusion channels are interrupted at the grain boundaries. Low activation energy here can be roughly considered E_a ~0.15 eV (E_a = 0.14 eV for $Cu_{1.75}Se$), for high conditionally, we can assume E_a ~0.40 eV and higher (E_a = 0.39 eV for $Li_{0.25}Cu_{1.75}S$). E_a value for the grain boundary diffusion in studied copper chalcogenides obviously lies between the above determined low and high activation energy of diffusion.

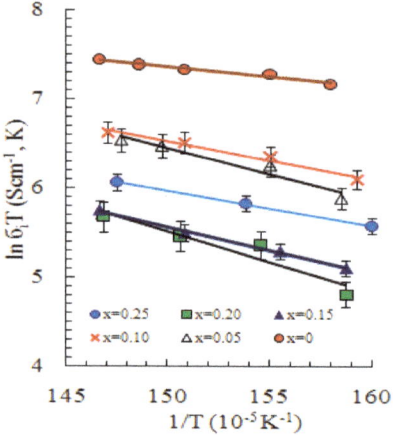

Figure 5. Temperature dependences of the ionic conductivity of Cu_2S and $Li_xCu_{2-x}S$ samples. Reprinted from reference [24]. Copyright 2003 John Wiley and Sons.

Thus, copper sulfide and selenide can be attributed to special superionic conductors, in which mobile ions move along the volume more easily than along grain boundaries. This leads to the conclusion that for thermoelectric applications at temperatures lower than ~200 °C, it is more advantageous to use copper sulfide and selenide with nanosized grains in order to reduce unwanted ionic conductivity.

Ionic conductivity and diffusion in sodium-doped copper chalcogenides has been poorly studied, if we do not take into account the works on the study of copper sulfide and selenide as active electrodes of sodium-ion batteries. In the work of M. Balapanov et al. [38], it is reported that the ionic conductivity of $Na_{0.2}Cu_{1.8}S$ is about 2 S cm^{-1} at 400 °C (the activation energy $E_a \approx 0.21$ eV). The Seebeck ionic coefficient has high values of $0.3 \div 0.4$ mV K^{-1}. The thermal conductivity of superionic $Na_{0.2}Cu_{1.8}S$ is low and provides high values of the dimensionless thermoelectric figure of merit ZT from 0.4 to 1 at temperatures from 150 °C to 340 °C.

High ionic and electronic conductivity of copper chalcogenides are very attractive to apply in chemical current sources. Sodium ion batteries (NIB) can become an alternative to lithium-ion batteries (LIB). Sodium is the fifth most abundant metal in the earth's crust (2.27%). World prices for the main source of sodium—sodium carbonate—are ~20–30 times lower than the prices for lithium carbonate, the main raw material for the production of LIB components [155,156].

When the battery is operating, sodium ions are extracted from the material of the negative electrode and are embedded in the matrix of the positive electrode; during charging, the directions of the processes change. In works [157,158], the excellent properties of copper selenide Cu_2Se as a cathode material for NIB are shown. The charge plateau is about 2.02 V during the charging process. The potential gap between the discharge and charge curves is only 0.1 V, which indicates very low polarization. The initial discharge capacity of the thin-film electrode is 253.0 mAh/g, and the charging capacity is 196.6 mAh/g. After 100 cycles, the discharge capacity was 113.6 mAh/g. The discharge power at 0.1 C, 0.5 C, and 2 C is 251.4, 122.8, and 90.8 mAh/g, respectively. These results indicate that Cu_2Se is suitable for NIB use in fast charge/discharge mode. The electrochemical system sodium/copper sulfide (Na/Cu_2S) was investigated by Kim J.-S. [159] using the 1M $NaCF_3SO_3$-TEGDME electrolyte. The first discharge curve of Na/Cu_2S cells shows an oblique shape without a plateau potential region. The first discharge capacity is 294 mAh/g and decreases to 220 mAh/g after 20 cycles. The discharge process is explained by the intercalation of sodium into the Cu_2S phase without separation of the Cu_2S phases. Perhaps, high diffusion coefficients of copper sulfide and copper selenide will help to solve the serious problem of the sodium-ion electrochemical system, which is its very long charge/discharge time, because NIBs cannot yet operate at high current densities.

3.2. Seebeck Coefficient and Thermal Conductivity

3.2.1. Seebeck Effect

In the Seebeck effect, the action of a temperature gradient in a material creates an electromotive force (e.m.f.). The ratio of this e.m.f. to the applied temperature difference is the Seebeck coefficient of the material. From a physical point of view, the Seebeck coefficient is the entropy transferred by a charge carrier during isothermal current flow, divided by the charge of the carrier [160]. Thus, the electronic Seebeck coefficient is directly related to the Fermi level of electrons. Usually, reduced Fermi level $\mu^* = (\mu - E_c)/k_B T$ and reduced energy of electrons $\varepsilon^* = (E - E_c)/k_B T$ are used in semiconductor physics, where E_c is the carrier energy corresponding to bottom of conductivity band. For nondegenerate semiconductors ($\mu^* \ll 1$), the Pisarenko formula for the Seebeck coefficient can be used [161]. In the case of carrier scattering by acoustic phonons in semiconductor with parabolic band, it looks as:

$$\alpha_L = \frac{k_B}{e}(2 - \mu^*). \tag{20}$$

For carrier scattering by impurity ions, the Pisarenko formula is written as:

$$\alpha_i = \frac{k_B}{e}(4 - \mu^*). \tag{21}$$

For degenerate semiconductors ($\mu^* > 1$) with a parabolic band, the Seebeck coefficient can be calculated as [161]:

$$\alpha_L = \frac{k_B}{e}\left(\frac{2F_1(\mu^*)}{F_0(\mu^*)} - \mu^*\right); \alpha_i = \frac{k_B}{e}\left(\frac{4F_3(\mu^*)}{3F_2(\mu^*)} - \mu^*\right). \tag{22}$$

In Equation (22), F_0, F_1, F_2, and F_3 are Fermi integrals, determined by general formula:

$$F_n = \int_0^\infty \frac{\varepsilon^{*n} d\varepsilon^*}{1 + \exp(\varepsilon^* - \mu^*)}. \tag{23}$$

For metallic systems ($\mu^* \gg 1$) with a parabolic band, the Seebeck coefficient can be calculated as [161]:

$$\alpha_L = \frac{k_B}{e}\frac{\pi^2}{3}\frac{1}{\mu^*}; \alpha_i = \frac{k_B}{e}\frac{\pi^2 \mu^*}{\pi^2 + \mu^{*2}}. \tag{24}$$

The Seebeck coefficient in an impurity semiconductor with a parabolic band is proportional to the effective carrier mass and temperature and decreases with increasing carrier concentration (see Equation (15) above).

The Seebeck coefficient, thus, directly depends on the position of the Fermi level and on the concentration of charge carriers. The electron gas in metals is in a degenerate state; therefore, the Fermi level, the energy, and the electron velocity are weakly dependent on temperature. As a consequence, the values of the Seebeck coefficient are small. The Seebeck coefficient reaches relatively large values in semimetals and their alloys, where the carrier concentration is lower and depends on temperature. In this case, the concentration of electrons is high, nevertheless, the Seebeck coefficient is large due to the fact that the average energy of conduction electrons differs from the Fermi energy. Sometimes, fast electrons have a lower diffusion capacity than slow ones, and the Seebeck coefficient changes sign accordingly. The magnitude and sign of the Seebeck coefficient also depend on the shape of the Fermi surface. In some metals and alloys with a complex Fermi surface, various parts of the latter can give opposite sign to the thermo-e.m.f., and the Seebeck coefficient can be equal to or close to zero [162,163].

The Seebeck coefficient in semiconductors strongly depends on the scattering mechanism of charge carriers; it is seen from comparing Equations (21) and (22), for instance. In real crystals, several scattering mechanisms usually operate simultaneously. The contribution of each type of scattering can vary greatly depending on the temperature and concentration of impurities in the sample [162,163]. Accordingly, Formulas (21) and (22) heavy doping could lead to increasing the Seebeck coefficient at the same carrier's concentration.

Binary copper and silver chalcogenides have excellent thermoelectric characteristics, which continue to be the subjects of intense research [2,4,82,83,109,139,164–167]. Over the past 5 years, as examples, we can mention the works [12,14,17,31,39,99,106,129,168,169] on $Cu_{2-\delta}S$, the works [15,16,165,167,170–178] on $Cu_{2-\delta}Se$, the works [110,179–185] on $Cu_{2-\delta}Te$, the works [186–190] on $Ag_{2+\delta}Se$, the works [191–194] on $Ag_{2+\delta}S$, and the works [195–199] on $Ag_{2\pm\delta}Te$.

Some high achievements in thermoelectric performance of superionic chalcogenides and a number of other thermoelectric materials are shown in Figure 2 above and in Table 2 below.

Table 2. State-of-art thermoelectric properties of some perspective materials.

Materials	Year	Synthesis	α, mVK^{-1}	σ, S cm^{-1}	$PF = \alpha^2\sigma$, μWcm^{-1}K^{-2}	k, Wm^{-1}K^{-1}	T, °C	ZT_{max}	Ref.
$Cu_{1.98}S_{1/3}Se_{1/3}Te_{1/3}$ (p-type)	2017	Melting + annealing + SPS	0.243	182	10.7	0.57	727	1.9	[179]
$Cu_{1.97}S$ (p-type)	2014	Melting + annealing + SPS	0.3	100	8.2	0.48	727	1.7	[84]
$Na_{0.01}Cu_{1.80}S$ (p-type)	2016	Mechanical Alloying + SPS	0.110	850	10.5	0.7	500	1.1	[31]
$Na_{0.04}Cu_{1.96}Se$ + micropores (p-type)	2019	Hydrothermal method + HP	~0.29	~130	~11	~0.54	700	2.1	[53]
$Cu_{1.98}Li_{0.02}S$ + nanopores	2018	Hydrothermal method + HP	~0.27	~145	~10.6	0.48	700	2.14	[200]
$Cu_{1.94}Al_{0.02}Se$	2014	Melting + BM + SPS	0.246	261	15.8	0.611	756	2.62	[177]
Cu_2Se + 1 mol% $CuInSe_2$ (p-type)	2017	BM + SPS	0.15	550	12.4	0.4	577	2.63	[85]
$Cu_{1.94}Se_{0.5}S_{0.5}$ (p-type)	2017	Melting + annealing + SPS + HP	0.37	~96	13.2	~0.6	727	2.3	[86]
$Cu_{2-x}S$ + 0.75 wt% Grapheme (p-type)	2018	BM+SPS + annealing in 95 vol% Ar and 5 vol% H_2	~0.16	~450	~12	0.67	600	~1.5	[169]
$Ag_2Sb_{0.02}Te_{0.98}$ (n-type)	2020	Vacuum melting + +annealing at 1000 °C and 400 °C + cooling at −173 °C	~0.106	~870	~9.8	~0.29	137	1.4	[166]
Cu_2Se + 0.3 wt.% carbon fiber (p-type)	2017	Solid state Cu_2Se synthesis + BM + CP + annealing	0.175	375	11.5	0.4	577	2.4	[201]
$Bi_{0.5}Sb_{1.5}Te_3$ + x% Te (p-type)	2015	Liquid-phase compacting + SPS	~0.24	~650	~37	~0.65	47	1.86	[87]
$Pb_{0.940}Mg_{0.020}Na_{0.040}Te$ (p-type)	2016	Solid state synthesis from elements + SPS	~0.24	~400	~23	~1	577	1.8	[89]
$PbTe$ + 0.2% PbI_2 (n-type)	2016	Solid state synthesis from elements + SPS	~0.23	~330	~20	~1.2	477	1.4	[89]

Table 2. Cont.

Materials	Year	Synthesis	α, mVK^{-1}	σ, S cm^{-1}	$PF = \alpha^2\sigma$, μWcm^{-1}K^{-2}	k, Wm^{-1}K^{-1}	T, °C	ZT_{max}	Ref.
Ta$_{0.74}$V$_{0.1}$Ti$_{0.16}$FeSb (p-type)	2019	BM + HP	~0.225	~1040	~52	~3.35	700	1.52	[88]
MgAg$_{0.965}$Ni$_{0.005}$Sb$_{0.99}$ (p-type)	2015	BM + HP	~0.235	~450	~25	~1.1	245	~1.15	[90]
SnSe$_{1-x}$Br$_x$ (n-type)	2018	The temperature gradient method and bromine doping	~0.48	~38	~9	~0.245	500	2.8	[91]
Bi$_2$Te$_{2.79}$Se$_{0.21}$ (n-type)	2015	Zone melting + BM + HP + hot deformation	~0.192	~970	~36	~1.1	84	1.2	[92]

BM is ball milling, SPS is spark plasma sintering, HP is hot pressing, CP is cold pressing, T is temperature, α is Seebeck coefficient, σ is electrical conductivity, k is thermal conductivity, PF is power factor, and ZT_{max} is the peak value of figure of merit.

Cu$_2$Se

Resent achievements on Cu$_2$Se-based thermoelectrics are carefully described in the review of Liu W.-D. et al., published in 2020 [165]. They state that very high ZT values close to 2 or higher have been achieved for Cu$_2$Se thermoelectric materials within the temperature range from ~600 to ~700 °C [165]. The most significant achievement noted in their review is $ZT = 2.6$ at 577 °C for copper selenide doped with 1% CuInSe$_2$ in 2017 [85]. In addition to describing thermoelectric performance, Liu W.-D. et al. summarize fundamentals of Cu$_2$Se, including crystal structure, band structure, phase transition, and other aspects. Considering the abundance of Cu and Se and their high ZT values, Cu$_2$Se-based thermoelectric materials are highly promising alternatives for the toxic PbTe thermoelectric materials, presume Liu W.-D. et al. [165].

It is well known that Cu$_2$Se has an excellent intrinsic TE performance, however, practice shows that alloying with other elements can greatly improve the thermoelectric characteristics of copper selenide. Doping effect on Cu$_2$Se thermoelectric performance was reviewed in 2020 by Y. Qin et al. [202]. The authors summed that through doping, many elements, such as Al, Li, Na, In, etc., have caused the ZT value of Cu$_2$Se to exceed 2, and almost break through the minimum requirement of commercial use ($ZT > 3$). It is found that the high TE performance mainly originates from the fact that dopants can form point defects and dislocations, bring the mass fluctuation and strain fluctuation into the lattice, shift the microstructures, and so on, all of which play an important role in significantly scattering phonons and carriers (holes for Cu$_2$Se, specifically), reducing the thermal conductivity below the glassy limit. Further, Y. Qin et al. suggest that, when doping Cu$_2$Se, researchers can pay attention to the manipulation of microstructures, the anomaly at phase transition, and the avoidance of Cu deficiency, with which higher ZT values may be produced. In our opinion, the deficiency of copper in copper selenide should not be completely avoided, but it should be optimized, otherwise, the conductivity of the material may be insufficient to obtain the required thermoelectric power $\alpha^2 \sigma$.

In recent years, copper selenide has literally become a testing ground for many new ideas of researchers in thermoelectric materials. For example, D. Byeon et al. [175] in 2019 discovered the colossal Seebeck effect at the superionic phase transition in copper selenide, which was placed in a thermal field with two perpendicular temperature gradients (horizontal and vertical). The authors observed that Cu$_2$Se shows two sign reversals and colossal values of α exceeding ± 2 mV K^{-1} in a narrow temperature range, 67 °C < t < 127 °C, where a structure phase transition takes place. The metallic behavior of σ possessing larger magnitude exceeding 600 S cm^{-1} leads to a colossal value of power factor $\alpha^2 \sigma = 2.3$ Wm^{-1} K^{-2}. The small thermal conductivity less than 2 Wm^{-1} K^{-1} results in a huge dimensionless figure of merit exceeding 400.

For interpreting the unusual Seebeck coefficient, D. Byeon et al. [175] assumed that the chemical potentials of copper ions and conduction electrons in the low-temperature phase could be different from those in the high-temperature phase. In such a case, copper ions and electrons slightly move from one of the phases to the other so as to reach the energy equilibrium between both phases. The effect of the chemical potentials should change the carrier concentration of each phase. With an increasing temperature, the sample bottom starts transforming to the high-temperature phase, leading to an increase in the electron concentration in the low-temperature phase situated above due to the difference between the chemical potentials of electrons and/or copper ions. The number of electrons in the low-temperature phase in the top surface is further enlarged with the increased volume fraction of the high-temperature phase. Under the effect of the chemical potentials on the carrier concentration, the positive Seebeck coefficient of the low-temperature phase below 47 °C becomes negative at around 57 °C and reaches -4347 µV K^{-1} at 74 °C. In the higher temperature range above 74 °C, the chemical potential of either electrons or copper ions leads to a reduction in the electron concentration in the low-temperature phase, and the positive peak of the Seebeck coefficient of 1982 µV K^{-1} at 76 °C is created. In addition, one of the factors would be sustained in the temperature range of 77–97 °C, forming the plateau of the Seebeck coefficient (+220 µV K^{-1}). The rather large value of $\sigma(T)$ together with the large magnitude of the Seebeck coefficient naturally leads to surprisingly large values of power factor at the peaks: 0.18–2.3 W m^{-1} K^{-2} and 0.06–0.5 W m^{-1} K^{-2} for n-type and p-type, respectively. These values are definitely much larger than a few mW m^{-1} K^{-2} of typical thermoelectric materials.

The Equation (1) leading to the efficiency of energy conversion in a thermoelectric generator, η, that increases with an increasing ZT, was derived on the basis of a model involving π-type junctions comprised of two thermoelectric materials. In this scenario, a temperature gradient is applied to the thermoelectric materials simply along the direction of electrical current. The temperature distribution of the samples, in the setup introduced in work of D. Byeon et al. [175], was certainly different from the case of the π-type module. Therefore, ZT is no longer valid to estimate η in the case of two perpendicular temperature gradients, D. Byeon et al. [175] note. Nevertheless, it should be emphasized that the large $\alpha(T)$, large $\sigma(T)$, and consequently obtained colossal PF, as measured using the same electrodes, can be certainly applicable. Of course, the narrowness of the temperature range in which the colossal Seebeck effect occurs and the condition of the presence of two cross-over temperature gradients will complicate the practical application of this interesting effect in thermoelectric devices. However, sometime technical problems will be solved, as, for example, it was with the use of liquid crystals in monitors, and, perhaps, a thermoelectric device based on this effect will be equally widespread.

$Cu_{2-\delta}S$

In 2014, He et al. [84] investigated the TE performances of Cu_xS (x = 1.97, 1.98, and 2) and achieved very high ZTs of 1.4–1.7 at 727 °C. Similar with the β-Cu_2Se phase with the character of "phonon-liquid electron-crystal", [1] such high ZTs are mainly contributed by the ultralow lattice thermal conductivities (about 0.3–0.5 W m^{-1} K^{-1}) caused by the liquid-like Cu ions. TE properties of copper sulfides with the Cu/S atomic ratios between 1.8 and 1.97 were studied in 2016 by P. Qiu et al. [203]. They reported results of investigations on crystal structures, valence states of elements, and thermoelectric properties of the compounds prepared by melting from elements at 1000 °C followed by hot pressing from fine powder. It was shown that the valence state of copper in these binary compounds does not change, and the thermoelectric properties were found to be very sensitive to copper deficiency. In addition, they reveal that the arrangement of sulfur in the crystal structure also plays an important role in electrical transfer. The optimal compositions of $Cu_{2-x}S$ were determined to obtain a high power factor and thermoelectric figure of merit. The crystal structure of the investigated compositions, namely the valence states of copper and sulfur in all compositions, had similar values. At 473 °C, the maximum power factor was 12.5 µW cm^{-1} K^{-2} in $Cu_{2-x}S$ (2−x = 1.90–1.92), and the maximum ZT = 0.8 was observed for the $Cu_{1.96}S$ composition with total thermal conductivity $\kappa \approx 1$ W m^{-1} K^{-1}. Due to the lower κ, $Cu_{1.97}S$ and $Cu_{1.98}S$ reported by He et al. [84] possess higher ZT than $Cu_{1.96}S$ at 477 °C in [203].

Taking into account its own ultra-low lattice thermal conductivity, an increase in the performance of TE in Cu_2S can be achieved by improving its electrical transport properties. Nanosized forms of copper sulfides (platelets, discs, rods, and others) promote their use at a more advanced level, adjusting their properties depending on the shape and size of the particles of materials. Large-scale Cu_2S tetradecahedrons microcrystals and sheet-like Cu_2S nanocrystals were synthesized by employing a hydrothermal synthesis (HS) method and wet chemistry method (WCM), respectively, by Yun-Qiao Tang et al. [14] in 2017. The polycrystalline copper sulfides bulk materials were obtained by densifying the as-prepared powders using the spark plasma sintering (SPS) technique. The pure Cu_2S bulk samples sintered by using the powders prepared via HS reached the significant thermoelectric figure of merit (ZT) value of 0.38 at 300 °C. Contrary to hydrothermal sintered Cu_2S bulk samples, the highly dense $Cu_{1.97}S$ bulks, fabricated by a melt-solidification technique by L. Zhao et al. [17], showed higher thermoelectric performance with ZT of 1.9 at 697 °C.

Often, introducing suitable impurities helps to enhance useful properties of the TE material. Guan M. et al. [129] carried out doping of Cu_2S with lithium. Series of $Cu_{2-x}Li_xS$ samples with various Li contents (x = 0, 0.005, 0.010, 0.050, and 0.100) were synthesized by fusion with subsequent annealing. At the composition of $x = 0.05$, the sample was more stable, and homogeneous phases with the same monoclinic structure as Cu_2S at room temperature were found. It was revealed that the electrical conductivity in the $Cu_{2-x}Li_xS$ samples is significantly improved by the introduction of Li; the electrical conductivity increases due to the increased concentration of carriers. The maximal figure of merit $ZT = 0.84$ was achieved for $Cu_{1.99}Li_{0.01}S$ composition at 627 °C, about a 133% improvement as compared with that in Cu_2S matrix.

Work on doping copper sulfide and copper selenide with lithium should be continued with the choice of other compositions of the initial chalcogenide, since the possibilities of increasing the thermoelectric power of the material seem to be higher, in our opinion, than was achieved in the described works. This was also shown in 2018 by the subsequent work of Hu et al. [200], in which for the $Li_{0.02}Cu_{1.98}S$ composition, the value $ZT = 2.14$ was achieved, which exceeds the indices of pure copper sulfide.

The use of the electrochemical method of doping with lithium made it possible to obtain and study solid solutions $Li_xCu_{2-x}S$ ($0 < x < 0.25$) and $Li_xCu_{(2-x)-\delta}Se$ ($x \leq 0.25$), which are promising for thermoelectric applications [23,24]. The authors of [149] investigated a semiconductor alloy with the composition $Li_{0.15}Cu_{1.85}S$ with an ionic conductivity 8–10 times lower than that of pure copper sulfide, characterized in that it was obtained by cold pressing from $Li_{0.15}Cu_{1.85}S$ nanopowder. At room temperature, the alloy is heterophase, consisting of an orthorhombic $Cu_{1.75}S$ phase, a tetragonal $Cu_{1.96}S$ phase, a hexagonal Cu_2S phase, and a cubic Cu_2S phase.

$Cu_{2-\delta}Te$

Pure copper telluride demonstrates more modest thermoelectric characteristics compared to copper selenide and sulfide. It may be caused the fact that after the same SPS sintering, copper telluride $Cu_{2-\delta}Te$ usually has higher electrical conductivity and lower thermopower in comparison with $Cu_{2-\delta}Se$ or $Cu_{2-\delta}S$, owing to its greater copper deficiency [110].

He Y. at al. [110] in 2015 investigated high-density Cu_2Te samples obtained using direct annealing without a sintering process (SPS). In the absence of sintering processes, the samples' compositions could be well controlled, leading to substantially reduced carrier concentrations that are close to the optimal value. The electrical transports were optimized, and the best power factor was achieved near 1300 $\mu W\ m^{-1}\ K^{-2}$, which is more than a 30% enhancement when compared with the sample sintered using SPS. The minimal thermal conductivity at 327 °C was reduced to ~0.7 $W\ m^{-1}\ K^{-1}$ for direct annealing sample instead of ~1.2 $W\ m^{-1}\ K^{-1}$ for SPS sintered sample. The ZT values were significantly improved by He Y. at al. [110] to 1.1 at 727 °C, which is nearly a 100% improvement compared with SPS sintered sample. Furthermore, He Y. at al. note that this method saves substantial time and cost during the sample's growth.

Cu_2Te/Te nanorod composites were fabricated by ultrasonication in ethanol from a mixture of Te and Cu_2Te nanorods and their thermoelectric properties were investigated by D. Park et al. [181]. The thermoelectric power factor of the Cu_2Te/Te nanorod composites at room temperature was the highest (431 $\mu V\ K^{-1}$) at 10 wt% Te, and the enhancement in power factor is achieved to ~440 $\mu W\ m^{-1}\ K^{-2}$ for this composition. The authors note that a great reduction in the total thermal conductivity κ was caused by the strong phonon scattering effect owing to a 1D nanostructure of the composites. Because of the enhancement in power factor and decrease in thermal conductivity, the composite samples showed an enhanced thermoelectric figure of merit $ZT = 0.22$ at room temperature, which was achieved for a Te content of 10 wt% and was ~4.5 times larger than that of the pristine Cu_2Te nanorods. D. Park et al. highlight the fact that the composites with the two types of nanorods form Cu_2Te/Te homo-interfaces and show a lower κ value than the two types of pristine nanorods, because they form an effective phonon-scattering center. This is one of many results that support our thesis about the desirability of greater attention of researchers to composite thermoelectric materials, including due to possible positive synergetic effects.

Mukherjee S. et al. [182] studied thermoelectric properties of $Cu_{2-x}Fe_xTe$ solid solutions until the solubility limit of Fe ~x = 0.05 and obtained a maximum figure of merit ZT ~0.16 at 477 °C for $Cu_{1.97}Fe_{0.03}Te$ composition. The authors revealed that the specific heat capacity C_p of Fe alloyed samples higher than the Dulong–Petit limit of 3 Nk_B, and the minimum thermal conductivity (~2.19 W m^{-1} K^{-1}) is high for superionic conductor. Mukherjee S. et al. [182] supposed that the superionic "liquid-like" behavior of the Cu ions is suppressed by the presence of Fe, which is rather probable. On the contrary, doping with a more pronounced superionic nanostructured Ag_2Te conductor made it possible for Mukherjee S. et al., in their next work [183], to sharply reduce the thermal conductivity of bulk copper telluride from ~3.5 W m^{-1} K^{-1} to ~0.26 W m^{-1} K^{-1} and achieve the high ZT ~0.99 at 315 °C for $(Cu_2Te)_{50.00}$-$(Ag_2Te)_{50.00}$ composition. The authors explain the achievement as due to the additional scattering of carriers as well as phonons by the Ag_2Te nanostructures. It can be clarified here that, possibly, the thermal conductivity strongly decreases both due to the liquid-like state of the lattice, since silver telluride is a fast ionic conductor (the ionic conductivity ~1.6 S cm^{-1} at 300 °C, the activation energy of the ionic conductivity E_a = 0.13 eV) [184], and due to additional scattering of phonons at interphase boundaries, the area of which greatly increases in the presence of silver telluride nanoparticles.

After long attempts, ZT was also brought to the level of 1.5 for copper tellurides. This was achieved by the group of Zhao K. et al. [185]. Doping with silver telluride also brought success. It is demonstrated that the Cu_2Te-based compounds are also excellent TE materials if Cu deficiency is sufficiently suppressed. By introducing Ag_2Te into Cu_2Te, the carrier concentration is substantially reduced to significantly improve the ZT with a record-high value of 1.8, a 323% improvement over Cu_2Te, and outperforming any other Cu_2Te-based materials. The single parabolic band model is used in [185] to prove that all Cu_2X-based compounds are excellent TE materials.

The highest ZT reported by Zhao et al. [185] for Cu_2Te 50% Ag_2Te is ~1.8 at 727 °C. Similarly, inclusion of nanosized Ag_2Se into Cu_2Se increased the ZT to ~0.9 at 315 °C and to ~1.85 at 527 °C, according to report of Ballikaya et al. [176]. In the work of Mukherjee S. [183], the incorporation of nanostructured Ag_2Te into bulk Cu_2Te obtained by the melting route also improved the thermoelectric properties of Cu_2Te significantly. The value ZT = 0.99 at 315 °C obtained in [183] for $(Cu_2Te)_{100-x}$-$(Ag_2Te)_x$ composites (where x varies from 0 to 50) is sufficiently higher than that reported for the alloy Cu_2Te 50% Ag_2Te by Zhao et al. (ZT ~0.79) at the same temperature [185]. Thus, the mechanical alloying of Cu_2Te with nanostructured Ag_2Te, as followed in the [183], is an effective method in enhancing the TE performance of Cu_2Te compared to the conventional vacuum melting route. Given the results of Zhao et al. [185], Ballikaya et al. [176], and Mukherjee S. [183], one can conclude that Cu_2X-Ag_2X (X = S, Se, and Te) composites are promising thermoelectric materials for mid-temperature applications.

Sodium doping in copper chalcogenides was studied in works [31,36–40,53,138]. Copper chalcogenides with a high sodium content are not homogeneous and form a mixture of phases with different conducting and thermoelectric properties, but deserve attention as nanocomposite thermoelectric materials with low thermal conductivity. The properties of such mixtures are poorly understood; however, we expect new effects here associated with the presence of numerous barrier layers and pores.

In alloys of the Na-Cu-S system, the results show that Na plays two important roles: the first reduces the concentration of carriers, thereby improving the Seebeck coefficient, the other reduces the thermal conductivity, which is generally favorable, improving the characteristics of the thermoelectric figure of merit [62,138].

Potassium doping in copper selenide was studied by Z. Zhu et al. [171] using hydrothermal synthesis and hot pressing. Numerous micro-pores were introduced by K doping, together with reduced electronic conductivity that result in low thermal conductivity. For the nominal component $Cu_{1.97}K_{0.03}Se$ (EPMA measured composition $Cu_{1.99}K_{0.01}Se$), the peak value of ZT reaches 1.19 at 500 °C, which is 47% larger than that of pure Cu_2Se (ZT_{max} = 0.81).

For comparison, we present several works on other groups of thermoelectric materials. Bismuth telluride-based solid solutions and bismuth telluride nanocrystals also remain in field of attention [92,98].

Half-Heusler alloys are intensively studied to improve thermoelectric performance [111–113]. Half-Heusler phases (space group $\bar{4}3m$, C1b) have recently captured much attention as promising thermoelectric materials for heat-to-electric power conversion in the mid-to-high temperature range. The most studied ones are the RNiSn-type half-Heusler compounds, where R represents refractory metals Hf, Zr, and Ti. These compounds have shown a high-power factor and high-power density, as well as good material stability and scalability. Due to their high thermal conductivity, however, the dimensionless figure of merit (ZT) of these materials has stagnated near 1 for a long time. Since 2013, the verifiable ZT of half-Heusler compounds has risen from 1 to near 1.5 for both n- and p-type compounds in the temperature range of 500–900 °C [111].

Skutterudites represent a promising group of thermoelectric materials [83,114,116,117]. Skutterudites are bulk TE materials with the crystal structure belonging to cubic space group $Im3$. They contain vacancies into which low-coordination ions (usually rare earth elements) can be inserted to decrease the thermal conductivity by enhancing phonon scattering without reducing electrical conductivity. Such a structure makes them exhibit PGEC (phonon glass electron crystal) behavior. ZT of skutterudites can be significantly improved by double, triple, and multiple filling of elements into the vacancies in their structure. Usually, alkali metals, alkaline earth metals, lanthanides, and similar elements are selected as "fillers" or "dopants" for skutterudites because of their moderate atom size. Hence, the selection and combination of different "fillers" is very crucial for achieving high TE performance. Replacement of host atoms in lattice sites, such as substitution for Co and Fe, has also been identified as an effective way to reduce thermal conductivity [164].

J. Chen et al. [190] in 2020 wrote, owing to the intrinsically good near-room-temperature thermoelectric performance, orthorhombic β-Ag_2Se has been considered as a promising alternative to n-type Bi_2Te_3 thermoelectric materials. Herein, we develop an energy- and time-efficient wet mechanical alloying and spark plasma sintering method to prepare porous β-Ag_2Se with hierarchical structures including high-density pores, a metastable phase, nanosized grains, semi-coherent grain boundaries, high-density dislocations, and localized strains, leading to an ultralow lattice thermal conductivity of ~0.35 W m^{-1} K^{-1} at 27 °C. Relatively high carrier mobility is obtained by adjusting the sintering temperature to obtain pores with an average size of ~260 nm, therefore resulting in a figure of merit, ZT, of ~0.7 at 27 °C and ~0.9 at 117 °C. The single parabolic band model predicts that ZT of such porous β-Ag_2Se can reach ~1.1 at 27 °C if the carrier concentration can be tuned to ~1 × 10^{18} cm^{-3}, suggesting that β-Ag_2Se can be a competitive candidate for room-temperature thermoelectric applications.

In the end of this section, let us mention the unusual thermoelectric studies that were published in recent years. G. Kim et al. reported in [196] about a new interesting method of control of the thermoelectric properties of polycrystalline Ag_2S by spatial-phase separation into low- and high-temperature phases using a configuration based on bottom heating and top measurement. The authors [196] experimentally confirmed that the Seebeck coefficient was determined by the low-temperature phase at the top surface but the electrical resistivity was dominated by the high-temperature phase lying below the low-temperature phase. As a result, a high Seebeck coefficient ~−0.650 mV K^{-1} and high conductivity ~500 S cm^{-1} were simultaneously observed over a broad temperature range (17–167 °C). The authors [196] hope that this idea suggests a new concept for thermoelectric materials and devices. Of course, the technical implementation of such a thermoelectric device is a big problem, but the observed effect is impressive and deserves attention. There is some similarity of this new approach to increasing the thermoelectric efficiency of materials with another original work by Japanese researchers [175], in which the use of a second transverse temperature gradient allows one to obtain a colossal thermoelectric effect near the temperature of the superionic phase transition in copper selenide.

In addition to ordinary rigid TE, the flexible thermoelectric materials are objects of greatly increasing attention in the past decade [167], and a variety of conductive polymers, carbon materials, nano-sized inorganic semiconductors, and metals have been applied to fabricate flexible thermoelectric films. So far, in spite of their relatively poor flexibility, the nano-sized inorganic semiconductors are superior to their counterparts because of their excellent thermo-electric performance [186]. For example, silver selenide is considered as a promising room-temperature thermoelectric material due to its excellent performance and high abundance. However, the silver selenide-based flexible film is still behind in thermoelectric performance compared with its bulk counterpart. In work by J. Gao et al. [186], the composition of paper-supported silver selenide film was successfully modulated through changing reactant ratio and annealing treatment. In consequence, the power factor value of 2450.9 ± 364.4 mW/(mK^2) at 30 °C, which is close to that of state-of-the-art bulk Ag$_2$Se, has been achieved. On base of the film, the thermoelectric device was established. At a temperature difference of 25 °C, the maximum power density of this device reaches 5.80 W/m^2, which is superior to that of previous film thermoelectric devices. Similar investigations of silver selenide were performed by Ding Y. et al. [187] and Perez-Taborda J.A. et al. [189]. Recently, Liang J. et al. [204] reported high intrinsic flexibility and state-of-the-art figures of merit (up to 0.44 at 27 °C and 0.63 at 177 °C) in Ag$_2$S-based inorganic materials, opening a new avenue of flexible thermoelectrics. In the flexible full-inorganic devices composed of such Ag$_2$S-based materials, high electrical mobility yielded a normalized maximum power density up to 0.08 W m^{-1} under a temperature difference of 20 °C near room temperature, orders of magnitude higher than organic devices and organic–inorganic hybrid devices.

Theoretically and experimentally, these works exerted a significant effort to achieve silver chalcogenide-based flexible thermoelectric films and devices with high performance to provide electricity to electronic devices for an individual's timely healthcare, real-time safety monitoring, and life improvement by utilizing the temperature difference between the skin and the ambient environment, all-weather, regardless of the motion state of the human body.

3.2.2. Thermal Conductivity

The corpuscular or phonon approach to the consideration of lattice vibrations is especially convenient in the study of energy conversion processes. These processes include the processes of phonon creation and annihilation. It is most convenient to describe thermal conductivity in terms of phonon scattering by other phonons, static imperfections of the lattice, or by electrons [124]. The electronic structure and dynamics of the lattice are of particular importance for thermoelectric materials. The work of Bikkulova N.N. [205] first studied the lattice dynamics in the superionic and nonsuperionic states for solid solutions of copper chalcogenides. It was found that in superionic conductors, the main vibration modes are of an acoustic nature. Resonant interaction of the d-p state of anions in the valence band leads to screening of the electronic field of cations, lowers the activation barrier, and promotes disordering of the cation sublattice. Considering that the thermal conductivity of superionic copper chalcogenides is low and does not depend too much on the composition, it can be predicted that at temperatures 30–127 °C, materials based on lightly doped copper will be the most effective in thermoelectric devices. Nanostructuring of a material can further reduce thermal conductivity due to phonon scattering on grain boundary inhomogeneities and increase thermoelectric efficiency.

The enhancement of the effect of increasing the thermoelectric figure of merit in copper chalcogenides is achieved with nanostructuring, and the thermal conductivity decreases significantly more than the conductivity. This is due to the fact that in the formation of the optimal structure of the material, it is necessary to create conditions for phonons to be scattered more strongly by structure inhomogeneities than electrons. Since the wavelengths of electrons and phonons are different, one of the factors that can be used here is the size factor, since scattering is enhanced when the de Broglie wavelength becomes comparable to the size of the inhomogeneities in the medium.

Debye introduced the concept of the mean free path l and obtained a formula for a thermal conductivity similar to the formula following from the kinetic theory of gases:

$$k_L = \frac{1}{3} c_V \bar{v} \bar{l} \tag{25}$$

where c_V is the heat capacity of 1 cm^3 of the crystal, $\bar{v}$ is the average sound speed in the crystal, and $\bar{l}$ is the phonon mean free path.

Peierls changed Debye's theory, showing that the establishment of thermal equilibrium in a system of phonons can be influenced by processes for which the energy conservation law is satisfied:

$$\hbar\omega_1 + \hbar\omega_2 = \hbar\omega_3 \tag{26}$$

$$\hbar k_1 + \hbar k_2 = \hbar k_3 \tag{27}$$

The wave vectors are related by:

$$k_1 + k_2 = k_3 + G \tag{28}$$

These are the laws of conservation of energy and momentum (N-processes). Here, G is the reciprocal lattice vector. Since a phonon with a wave vector (+G) in a periodic lattice is indistinguishable from a phonon with a wave vector, Peierls called such processes Umklapp or U-processes. A feature of U-processes is that they change the momentum and direction of the energy flow. Thus, the U and N processes are responsible for the formation of thermal resistance to the phonon flux and ensure the establishment of thermal equilibrium in the phonon distribution [201].

For a known Fermi level and carrier scattering mechanism, the electronic component of thermal conductivity k_e can be calculated through the Fermi integrals using the formulas [206,207]

$$K_{e_a} = \left(\frac{k_B}{e}\right)^2 \left(\frac{3F_0(\mu^*)F_2(\mu^*) - 4F_1^2(\mu^*)}{F_0^2(\mu^*)}\right); K_{e_i} = \left(\frac{k_B}{e}\right)^2 \left(\frac{15F_2(\mu^*)F_4(\mu^*) - 16F_3^2(\mu^*)}{9F_2^2(\mu^*)}\right). \tag{29}$$

where μ^* is the reduced Fermi level, K_{e_a} and K_{e_i} are electronic thermal conductivities for scattering by acoustic phonons and by impurity ions, respectively, and the Fermi integrals $F_n(\mu^*)$ are given by Equation (23) above. To determine the Fermi level, one usually uses the experimental values of Seebeck coefficient and theoretical expressions for the Seebeck coefficient (see Equation (22)).

As already noted, the interaction of the electronic, ionic, and phonon subsystems of the crystal is well manifested in copper chalcogenides [20,144]. Thus, the "liquid-like" cationic sublattice is responsible for very low thermal conductivity and nonstoichiometric defects determine high electronic conductivity, which contributes to the high thermoelectric figure of merit of copper chalcogenides [1,144,173].

Lattice vibrations in crystalline materials generate phonons as heat carriers for heat conduction, and the phonon dispersion (energy versus momentum) is fundamentally determined by the mass of lattice vibrators (atoms) and the interaction force between atoms. A significant manipulation of lattice thermal conductivity through a change in atomic mass usually requires a large variation in chemical composition, which is not always valid thermodynamically or may risk the resultant detriment of other functionalities (e.g., carrier mobility). Here, Wu et al. [208] show a strategy of alternatively manipulating the interaction force between atoms through lattice strains without changing the composition, for remarkably reducing the lattice thermal conductivity without reducing carrier mobility, in Na$_{0.03}$Eu$_{0.03}$Sn$_{0.02}$Pb$_{0.92}$Te with stable lattice dislocations. This successfully leads to an extraordinarily high thermoelectric figure of merit ZT = 2.4 at 527 °C, with the help of valence band convergence. This work offers both insights and solutions on lattice strain engineering for reducing lattice thermal conductivity, thus advancing thermoelectrics.

In [209], Liu et al. reported results showing that binary ordered Cu$_{2-\delta}$Se has an extremely low lattice thermal conductivity at low temperatures. The low energy multi-Einstein optic modes are the dominant approach to obtaining such an extremely low lattice thermal conductivity. It is indicated that the damped vibrations of copper ions could contribute to the low energy multi-Einstein optic modes, especially for those low energy branches at 2–4 meV. Recent work on Cu$_{2-\delta}$X (X = S, Se, or Te) thermoelectric materials [15] reveals the ultralow lattice thermal conductivity (0.3–0.6 W m^{-1} K^{-1}) with the reduced specific heat down to the limit of a solid material at high temperatures. The "liquid-like" diffusion of disordered copper ions is believed to be the origin of the abnormal and interesting thermal transport phenomenon. In addition, these materials also exhibit extremely low thermal conductivity at low temperatures at which the copper ions display highly ordered distributions. The ultralow lattice thermal conductivity in the well-ordered simple binary Cu$_{2-\delta}$Se compounds below room temperature is quite abnormal and special, but the mechanism is still unknown. The lattice dynamics of several copper based thermoelectric materials were performed by studying the heat capacity and using inelastic neutron scattering techniques [210,211], demonstrating that those low energy localized vibrational modes mainly form the motion of copper atoms.

In 2021, Dutta M. et al. [212] presented lattice dynamics associated with the local chemical bonding hierarchy in Zintl compound TlInTe$_2$, which cause intriguing phonon excitations and strongly suppress the lattice thermal conductivity to an ultralow value (0.46–0.31 W m^{-1} K^{-1}) in the range 30–400 °C. They established an intrinsic rattling nature in TlInTe$_2$ by studying the local structure and phonon vibrations using synchrotron X-ray pair distribution function (PDF) (−173–230 °C) and inelastic neutron scattering (INS) (−268–177 °C), respectively. They showed that during 1D chain of covalently bonded $[InTe_2]_n^{-n}$ transport heat with Debye type phonon excitation, ionically bonded Tl rattles with frequency ~30 cm^{-1} inside a distorted Thompson cage formed by $[InTe_2]_n^{-n}$. This highly anharmonic Tl rattling causes strong phonon scattering and consequently, phonon lifetime reduces to an ultralow value of ~0.66(6) ps, resulting in ultralow thermal conductivity in TlInTe$_2$. The temperature-dependent X-ray PDF and INS investigations provided conclusive evidence the origin of low thermal conductivity in Zintl TlInTe$_2$. Thus, critical examination of chemical bonding, local structure, and experimental determination of phonon DOS should be the way forward to explore the fundamental origin of low thermal conductivity in crystalline material.

In the recent works of Bulat et al. [171,172], it was found that the experimental values of thermal conductivity in nanostructured samples of copper selenide are significantly lower than those given by theoretical calculations. According to the results of electron microscopic (high resolution) studies [172], a large number of nanosized grains and defects are observed in nanostructured copper selenide. This allowed the authors to conclude that the extremely low lattice thermal conductivity is due to phonon scattering at these intergrain nanoboundaries and nanodefects. This conclusion is confirmed by the results of studies where in nanostructured samples synthesized by a chemical method followed by SPS, the lattice thermal conductivity was 0.2 [173] and 0.23 W m^{-1} K^{-1} [174] at 577−627 °C. The presence of such nanodefects was not taken into account in the calculations of thermal conductivity. This is precisely the reason for the significant difference between the calculated and experimental values of the lattice thermal conductivity at temperatures above 507 °C.

In [213], Matthias Agne et al. expressed his concerns about the use of the well-known formula for calculating thermal conductivity. The ease and access of thermal diffusivity D measurements allows for the calculation of κ when the volumetric heat capacity, ρc_p, of the material is known. However, in the relation $\kappa = \rho c_p D$, there is some confusion as to what value of c_p should be used in materials undergoing phase transformations. In this paper, it is demonstrated that the Dulong–Petit estimate of c_p at high temperature is not appropriate for materials having phase transformations with kinetic timescales relevant to thermal transport. In these materials, there is an additional capacity to store heat in the material through the enthalpy of transformation ΔH. This can be described using a generalized model for the total heat capacity for a material $\rho c_p = C_{p\varphi} + \Delta H \cdot (\partial \varphi / \partial T)_p$, where φ is an order parameter that describes how much latent heat responds "instantly" to temperature changes. Here, $C_{p\varphi}$ is the intrinsic heat capacity (e.g., approximately the Dulong–Petit heat capacity at high temperature). It is shown experimentally in Zn_4Sb_3 that the decrease in D through the phase transition at $-23\,°C$ is fully accounted for by the increase in c_p, while κ changes smoothly through the phase transition. Consequently, conclude Agne et al. [213], reports of κ dropping near phase transitions in widely studied materials such as PbTe and SnSe have likely overlooked the effects of excess heat capacity and overestimated the thermoelectric efficiency, ZT.

An incomplete understanding of heat capacity measurements and models can lead to inaccurate estimations of κ in some systems, especially those having substantial latent heats (e.g., during phase transitions). The recent debate surrounding the thermoelectric material Cu_2Se is an excellent example [214–218]. In this material and others [91,219], the thermal diffusivity drops markedly as the material undergoes a phase transition. Depending on the heat capacity used to calculate κ, a maximum ZT between 0.6 [220] and 2.3 [214] has been reported due to the superionic phase transition in Cu_2Se. It is exemplified in the paper [213] that the choice of heat capacity can have a drastic impact on ZT values. Recognizing that the total capacity of a material to absorb heat includes both the intrinsic heat capacity of the phases present and the enthalpy (heat) of transformation ΔH that is required to maintain equilibrium (characterized by the order parameter φ), the temperature is changed as:

$$\rho c_p = \left(\frac{\partial H}{\partial T}\right)_p = C_{p\varphi} + \Delta H \cdot \left(\frac{\partial \varphi}{\partial T}\right)_p. \tag{30}$$

This is understandable, as one could argue that $(\partial \varphi/\partial T)_p$ should be zero in the steady-state measurement of κ. However, it is shown theoretically in the article by Agne et al. [214] why this term is non-zero and should be included in the calculation of κ when transformation kinetics are fast on the timescale of thermal transport. Not including the enthalpy of transformation can lead to significantly underestimated values of κ in the region of peak ZT for many important cases, such as Cu_2Se, PbTe, and SnSe, where exceptional ZT > 2 has been reported. In many good thermoelectric materials at their operation temperatures, atomic rearrangement may be fast enough that latent heats will suppress the thermal diffusivity. An apparent increase in ZT will result if a heat capacity is used that does not account for the latent heats. In particular, any discontinuity, spike, or sharp decrease that is found in thermal diffusivity measurements should be scrutinized before the same features are ascribed to the thermal conductivity. Even estimated values of lattice thermal conductivity κ_L which are substantially below estimates for the lower limit of thermal conductivity [221] should be scrutinized as they may indicate that κ is underestimated, as was found in cases of dynamic doping. Substantial underestimates of κ and overestimates of ZT, as demonstrated in Cu_2Se, are likely prevalent in other systems such as SnSe [213].

The crystal structure and various physical properties of semiconductor compounds of copper sulfides have been studied for a long period of time, starting from the 1930s; therefore, there is a need for new research in this area using modern research methods, since equipment and methods have emerged that make it possible to take a new look at the already established facts and discover new sides of known structures and phenomena. Differential thermal analysis (DTA) and differential scanning calorimetry (DSC) are among the most sensitive and reliable methods for studying the thermal properties of solids, which are often used in practice, as they allow one to determine all the main properties of heat transfer, specific heat, enthalpy, and phase equilibria of structural transition.

The thermal properties of copper sulfides have been less studied than other physical properties. The earliest studies of the thermal and thermoelectric properties of $Cu_{2-x}S$ were carried out on polycrystalline and single-crystal samples [47,222]. It was shown that the properties of $Cu_{2-x}S$ strongly depend on the nonstoichiometry x of the composition. In addition, the thermal properties were found to exhibit anomalous behavior near the temperatures of the phase transformation in $Cu_{2-x}S$. The results show that the mechanism of heat transfer is mainly due to phonons, while the contribution of electrons and dipoles was indeed very small [222].

The heat capacity and thermodynamic properties of Cu_2S in the temperature range 268–677 °C were studied in detail by F. Gronvold [223]. The phase transitions were recorded at about 103 and 437 °C. Numerical results on the enthalpies in the transitions ($\gamma \rightarrow \beta$) and ($\beta \rightarrow \alpha$) in Cu_2S are given [223]. Significantly higher values of heat capacity $C_p \approx 11.9\ R$ were obtained under conditions close to equilibrium. In the region between transitions, the heat capacity decreases significantly with increasing temperature. F. Gronvold notes that for copper sulfide, for virtually every temperature, a thermodynamically equilibrium arrangement of copper atoms is formed in various voids of the crystal structure.

In work [36], phase transitions and thermal effects of solid $Na_xCu_{2-x}S$ samples were investigated by DSC in an argon atmosphere in the temperature range (30–430 °C). DSC was detected at 103 °C with an enthalpy area of 5.234 µW mg^{-1}. The beginning of the effect is about 69 ± 3 °C. The end of the thermal effect is about 122 °C. The heat capacity varies within 150–480 J kg^{-1} K^{-1}. In the series of compositions $Na_{0.15}Cu_{1.85}S$, $Na_{0.17}Cu_{1.80}S$ and $Na_{0.20}Cu_{1.77}S$, prepared by an exchange reaction in a melt mixture of sodium and potassium hydroxides and cold pressing, the heat capacity at the point of phase transition increases with increasing sodium content. This can be interpreted as an indirect evidence of sodium participation in the formation of the crystal lattice of copper sulfide, as with increasing sodium concentration, energy expenditures on mixing of cations (copper and sodium) should increase. Enthalpy of the transition increases with increasing sodium content in the $Na_{0.15}Cu_{1.85}S$, $Na_{0.17}Cu_{1.80}S$, and $Na_{0.20}Cu_{1.77}S$ chain (5234, 6923, and 11,720 J kg^{-1}K^{-1}, respectively). Extremely low thermal conductivities 0.3–0.1 W m^{-1} K^{-1} were obtained for $Na_{0.15}Cu_{1.85}S$ composition. The ZT = 0.28 value achieved for the alloys studied is good at 320 °C for copper sulfides, but it can be greatly increased, in our view, by lowering the sodium content in non-stoichiometric copper sulfide ($Cu_{1.75}S \div Cu_{1.85}S$) to the optimum so that the electrical conductivity remains high enough.

In following work [39], X-ray phase analysis showed that at 20 °C, the samples $Na_{0.3}Cu_{1.6}S$, $Na_{0.35}Cu_{1.5}S$, and $Na_{0.4}Cu_{1.55}S$ are a mixture of phases: monoclinic jarleite $Cu_{1.93}S$, rhombohedral digenite $Cu_{1.8}S$, and hexagonal Na_2S_2. Synthesized mixtures were compacted by cool pressing. The crystallite sizes obtained from the analysis of X-ray were 32–67 nm, 41–96 nm, and 15–37 nm for $Na_{0.3}Cu_{1.6}S$, $Na_{0.35}Cu_{1.5}S$, and $Na_{0.4}Cu_{1.55}S$ samples, respectively. For all samples, there is a sharp increase in the Seebeck coefficient after 250 °C. The energy dependence of the transparency of additional barriers for current carriers arising due to the presence of inclusions of the dielectric phase (Na_2S_2), the content of which is proportional to the sodium concentration in the sample, can lead to an increase in the Seebeck coefficient. For the $Na_{0.4}Cu_{1.55}S$ sample, a high value of ZT = 0.84 was obtained at 358 °C. Above 350 °C, the conductivity of all samples sharply decreases and the Seebeck coefficient rises abruptly, since the fusion of individual phases with the formation of a new phase on the basis of the cubic modification of copper sulfide (chalcocite) is possible, which is characterized by low electronic conductivity at the level of units S cm^{-1} [46,122].

In [138], the results of thermal conductivity in $Cu_{1.8}S$ compounds with the addition of Na_2S are shown. It was found that after the introduction of Na_2S into the $Cu_{1.8}S$ matrix, the thermal conductivity values decreased. Y. Zhang et al. [138] explain this behavior of thermal conductivity by enhanced phonon scattering due to an increase in the number of micropores, which have a great influence on the decrease in thermal conductivity; in addition, the acoustic velocity has decreased, which is considered as an effective thermal barrier, and, as a consequence, leads to a decrease in thermal conductivity.

3.2.3. Possibilities of Practical Application of Copper Chalcogenides and Its Alloys

After the sobering article by Dennler [18] with doubts about the usefulness of thermoelements unstable to degradation based on copper chalcogenides, the boom of studies of thermoelectric properties of these materials did not stop, and the number of studies did not decrease. Against the background of a further gradual increase in ZT, we would like to note one message [224]. The message concerns the technical possibility of avoiding the release of copper from the chalcogenide during the operation of the thermoelement. In the work [224] by Qiu P. et al., through systematically investigating electromigration in copper sulfide/selenide thermoelectric materials, the mechanism for atom migration and deposition based on a critical chemical potential difference was revealed. The authors have shown that the release of copper from the sample begins when the electrical potential difference applied to the sample exceeds the critical voltage determined by equality

$$V_c = -\frac{1}{FZ_e}\Delta\mu_{Cu}^{crit} - S^*\Delta T, \tag{31}$$

where Z_e defines the charge (-1 for electrons or $+1$ for holes), F is Faraday's constant, $\Delta\mu_{Cu}^{crit}$ is some critical chemical potential, and S^* accounts for the net effect of thermodiffusion. From this analysis, it is expected that a voltage difference, not current density, is the critical parameter for Cu deposition. The microscopic defect model of Yokota and Korte [225,226] allows one to determine the critical chemical potential by using the term off-stoichiometry (nonstoichiometry degree), δ, (in $Cu_{2-\delta}X$, X = S, Se) and to calculate V_c as:

$$V_c = -\frac{RT}{F}\left[Arsinh\left(\frac{\delta_C}{2\sqrt{K_e}}\right) - Arsinh\left(\frac{2\delta - \delta_C}{2\sqrt{K_e}}\right)\right]. \tag{32}$$

A parameter named the critical off-stoichiometry (δ_c) is introduced here, corresponding to the "solubility limit" of Cu concentration at the cathode of the cell with the mixed ionic–electronic conductor. K_e is the equilibrium constant for electrons and holes that is independent of stoichiometry, R is the gas constant, and T is the temperature. The thermodynamic theory developed in the work predicts that a given off-stoichiometry and temperature difference will result in limitations on the electrical potential difference that is stable across the material.

On basis of the analysis of copper release, a strategy for stable use is proposed by the authors of paper [224]: constructing a series of electronically conducting, but ion-blocking, barriers to reset the chemical potential of such conductors to keep it below the threshold for decomposition, even if it is used with high electric currents and/or large temperature differences. This strategy opens the possibility of using such conductors in thermoelectric applications and may also provide approaches to engineer perovskite photovoltaic materials.

Despite the importance of studying the physical side of the phenomena and progress in synthesis, the practical application of copper sulfide and its alloys is no less important. Today, materials with ZT ~2 and higher are considered promising for use on an industrial scale. In addition to this main criterion for the development of thermoelectric devices, important conditions are cheapness and availability of raw materials, ease of synthesis, material stability and compatibility with other components of the thermoelectric module, and others [4,82,83]. Doping with lithium and sodium makes it possible to enhance the useful properties of these materials, in particular, to increase the thermoelectric effect, and the creation of composites with nanoscale carbon blocks and numerous pores makes it possible to achieve thermal conductivity up to 0.1 W m^{-1} K^{-1}.

4. Conclusions and Suggestions

Copper chalcogenides and their alloys are promising thermoelectric materials that have recently attracted increased attention of researchers. They clearly show the interaction of the electronic, ionic, and phonon subsystems of the crystal [1,20,144]. Thus, the "liquid-like" cationic sublattice is responsible for very low thermal conductivity, while nonstoichiometric defects provide high electronic conductivity, which contributes to the high thermoelectric figure of merit of copper chalcogenides [1]. Of course, high mobility of copper ions creates a problem of copper release in a long operation at high temperatures; however, this problem is surmountable. On the basis of the analysis of copper release, a strategy for stable use was proposed by P. Qiu et al. [224]: constructing a series of electronically conducting, but ion-blocking, barriers to reset the chemical potential of such conductors to keep it below the threshold for decomposition, even if it is used with high electric currents and/or large temperature differences. This strategy opens the possibility of using such conductors in thermoelectric applications.

Meanwhile, the ZT of alloys based on copper chalcogenides approached the value 3. Thus, $ZT = 2.62$ was obtained for the composition $Cu_{1.94}Al_{0.02}Se$ at a temperature of 756 °C by B. Ghong et al. [177] and $ZT = 2.63$ was achieved for Cu_2Se + 1 mol% $CuInSe_2$ at 577 °C by A. Olvera et al. [85]. Doping, as well as the improvement of methods and conditions of synthesis, remain the main tools of researchers.

Doping of copper chalcogenides with lithium is convenient due to the proximity of the ionic radii of copper and lithium, which favors the formation of solid solutions. For the composition $Li_{0.09}Cu_{1.9}Se$, the maximum $ZT \approx 1.4$ at 727 °C was obtained [30]. Due to the large difference in the ionic radii of copper and sodium, the solid solubility of sodium does not exceed a few percentage points, but this turned out to be enough to reach $ZT = 2.1$ at 700 °C for microporous $Na_{0.04}Cu_{1.96}Se$ obtained by hydrothermal method and hot pressing [53]. It is possible that the result would have been higher if a more non-stoichiometric composition was chosen as the matrix for doping with sodium, e.g., $Cu_{1.94}Se$. Alloys of copper chalcogenides with high sodium content are not homogeneous and form a mixture of phases with different conducting and thermoelectric properties, but deserve attention as nanocomposite thermoelectric materials with low thermal conductivity. The properties of such mixtures are poorly understood, but we expect good thermoelectric performance here associated with the presence of numerous barrier layers and pores.

Not only superionic conductors have low lattice thermal conductivity, but also other classes of materials, such as multi-filled skutterudites [117], for example. The multiple-filled skutterudite TE material $Ba_uLa_vYb_wCo_4Sb_{12}$ prepared by spark plasma sintering [7] shows $ZT \sim 1.7$ at 577 °C [115]. The very high ZT value benefits from the following two aspects: (a) controlling the filling fraction of multiple filled atoms to optimize the carrier concentration, leading to a higher power factor and (b) strong scattering of wide-band phonons can be achieved by different rattling frequencies of the multiple atomic filling in the icosahedral voids of $CoSb_3$ TE material, making the lattice thermal conductivity close to the theoretical minimum.

The results of recent years show that the decrease in thermal conductivity due to nanostructuring and nanoinclusions of another phase is comparable in magnitude with the contribution of "melting" of the sublattice of mobile ions in a super-ionic conductor.

Moreover, do not underestimate heterogeneous thermoelectrics and composites. The recent works reveal that inclusions of the second phase, such as the addition of graphene or carbon fiber, make it possible to achieve a sharp increase in ZT. Thus, L. Zhao et al. [178] achieved $ZT = 2.4$ for composition Cu_2Se + 0.3 wt.% carbon fiber at 577 °C.

In our opinion: bulk materials soon will be able to exceed $ZT = 3$. Recently, the highest $ZT = 2.8$ at 500 °C was achieved for $SnSe_{1-x}Br_x$ composition by C. Chang et al. [91], who maintained that a continuous phase transition increases the symmetry and diverges two converged conduction bands. These two factors improve carrier mobility, while preserving a large Seebeck coefficient.

Funding: This research was funded by the Science Committee of the Ministry of Education and Science of the Republic of Kazakhstan (No. AP08856636).

Conflicts of Interest: The authors declare no conflict of interest.

References

1. Liu, H.; Shi, X.; Xu, F.; Zhang, L.; Zhang, W.; Chen, L.; Li, Q.; Uher, C.; Day, T.; Snyder, J. Copper ion liquid-like thermoelectrics. *Nat. Mater.* **2012**, *11*, 422–425. [CrossRef]
2. Qiu, P.; Shi, X.; Chen, L. Cu-based thermoelectric materials. *Energy Storage Mater.* **2016**, *3*, 85–97. [CrossRef]
3. Dmitriev, A.V.; Zvyagin, I.P. Current trends in the physics of thermoelectric materials. *Physics-Uspekhi* **2010**, *53*, 789–803. [CrossRef]
4. Mao, J.; Liu, Z.; Zhou, J.; Zhu, H.; Zhang, Q.; Chen, G.; Ren, Z. Advances in thermoelectrics. *Adv. Phys.* **2018**, *67*, 69–147. [CrossRef]
5. Gorbachev, V.V. *Poluprovodnikovy'e Soedineniya*; Metallurgiya: Moscow, Russia, 1980; 132p. (In Russian)
6. El Akkad, F.; Mansour, B.; Hendeya, T. Electrical and thermoelectric properties of Cu_2Se and Cu_2S. *Mater. Res. Bull.* **1981**, *16*, 535–539. [CrossRef]
7. Konev, V.N.; Bikkin, K.M.; Fomenkov, S.A. Thermo-e.m.f. of $Cu_{2-\delta}X$ (X-S, Se). *Inorg. Mater.* **1983**, *19*, 1066–1069.
8. Brown, D.R.; Day, T.; Caillat, T.; Snyder, J. Chemical Stability of (Ag, $Cu)_2Se$: A Historical Overview. *J. Electron. Mater.* **2013**, *42*, 2014–2019. [CrossRef]
9. Korzhuev, M.A.; Laptev, A.V. Effekty izmeneniya sostava obrazczov superionnogo $Cu_{2-x}Se$ pod dejstviem elektricheskogo toka. *Zhurnal Tekhnicheskoj Fiz. (USSR)* **1989**, *59*, 62–66. (In Russian)
10. Korzhuev, M.A. Inhibition of the growth of excrescences in mixed electronic-ionic conductors. *Tech. Phys.* **1998**, *43*, 1333–1337. [CrossRef]
11. Slack, G.A. *CRC Handbook of Thermoelectricity*; CRC Press: Cardiff, UK, 1995; p. 157.
12. Qin, P.; Qian, X.; Ge, Z.-H.; Zheng, L.; Feng, J.; Zhao, L.-D. Improvements of thermoelectric properties for p-type $Cu_{1.8}S$ bulk materials via optimizing the mechanical alloying process. *Inorg. Chem. Front.* **2017**, *4*, 1192–1199. [CrossRef]
13. Zhao, L.; Fei, F.Y.; Wang, J.; Wang, F.; Wang, C.; Li, J.; Wang, J.; Cheng, Z.; Dou, S.X.; Wang, X. Improvement of thermoelectric properties and their correlations with electron effective mass in $Cu_{1.98}S_xSe_{1-x}$. *Sci. Rep.* **2017**, *7*, 40436. [CrossRef] [PubMed]
14. Tang, Y.-Q.; Ge, Z.-H.; Feng, J. Synthesis and Thermoelectric Properties of Copper Sulfides via Solution Phase Methods and Spark Plasma Sintering. *Crystals* **2017**, *7*, 141. [CrossRef]
15. Zhao, L.-L.; Wang, X.-L.; Wang, J.-Y.; Cheng, Z.; Dou, S.X.; Wang, J.; Liu, L.-Q. Superior intrinsic thermoelectric performance with zT of 1.8 in single-crystal and melt-quenched highly dense $Cu_{2-x}Se$ bulks. *Sci. Rep.* **2015**, *5*, 7671. [CrossRef]
16. Gahtori, B.; Bathula, S.; Tyagi, K.; Jayasimhadri, M.; Srivastava, A.; Singh, S.; Budhani, R.; Dhar, A. Giant enhancement in thermoelectric performance of copper selenide by incorporation of different nanoscale dimensional defect features. *Nano Energy* **2015**, *13*, 36–46. [CrossRef]
17. Zhao, L.; Wang, X.; Fei, F.Y.; Wang, J.; Cheng, Z.; Dou, S.X.; Wang, J.; Snyder, J. High thermoelectric and mechanical performance in highly dense $Cu_{2-x}S$ bulks prepared by a melt-solidification technique. *J. Mater. Chem. A* **2015**, *3*, 9432–9437. [CrossRef]
18. Dennler, G.; Chmielowski, R.; Jacob, S.; Capet, F.; Roussel, P.; Zastrow, S.; Nielsch, K.; Opahle, I.; Madsen, G.K.H. Are Binary Copper Sulfides/Selenides Really New and Promising Thermoelectric Materials? *Adv. Energy Mater.* **2014**, *4*, 1301581. [CrossRef]
19. Ivanov-Shicz, A.K.; Murin, I.V. *Ionika Tverdogo Tela*; Izdatelstvo Sankt-Peterburgskogo Universiteta: St. Petersburg, Russia, 2000; Volume 1, 616p. (In Russian)
20. Berezin, V.M.; Vyatkin, G.P. *Superionnye Poluprovodnikovye Khalkogenidy*; Yu.-UrGU: Chelyabinsk, Russia, 2001; 135p.
21. Shahi, K. Transport studies on superionic conductors. *Phys. Status Solidi* **1977**, *41*, 11–44. [CrossRef]
22. Balapanov, M.K.; Nadejzdina, A.F.; Yakshibayev, R.A.; Lukmanov, D.R.; Gabitova, R.J. Ionic conductivity and chemical diffusion in $LixCu_{2-x}Se$ superionic alloys. *Ionics* **1999**, *5*, 20–22. [CrossRef]
23. Balapanov, M.K.; Bikkulova, N.N.; Mukhamedyanov, U.K.; Asilguschina, G.N.; Musalimov, R.S.; Zeleev, M.K. Phase transitions and transport phenomena in $Li_{0.25}Cu_{1.75}Se$ superionic compound. *Phys. Status Solidi* **2004**, *241*, 3517–3524. [CrossRef]
24. Balapanov, M.K.; Gafurov, I.G.; Mukhamed'Yanov, U.K.; Yakshibaev, R.A.; Ishembetov, R.K. Ionic conductivity and chemical diffusion in superionic $LixCu_{2-x}Se$ ($0 \leq x \leq 0.25$). *Phys. Stat. Sol.* **2004**, *241*, 114–119. [CrossRef]
25. Balapanov, M.K.; Yakshibaev, R.A.; Gafurov, I.G.; Ishembetov, R.K.; Kagarmanov, S.M. Superionic conductivity and crystal structure of $LixCu_{2-x}S$ alloys. *Bull. Russ. Acad. Sci. Phys.* **2005**, *69*, 623–626.
26. Balapanov, M.K.; Zinnurov, I.B.; Mukhamed'Yanov, U.K. Ionic conduction and chemical diffusion in solid solutions of superionic conductors Cu_2X-Me_2X (Me = Ag, Li; X = S, Se). *Russ. J. Electrochem.* **2007**, *43*, 585–589. [CrossRef]
27. Balapanov, M.K. Grain size effect on diffusion processes in superionic phases $Cu_{1.75}S$, $Li_{0.25}Cu_{1.75}Se$, and $Li_{0.25}Cu_{1.75}S$. *Russ. J. Electrochem.* **2007**, *43*, 590–594. [CrossRef]
28. Ishembetov, R.K.; Balapanov, M.K.; Yulaeva, Y.K. Electronic Peltier effect in $Li_xCu_{(2-x)-\delta}S$. *Russ. J. Electrochem.* **2011**, *47*, 416–419. [CrossRef]
29. Balapanov, M.K.; Ishembetov, R.K.; Kuterbekov, K.A.; Nurakhmetov, T.N.; Urazaeva, E.K.; Yakshibaev, R.A. Influence of the cation sublattice defectness on the electronic thermoelectric power of $Li_xCu_{(2-x)-\delta}S(x \leq 0.25)$. *Inorg. Mater.* **2014**, *50*, 930–933. [CrossRef]
30. Kang, S.D.; Pöhls, J.-H.; Aydemir, U.; Qiu, P.; Stoumpos, C.; Hanus, R.; White, M.A.; Shi, X.; Chen, L.; Kanatzidis, M.G.; et al. Enhanced stability and thermoelectric figure-of-merit in copper selenide by lithium doping. *Mater. Today Phys.* **2017**, *1*, 7–13. [CrossRef]
31. Ge, Z.-H.; Liu, X.; Feng, D.; Lin, J.; He, J. High-Performance Thermoelectricity in Nanostructured Earth-Abundant Copper Sulfides Bulk Materials. *Adv. Energy Mater.* **2016**, *6*, 1600607. [CrossRef]

32. Xiao, X.-X.; Xie, W.-J.; Tang, X.-F.; Zhang, Q.-J. Phase transition and high temperature thermoelectric properties of copper selenide $Cu_{2-x}Se$ ($0 \leq x \leq 0.25$). *Chin. Phys.* **2011**, *20*, 087201. [CrossRef]
33. Balapanov, M.; Zinnurov, I.; Akmanova, G. The ionic Seebeck effect and heat of cation transfer in $Cu_{2-\delta}Se$ superionic conductors. *Physics of the Solid State.* **2006**, *48*, 1868–1871. [CrossRef]
34. Liu, H.; Shi, X.; Kirkham, M.; Wang, H.; Li, Q.; Uher, C.; Zhang, W.; Chen, L. Structure-transformation-induced abnormal thermoelectric properties in semiconductor copper selenide. *Mater. Lett.* **2013**, *93*, 121–124. [CrossRef]
35. Balapanov, M.K.; Ishembetov, R.K.; Kuterbekov, K.A.; Kubenova, M.M.; Danilenko, V.N.; Nazarov, K.; Yakshibaev, R.A. Thermoelectric and thermal properties of superionic $Ag_xCu_{2-x}Se$ (x = 0.01, 0.02, 0.03, 0.04, 0.25) compounds. *Lett. Mater.* **2016**, *6*, 360–365. [CrossRef]
36. Balapanov, M.; Kubenova, M.; Kuterbekov, K.; Kozlovskiy, A.; Nurakov, S.; Ishembetov, R.; Yakshibaev, R. Phase analysis, thermal and thermoelectric properties of nanocrystalline $Na_{0.15}Cu_{1.85}S$, $Na_{0.17}Cu_{1.80}S$, $Na_{0.20}Cu_{1.77}S$ alloys. *Eurasian J. Phys. Funct. Mater.* **2018**, *2*, 231–241. [CrossRef]
37. Kubenova, M.M.; A Kuterbekov, K.; Abseitov, E.T.; Kabyshev, A.M.; Kozlovskiy, A.; Nurakov, S.N.; Ishembetov, R.K.; Balapanov, M.K. Electrophysical and thermal properties of $Na_xCu_{2-x}S$ (x = 0.05, 0.075, 0.10) and $Na_{0.125}Cu_{1.75}S$ semiconductor alloys. *IOP Conf. Series: Mater. Sci. Eng.* **2018**, *447*, 012031. [CrossRef]
38. Balapanov, M.K.; Ishembetov, R.K.; Kuterbekov, K.A.; Kubenova, M.M.; Almukhametov, R.F.; Yakshibaev, R.A. Transport phenomena in superionic $Na_xCu_{2-x}S$ (x= 0.05; 0.1; 0.15; 0.2) compounds. *Ionics* **2018**, *24*, 1349–1356. [CrossRef]
39. Balapanov, M.K.; Ishembetov, R.K.; Kabyshev, A.M.; Kubenova, M.M.; Kuterbekov, K.A.; Yulaeva, Y.K.; Yakshibaev, R.A. Influence of sodium doping on electron conductivity and electronic seebeck coefficient of copper sulfide. *Vestn. Bashkirskogo Univ.* **2019**, *24*, 823. [CrossRef]
40. Kubenova, M.; Balapanov, M.; Kuterbekov, K.; Ishembetov, R.; Kabyshev, A.; Yulaeva, Y. Phase composition and thermoelectric properties of the nanocomposite alloys $Na_xCu_{2-x-y}S$. *Eurasian J. Phys. Funct. Mater.* **2020**, *4*, 67–85. [CrossRef]
41. Villars, P.; Cenzual, K.; Daams, J.; Gladyshevskii, R.; Shcherban, O.; Dubenskyy, V.; Melnichenko-Koblyuk, N.; Pavlyuk, O.; Savysyuk, I.; Stoyko, S.; et al. $NaCu_4S_4$ Structure Types. Part 6: Space Groups (166) R-3m-(160) R3m · $NaCu_4S_4$. *Landolt-Börnstein Group III Condensed Matter.* **2008**, *43A6*. [CrossRef]
42. Pichanusakorn, P.; Bandaru, P. Nanostructured thermoelectrics. *Mater. Sci. Eng. R Rep.* **2010**, *67*, 19–63. [CrossRef]
43. Chakrabarti, D.J.; Laughlin, D.E. The Cu-S (Copper-Sulfur) system. *Bull. Alloy. Phase Diagr.* **1983**, *4*, 254–271. [CrossRef]
44. Madelung, O.; Rössler, U.; Schulz, M. *Non-Tetrahedrally Bonded Elements and Binary Compounds I.*; Springer: Berlin/Heidelberg, Germany, 1998; 185p.
45. Evans, H.T. The crystal structures of low chalcocite and djurleite. *Z. Krist.* **1979**, *150*, 299–320. [CrossRef]
46. Roseboom, E.H. An investigation of the system Cu-S and some natural copper sulfides between 25 and 700 °C. *Econ. Geol.* **1966**, *61*, 641–672. [CrossRef]
47. Will, G.; Hinze, E.; Abdelrahman, A.R.M. Crystal structure analysis and refinement of digenite, $Cu_{1.8}S$, in the temperature range 20 to 500 °C under controlled sulfur partial pressure. *Eur. J. Miner.* **2002**, *14*, 591–598. [CrossRef]
48. Yamamoto, K.; Kashida, S. X-ray study of the cation distribution in Cu_2Se, $Cu_{1.8}Se$ and $Cu_{1.8}S$; analysis by the maximum entropy method. *Solid State Ion.* **1991**, *48*, 241–248. [CrossRef]
49. Potter, R.W. An Electrochemical Investigation of the System Cu-S. *Econ. Geol.* **1977**, *72*, 1524–1542. [CrossRef]
50. Shah, D.; Khalafalla, S.E. Kinetics and mechanism of the conversion of covellite (CuS) to digenite ($Cu_{1.8}S$). *Metall. Trans.* **1971**, *2*, 2637–2643. [CrossRef]
51. Mumme, W.G.; Gable, R.W.; Petricek, V. THE CRYSTAL STRUCTURE OF ROXBYITE, $Cu_{58}S_{32}$. *Can. Miner.* **2012**, *50*, 423–430. [CrossRef]
52. Goble, R.G. Copper Sulfides From Alberta: Yarrowite Cu_9S_8 and Spionkopite $Cu_{39}S_{28}$. *Canad. Min.* **1980**, *18*, 511–518.
53. Zhu, Z.; Zhang, Y.; Song, H.; Li, X.-J. High thermoelectric performance and low thermal conductivity in $Cu_{2-x}Na_xSe$ bulk materials with micro-pores. *Appl. Phys. A* **2019**, *125*, 1–7. [CrossRef]
54. Bertheville, B.; Low, D.; Bill, H.; Kubel, F. Ionic conductivity of Na_2S single crystals between 295 and 1350 K experimental setup and first results. *J. Phys. Chem. Solids* **1997**, *58*, 1569–1577. [CrossRef]
55. Eithiraj, R.D.; Jaiganesh, G.; Kalpana, G.; Rajagopalan, M. First-principles study of electronic structure and ground-state properties of alkali-metal sulfides—Li_2S, Na_2S, K_2S and Rb_2S. *Phys. Status Solidi* **2007**, *244*, 1337–1346. [CrossRef]
56. Zhuravlev, Y.N.; Kosobutskii, A.B.; Poplavnoi, A.S.; Zhuravlev, Y. Energy Band Genesis from Sublattice States in Sulfides of Alkali Metals with an Antifluorite Lattice. *Russ. Phys. J.* **2004**, *48*, 138–142. [CrossRef]
57. Kizilyalli, M.; Bilgin, M.; Kizilyalli, H. Solid-state synthesis and X-ray diffraction studies of Na_2S. *J. Solid State Chem.* **1990**, *85*, 283–292. [CrossRef]
58. Savelsberg, G.; Schäfer, H. Zurkenntnis von. $Na_2Cu_4S_3$ und KCu_3Te_2. *Mater. Res. Bull.* **1981**, *16*, 1291–1297. [CrossRef]
59. Jain, A.; Ong, S.P.; Hautier, G.; Chen, W.; Richards, W.D.; Dacek, S.; Cholia, S.; Gunter, D.; Skinner, D.; Ceder, G.; et al. Commentary: The Materials Project: A materials genome approach to accelerating materials innovation. *APL Mater.* **2013**, *1*, 011002. [CrossRef]
60. Burschka, C.; Naturforsch, Z. $Na_3Cu_4S_4$-a thiocuprate with isolated $[Cu_4S_4]$-chains. *Z. Nat.* **1979**, *34*, 396–397.
61. Effenberger, H.; Pertlik, F. Crystal structure of $NaCu_5S_3$. *Mon. Für Chem. Chem. Mon.* **1985**, *116*, 921–926. [CrossRef]
62. Yong, W.; She, Y.; Qing, F.; Ao, W. Hydrothermal synthesis of K, Na doped Cu-S nanocrystalline and effect of doping on crystal structure and performance. *Acta Phys. Sin.* **2013**, *62*, 17802–17809.

63. Zhang, X.; Kanatzidis, M.G.; Hogan, T.; Kannewurf, C.R. NaCu$_4$S$_4$, a Simple New Low-Dimensional, Metallic Copper Polychalcogenide, Structurally Related to CuS. *J. Am. Chem. Soc.* **1996**, *118*, 693–694. [CrossRef]
64. Klepp, K.O.; Sing, M.; Boller, H. Preparation and crystal structure of Na$_4$Cu$_2$S$_3$, a thiocuprate with discrete anions. *J. Alloys Compd.* **1992**, *184*, 265–273. [CrossRef]
65. Klepp, K.O.; Sing, M.; Boller, H. Preparation and crystal structure of Na$_7$Cu$_{12}$S$_{10}$, a mixed valent thiocuprate with a pseudo-one-dimensional structure. *J. Alloys Compd.* **1993**, *198*, 25–30. [CrossRef]
66. Savelsberg, G.; Schafer, H. Preparation and crystal-structure of Na$_2$AgAs and KCuS. *Naturforsch* **1978**, *33*, 711–713. [CrossRef]
67. Kubel, F.; Bertheville, B.; Bill, H. Crystal structure of dilithiumsulfide, Li$_2$S. *Z. Kristallogr.* **1999**, *214*, 302. [CrossRef]
68. Buehrer, W.; Altorfer, F.; Mesot, J.; Bill, H.; Carron, P.; Smith, H.G. Lattice dynamics and the diffuse phase transition of lithium sulfide investigated by coherent neutron scattering. *J. Phys. Condens. Matter.* **1991**, *3*, 1055–1064. [CrossRef]
69. Altorfer, F.; Buhrer, W.; Anderson, I.; Scharpf, O.; Bill, H.; Carron, P.L. Fast ionic diffusion in Li$_2$S investigated by quasielastic neutron scattering. *J. Phys. Condens. Matter* **1994**, *6*, 9937–9947. [CrossRef]
70. Mjwara, P.M.; Comins, J.D.; E Ngoepe, P.; Buhrer, W.; Bill, H. Brillouin scattering investigation of the high temperature diffuse phase transition in Li$_2$S. *J. Phys. Condens. Matter* **1991**, *3*, 4289–4292. [CrossRef]
71. Tsuji, J.; Nakamatsu, H.; Mukoyama, T.; Kojima, K.; Ikeda, S.; Taniguchi, K. Lithium K-edge XANES spectra for lithium compounds. *X-ray Spectrum.* **2002**, *31*, 319–326. [CrossRef]
72. Ohtani, T.; Ogura, J.; Sakai, M.; Sano, Y. Phase transitions in new quasi-one-dimensional sulfides T$_1$Cu$_7$S$_4$ and KCu$_7$S$_4$. *Solid State Commun.* **1991**, *78*, 913–917. [CrossRef]
73. Ohtani, T.; Ogura, J.; Yoshihara, H.; Yokota, Y. Physical Properties and Successive Phase Transitions in Quasi-One-Dimensional Sulfides ACu$_7$S (A = Tl, K, Rb). *J. Solid State Chem.* **1995**, *115*, 379–389. [CrossRef]
74. Klepp, K.O.; Sing, M. Crystal structure of rubidium dithiotricuprate, RbCu$_3$S$_2$. *Z. Kristallogr. NCS* **2002**, *217*, 474.
75. Burschka, C.; Bronger, W. KCu$_3$S$_2$ ein neues Thiocuprat. *Z. Naturforsch.* **1977**, *32*, 11–14. [CrossRef]
76. Klepp, K.O.; Yvon, K. Thallium-dithio-tricuprate, TlCu$_3$S$_2$. *Acta Crystallogr.* **1980**, *36*, 2389–2391. [CrossRef]
77. Chen, E.M.; Poudeu, P. Thermal and electrochemical behavior of Cu$_{4-x}$Li$_x$S$_2$ (x= 1, 2, 3) phases. *J. Solid State Chem.* **2015**, *232*, 8–13. [CrossRef]
78. Bikkulova, N.N.; Danilkin, S.A.; Beskrovnyi, A.I.; Yadrovskii, E.L.; Semenov, V.A.; Asylguzhina, G.N.; Balapanov, M.K.; Sagdatkireeva, M.B.; Mukhamed'Yanov, U.K. Neutron diffraction study of phase transitions in the superionic conductor Li$_{0.25}$Cu$_{1.75}$Se. *Crystallogr. Rep.* **2003**, *48*, 457–460. [CrossRef]
79. Kieven, D.; Grimm, A.; Beleanu, A.; Blum, C.; Schmidt, J.; Rissom, T.; Lauermann, I.; Gruhn, T.; Felser, C.; Klenk, R. Preparation and properties of radio-frequency-sputtered half-Heusler films for use in solar cells. *Thin Solid Films* **2011**, *519*, 1866–1871. [CrossRef]
80. Beleanu, A.; Kiss, J.; Baenitz, M.; Majumder, M.; Senyshyn, A.; Kreiner, G.; Felser, C. LiCuS, an intermediate phase in the electrochemical conversion reaction of CuS with Li: A potential environment-friendly battery and solar cell material. *Solid State Sci.* **2016**, *55*, 83–87. [CrossRef]
81. Soliman, S. Theoretical investigation of Cu-containing materials with different valence structure types: BaCu$_2$S$_2$, Li$_2$CuSb, and LiCuS. *J. Phys. Chem. Solids* **2014**, *75*, 927–930. [CrossRef]
82. Tan, G.; Ohta, M.; Kanatzidis, M.G. Thermoelectric power generation: From new materials to devices. *Philos. Trans. R. Soc. A Math. Phys. Eng. Sci.* **2019**, *377*, 20180450. [CrossRef] [PubMed]
83. Jaldurgam, F.; Ahmad, Z.; Touati, F. Low-Toxic, Earth-Abundant Nanostructured Materials for Thermoelectric Applications. *Nanomaterials* **2021**, *11*, 895. [CrossRef] [PubMed]
84. He, Y.; Day, T.; Zhang, T.; Liu, H.; Shi, X.; Chen, L.; Snyder, G.J. High Thermoelectric Performance in Non-Toxic Earth-Abundant Copper Sulfide. *Adv. Mater.* **2014**, *26*, 3974–3978. [CrossRef] [PubMed]
85. Olvera, A.A.; Moroz, N.A.; Sahoo, P.; Ren, P.; Bailey, T.P.; Page, A.A.; Uher, C.; Poudeu, P.F.P. Partial indium solubility induces chemical stability and colossal thermoelectric figure of merit in Cu$_2$Se. *Energy Environ. Sci.* **2017**, *10*, 1668–1676. [CrossRef]
86. Zhao, K.; Qiu, P.; Song, Q.; Blichfeld, A.B.; Eikeland, E.; Ren, D.; Ge, B.; Iversen, B.B.; Shi, X.; Chen, L. Ultrahigh thermoelectric performance in Cu$_2$-ySe$_{0.5}$S$_{0.5}$ liquid-like materials. *Mater. Today Phys.* **2017**, *1*, 14–23. [CrossRef]
87. Kim, S.I.; Lee, K.H.; A Mun, H.; Kim, H.S.; Hwang, S.W.; Roh, J.W.; Yang, D.J.; Shin, W.H.; Li, X.S.; Lee, Y.H.; et al. Dense dislocation arrays embedded in grain boundaries for high-performance bulk thermoelectrics. *Science* **2015**, *348*, 109–114. [CrossRef]
88. Zhu, H.; Mao, J.; Li, Y.; Sun, J.; Wang, Y.; Zhu, Q.; Li, G.; Song, Q.; Zhou, J.; Fu, Y.; et al. Discovery of TaFeSb—Based half—Heuslers with high thermoelectric performance. *Nat. Commun.* **2019**, *10*, 270. [CrossRef]
89. Hu, X.; Jood, P.; Ohta, M.; Kunii, M.; Nagase, K.; Nishiate, H.; Kanatzidis, M.G.; Yamamoto, A. Power generation from nanostructured PbTe-based thermoelectrics: Comprehensive development from materials to modules. *Energy Environ. Sci.* **2016**, *9*, 517–529. [CrossRef]
90. Kraemer, D.; Sui, J.; McEnaney, K.; Zhao, H.; Jie, Q.; Ren, Z.F.; Chen, G. High thermoelectric conversion efficiency of MgAgSb-based material with hot-pressed contacts. *Energy Environ. Sci.* **2015**, *8*, 1299–1308. [CrossRef]
91. Chang, C.; Wu, M.; He, D.; Pei, Y.; Wu, C.F.; Wu, X.; Yu, H.; Zhu, F.; Wang, K.; Chen, Y.; et al. 3D charge and 2D phonon transports leading to high out-of-plane ZT in n-type SnSe crystals. *Science* **2018**, *360*, 778–783. [CrossRef]
92. Hu, L.; Wu, H.; Zhu, T.; Fu, C.; He, J.; Ying, P.; Zhao, X. Tuning Multiscale Microstructures to Enhance Thermoelectric Performance of n-Type Bismuth-Telluride-Based Solid Solutions. *Adv. Energy Mater.* **2015**, *5*, 1500411. [CrossRef]

93. Gao, M.-R.; Xu, Y.; Jiang, J.; Yu, S.-H. Nanostructured metal chalcogenides: Synthesis, modification, and applications in energy conversion and storage devices. *Chem. Soc. Rev.* **2013**, *42*, 2986–3017. [CrossRef]
94. Ding, Z.; Bux, S.K.; King, D.J.; Chang, F.L.; Chen, T.-H.; Huang, S.-C.; Kaner, R.B. Lithium intercalation and exfoliation of layered bismuth selenide and bismuth telluride. *J. Mater. Chem.* **2009**, *19*, 2588–2592. [CrossRef]
95. Wu, Y.; Wadia, C.; Ma, W.; Sadtler, B.; Alivisatos, P. Synthesis and Photovoltaic Application of Copper(I) Sulfide Nanocrystals. *Nano Lett.* **2008**, *8*, 2551–2555. [CrossRef] [PubMed]
96. Wang, C.; Zhang, D.; Xu, L.; Jiang, Y.; Dong, F.; Yang, B.; Yu, K.; Lin, Q. A Simple Reducing Approach Using Amine To Give Dual Functional EuSe Nanocrystals and Morphological Tuning. *Angew. Chem. Int. Ed.* **2011**, *50*, 7587–7591. [CrossRef] [PubMed]
97. Shen, H.B.; Wang, H.Z.; Yuan, H.; Ma, L.; Li, L.S. Size-, shape-, and assembly-controlled synthesis of $Cu_{2-x}Se$ nanocrystals via a non-injection phosphine-free colloidal method. *Cryst. Eng. Comm.* **2012**, *14*, 555–560. [CrossRef]
98. Fu, J.; Song, S.; Zhang, X.; Cao, F.; Zhou, L.; Li, X.; Zhang, H. Bi_2Te_3 nanoplates and nanoflowers: Synthesized by hydrothermal process and their enhanced thermoelectric properties. *Cryst. Eng. Comm.* **2012**, *14*, 2159–2165. [CrossRef]
99. Wu, Z.; Pan, C.; Yao, Z.; Zhao, Q.; Xie, Y. Large-Scale Synthesis of Single-Crystal Double-Fold Snowflake Cu_2S Dendrites. *Cryst. Growth Des.* **2006**, *6*, 1717–1719. [CrossRef]
100. Wang, X.; Zhuang, J.; Peng, Q.; Li, Y. A general strategy for nanocrystal synthesis. *Nat. Cell Biol.* **2005**, *437*, 121–124. [CrossRef]
101. Bilecka, I.; Niederberger, M. Microwave chemistry for inorganic nanomaterials synthesis. *Nanoscale* **2010**, *2*, 1358–1374. [CrossRef]
102. Li, B.; Xie, Y.; Huang, J.; Liu, Y.; Qian, Y. Sonochemical Synthesis of Nanocrystalline Copper Tellurides Cu_7Te_4 and Cu_4Te_3 at Room Temperature. *Chem. Mater.* **2000**, *12*, 2614–2616. [CrossRef]
103. She, G.; Zhang, X.; Shi, W.; Cai, Y.; Wang, N.; Liu, P.; Chen, D. Template-Free Electrochemical Synthesis of Single-Crystal CuTe Nanoribbons. *Cryst. Growth Des.* **2008**, *8*, 1789–1791. [CrossRef]
104. Lee, K.-J.; Song, H.; Lee, Y.-I.; Jung, H.; Zhang, M.; Choa, Y.-H.; Myung, N.V. Synthesis of ultra-long hollow chalcogenide nanofibers. *Chem. Commun.* **2011**, *47*, 9107–9109. [CrossRef]
105. Ng, C.H.B.; Tan, H.; Fan, W.Y. Formation of Ag_2Se Nanotubes and Dendrite-like Structures from UV Irradiation of a CSe_2/Ag Colloidal Solution. *Langmuir* **2006**, *22*, 9712–9717. [CrossRef]
106. Li, Z.; Yang, H.; Ding, Y.; Xiong, Y.; Xie, Y. Solution-phase template approach for the synthesis of Cu_2S nanoribbons. *Dalton Trans.* **2005**, 149–151. [CrossRef] [PubMed]
107. Zhao, Y.; Pan, H.; Lou, Y.; Qiu, X.; Zhu, J.; Burda, C. Plasmonic $Cu_{2−x}S$ Nanocrystals: Optical and Structural Properties of Copper-Deficient Copper(I) Sulfides. *J. Am. Chem. Soc.* **2009**, *131*, 4253–4261. [CrossRef] [PubMed]
108. Ishembetov, R.K.; Yulaeva, Y.K.; Balapanov, M.K.; Sharipov, T.; Yakshibayev, R. Electrophysical properties of nanostructured copper selenide ($Cu_{1.9}Li_{0.1}Se$). *Perspect. Mater.* **2011**, *12*, 55–59. (In Russian)
109. Yang, D.; Benton, A.; He, J.; Tang, X. Novel synthesis recipes boosting thermoelectric study of A_2Q (A = Cu, Ag; Q = S, Se, Te). *J. Phys. D Appl. Phys.* **2020**, *53*, 193001. [CrossRef]
110. He, Y.; Zhang, T.; Shi, X.; Wei, S.-H.; Chen, L. High thermoelectric performance in copper telluride. *NPG Asia Mater.* **2015**, *7*, e210. [CrossRef]
111. Poon, S.J. Recent Advances in Thermoelectric Performance of Half-Heusler Compounds. *Metals* **2018**, *8*, 989. [CrossRef]
112. Zhu, H.; He, R.; Mao, J.; Zhu, Q.; Li, C.; Sun, J.; Ren, W.; Wang, Y.; Liu, Z.; Tang, Z.; et al. Discovery of ZrCoBi based half Heuslers with high thermoelectric conversion efficiency. *Nat. Commun.* **2018**, *9*, 1–9. [CrossRef] [PubMed]
113. Rogl, G.; Ghosh, S.; Wang, L.; Bursik, J.; Grytsiv, A.; Kerber, M.; Bauer, E.; Mallik, R.C.; Chen, X.-Q.; Zehetbauer, M.; et al. Half-Heusler alloys: Enhancement of ZT after severe plastic deformation (ultra-low thermal conductivity). *Acta Mater.* **2020**, *183*, 285–300. [CrossRef]
114. Rogl, G.; Rogl, P. Skutterudites, a most promising group of thermoelectric materials. *Curr. Opin. Green Sustain. Chem.* **2017**, *4*, 50–57. [CrossRef]
115. Shi, X.; Yang, J.; Salvador, J.R.; Chi, M.; Cho, J.Y.; Wang, H.; Bai, S.; Yang, J.; Zhang, W.; Chen, L. Multiple-Filled Skutterudites: High Thermoelectric Figure of Merit through Separately Optimizing Electrical and Thermal Transports. *J. Am. Chem. Soc.* **2011**, *133*, 7837–7846. [CrossRef]
116. Liu, Z.-Y.; Zhu, J.-L.; Tong, X.; Niu, S.; Zhao, W.-Y. A review of $CoSb_3$-based skutterudite thermoelectric materials. *J. Adv. Ceram.* **2020**, *9*, 647–673. [CrossRef]
117. Zhang, S.; Xu, S.; Gao, H.; Lu, Q.; Lin, T.; He, P.; Geng, H. Characterization of multiple-filled skutterudites with high thermoelectric performance. *J. Alloys Compd.* **2020**, *814*, 152272. [CrossRef]
118. Mott, N.F.; Davis, E.A.; Weiser, K. Electronic Processes in Non-Crystalline Materials. *Phys. Today* **1972**, *25*, 55. [CrossRef]
119. Shklovsky, B.I.; Efros, A.A. *Electronic Properties of Doped Semiconductors*; Springer: New York, NY, USA, 1984.
120. Titov, A.; Yarmoshenko, Y.; Titova, S.; Krasavin, L.; Neumann, M. Localization of charge carriers in materials with high polaron concentration. *Phys. B Condens. Matter* **2003**, *328*, 108–110. [CrossRef]
121. Yarmoshenko, Y.M.; Shkvarin, A.; Yablonskikh, M.V.; Merentsov, A.I.; Titov, A. Localization of charge carriers in layered crystals Me_xTiSe_2 (Me = Cr, Mn, Cu) studied by the resonant photoemission. *J. Appl. Phys.* **2013**, *114*, 133704. [CrossRef]
122. Wagner, J.B.; Wagner, C. Investigations on Cuprous Sulfide. *J. Chem. Phys.* **1957**, *26*, 1602–1606. [CrossRef]
123. Wagner, J.B.; Wagner, C. Electrical Conductivity Measurements on Cuprous Halides. *J. Chem. Phys.* **1957**, *26*, 1597–1601. [CrossRef]
124. Blakemore, J.S. *Solid State Physics*, 2nd ed.; Saunders: Philadelphia, PA, USA, 1974; 506p.

125. Yokota, I. On the Theory of Mixed Conduction with Special Reference to Conduction in Silver Sulfide Group Semiconductors. *J. Phys. Soc. Jpn.* **1961**, *16*, 2213–2223. [CrossRef]
126. Ishikawa, I.; Miyatani, S. Electronic and Ionic Conduction in $Cu_{2-\delta}Se$, $Cu_{2-\delta}S$ and $Cu_{2-\delta}(S,Se)$. *J. Phys. Soc. Jpn.* **1977**, *42*, 159–167. [CrossRef]
127. Yokota, I.; Miyatani, S. Conduction and diffusion in ionic-electronic conductors. *Solid State Ion.* **1981**, *3–4*, 17–21. [CrossRef]
128. Gafurov, I.G. Transport phenomenon and structural features in superionic alloys $Cu_{2-x}LixS$ (0.05 < x < 0.25). Doctoral Dissertation, Bashkir State University, Ufa, Russia, 1998; 20p.
129. Guan, M.-J.; Qiu, P.-F.; Song, Q.-F.; Yang, J.; Ren, D.-D.; Shi, X.; Chen, L. Improved electrical transport properties and optimized thermoelectric figure of merit in lithium-doped copper sulfides. *Rare Met.* **2018**, *37*, 282–289. [CrossRef]
130. Ishembetov, R.K. *Yavleniya Perenosa v Superionnykh Khalkogenidakh Medi, Zamesennykh Serebrom i Litiem*; Bashkirski Gosudarstvennyi Universitet: Ufa, Russia, 2006; 20p, Available online: https://search.rsl.ru/ru/record/01000261087 (accessed on 11 July 2021). (In Russian)
131. Balapanov, M.K.; Ishembetov, R.K.; Ishembetov, S.R.; Kubenova, M.M.; Kuterbekov, K.A.; Nazarov, K.; Yakshibaev, R.A. Electronic and ionic zeebeck coefficients in mixed conductors of $Ag_{0.25-\delta}Cu_{1.75}Se$, $Ag_{1.2-\delta}Cu_{0.8}Se$. *Russ. J. Electrochem.* **2017**, *53*, 859–865. [CrossRef]
132. Ishiwata, S.; Shiomi, Y.; Lee, J.S.; Bahramy, M.S.; Suzuki, T.; Uchida, M.; Arita, R.; Taguchi, Y.; Tokura, Y. Extremely high electron mobility in a phonon-glass semimetal. *Nat. Mater.* **2013**, *12*, 512–517. [CrossRef]
133. Moroz, N.A.; Olvera, A.; Willis, G.M.; Poudeu, P.F.P. Rapid direct conversion of $Cu_{2-x}Se$ to CuAgSe nanoplatelets via ion exchange reactions at room temperature. *Nanoscale* **2015**, *7*, 9452–9456. [CrossRef] [PubMed]
134. Hong, A.; Li, L.; Zhu, H.; Zhou, X.; He, Q.; Liu, W.; Yan, Z.; Liu, J.-M.; Ren, Z. Anomalous transport and thermoelectric performances of CuAgSe compounds. *Solid State Ion.* **2014**, *261*, 21–25. [CrossRef]
135. Miyatani, S.; Toyota, Y.; Yanagihara, T.; Iida, K. α-Ag2Se as a Degenerate Semicanductor. *J. Phys. Soc. Jpn.* **1967**, *23*, 35–43. [CrossRef]
136. Peplinski, Z.; Brown, D.B.; Watt, T.; Hatfield, W.E.; Day, P. Electrical properties of sodium copper sulfide ($Na_3Cu_4S_4$), a mixed-valence one-dimensional metal. *Inorg. Chem.* **1982**, *21*, 1752–1755. [CrossRef]
137. Ge, Z.-H.; Zhang, B.-P.; Chen, Y.-X.; Yu, Z.-X.; Liu, Y.; Li, J.-F. Synthesis and transport property of $Cu_{1.8}S$ as a promising thermoelectric compound. *Chem. Commun.* **2011**, *47*, 12697–12699. [CrossRef]
138. Zhang, Y.-X.; Feng, J.; Ge, Z.-H. High thermoelectric performance realized in porous $Cu_{1.8}S$ based composites by Na_2S addition. *Mater. Sci. Semicond. Process.* **2019**, *107*, 104848. [CrossRef]
139. Qiu, P.; Shi, X.; Chen, L. Thermoelectric Properties of $Cu_{2-\delta}X$ (X = S, Se, and Te). In *Materials Aspect of Thermoelectricity*; Uher, C., Ed.; CRC Press: Boca Raton, FL, USA, 2016; 624p.
140. Slack, G. New Materials and Performance Limits for Thermoelectric Cooling. In *CRC Handbook of Thermoelectrics*; Pollock Industries, Inc.: White River, VT, USA, 1995; pp. 407–440.
141. He, X.; Zhu, Y.; Mo, Y. Origin of fast ion diffusion in super-ionic conductors. *Nat. Commun.* **2017**, *8*, 15893. [CrossRef]
142. Donati, C.; Douglas, J.F.; Kob, W.; Plimpton, S.J.; Poole, P.; Glotzer, S.C. Stringlike Cooperative Motion in a Supercooled Liquid. *Phys. Rev. Lett.* **1998**, *80*, 2338–2341. [CrossRef]
143. Keys, A.S.; Hedges, L.O.; Garrahan, J.; Glotzer, S.C.; Chandler, D. Excitations are localized and relaxation is hierarchical in glass-forming liquids. *Phys. Rev. X* **2011**, *1*, 021013. [CrossRef]
144. Wakamura, K. Interpretation of high ionic conduction in superionic conductors based on electronic and phonon properties. *Solid State Ion.* **2004**, *171*, 229–235. [CrossRef]
145. Kikuchi, H.; Iyetomi, H.; Hasegawa, A. Insight into the origin of superionic conductivity from electronic structure theory. *J. Phys. Condens. Matter* **1998**, *10*, 11439–11448. [CrossRef]
146. Lavrent'ev, A.A.; Nikiforov, I.Y.; Dubeiko, V.A.; Gabrel'yan, B.V.; Domashevskaya, E.P.; Rehr, J.J.; Ankudinov, A.L. d-p-rezonansnoe vozdeystvie v soedineniyah medi s razlichnymi kristallicheskimi strukturami. *Kondens. Sredy I Mezhfaznye Granitsy* **2001**, *3*, 107–121. (In Russian)
147. Domashevskaya, E.; Gorbachev, V.; Terekhov, V.; Kashkarov, V.; Panfilova, E.; Shchukarev, A. XPS and XES emission investigations of d–p resonance in some copper chalcogenides. *J. Electron. Spectrosc. Relat. Phenom.* **2001**, *114*, 901–908. [CrossRef]
148. Balapanov, M.K.; Urazaeva, E.K.; Zinnurov, I.B.; Musalimov, R.S.; Yakshibaev, R.A. Influence of grain sizes on the ionic conductivity and the chemical diffusion coefficient in copper selenide. *Ionics* **2006**, *12*, 205–209. [CrossRef]
149. Bruheim, I.; Cameron, J. Flowable Concentrated Phospholipid Krill Oil Composition. U.S. Patent 20170020928 Al, 26 January 2017.
150. Meyer, M.; Jaenisch, V.; Maass, P.; Bunde, A. Mixed Alkali Effect in Crystals of β- and β″-Alumina Structure. *Phys. Rev. Lett.* **1996**, *76*, 2338–2341. [CrossRef] [PubMed]
151. Kadrgulov, R.F.; Yakshibaev, R.A.; Khasanov, M.A. Phase Relations, Ionic Conductivity and Diffusion in the Alloys of Cu_2S and Ag_2S Mixed Conductors. *Ionics* **2001**, *7*, 156–160. [CrossRef]
152. Yakshibaev, R.A.; Balapanov, M.K.; Mukhamadeeva, N.N.; Akmanova, G.R. Partial Conductivity of Cations of Different Kinds in the Alloys of Cu_2X-Ag_2X (X = Se, Te) Mixed Conductors. *Phys. Stat. Sol.* **1989**, *112*, 97. [CrossRef]
153. Balapanov, M.K. Effect of Cationic Substitution on Ion Transfer Phenomena in Superionic Copper Chalcogenides. *Bull. Bashkir Univ.* **2006**, *2*, 33–36.

154. Balapanov, M.K.; Ishembetov, R.K.; Yakshibaev, R.A. Soret effect and heat of silver atom transport in $Ag_{(2-x)} + \delta Cu_xSe$ (x = 0.1, 0.2, 0.4) superionic solid solutions. *Inorg. Mater.* **2006**, *42*, 705–707. [CrossRef]
155. Yabuuchi, N.; Kubota, K.; Dahbi, M.; Komaba, S. Research development on sodium-ion batteries (Reviev). *Chem. Rev.* **2014**, *114*, 11636–11682. [CrossRef] [PubMed]
156. Kulova, T.L.; Skundin, A.M. From lithium-ion to sodium-ion batteries. *Electrochem. Power Eng.* **2016**, *16*, 122–150. [CrossRef]
157. Li, L.; Zheng, Y.; Zhang, S.; Yang, J.; Shao, Z.; Guo, Z. Recent progress on sodium ion batteries: Potential high-performance anodes. *Energy Environ. Sci.* **2018**, *11*, 2310–2340. [CrossRef]
158. Yue, J.-L.; Sun, Q.; Fu, Z.-W. Cu_2Se with facile synthesis as a cathode material for rechargeable sodium batteries. *Chem. Commun.* **2013**, *49*, 5868–5870. [CrossRef] [PubMed]
159. Kim, J.-S.; Kim, D.-Y.; Cho, G.-B.; Nam, T.-H.; Kim, K.-W.; Ryu, H.-S.; Ahn, J.-H.; Ahn, H.-J. The electrochemical properties of copper sulfide as cathode material for rechargeable sodium cell at room temperature. *J. Power Sources* **2009**, *189*, 864–868. [CrossRef]
160. Emin, D. Seebeck Effect. In *Wiley Encyclopedia of Electrical and Electronics Engineering*; Webster, J.G., Ed.; WILEY: Madison, WI, USA, 2002; p. 33.
161. Fistul, V.I. *Vvedenie v Fiziku Polyprovodnikov*; Vysshaya Shkola: Moscow, Russia, 1984; 352p.
162. Bonch-Bruevich, V.L.; Kalashnikov, S.G. *Semiconductor Physics*; Science: Moscow, Russia, 1977; Volume 672.
163. Anselm, A.I. *Introduction to the Theory of Semiconductors*; Science: Moscow, Russia, 1978; 616p.
164. Han, C.; Li, Z.; Dou, S. Recent progress in thermoelectric materials. *Chin. Sci. Bull.* **2014**, *59*, 2073–2091. [CrossRef]
165. Liu, W.-D.; Yanga, L.; Chen, Z.-G. Cu_2Se thermoelectrics: Property, methodology, and devices. *Nano Today* **2020**, *35*, 100938. [CrossRef]
166. Ma, Z.; Wei, J.; Song, P.; Zhang, M.; Yang, L.; Ma, J.; Liu, W.; Yang, F.; Wang, X. Review of experimental approaches for improving zT of thermoelectric materials. *Mater. Sci. Semicond. Process.* **2021**, *121*, 105303. [CrossRef]
167. Zhang, Z.; Zhao, K.; Wei, T.-R.; Qiu, P.; Chen, L.; Shi, X. Cu_2Se-Based liquid-like thermoelectric materials: Looking back and stepping forward. *Energy Environ. Sci.* **2020**, *13*, 3307–3329. [CrossRef]
168. Sun, Y.; Xi, L.; Yang, J.; Wu, L.; Shi, X.; Chen, L.; Snyder, J.; Yang, J.; Zhang, W. The "electron crystal" behavior in copper chalcogenides Cu_2X (X = Se, S). *J. Mater. Chem. A* **2017**, *5*, 5098–5105. [CrossRef]
169. Tang, H.; Sun, F.-H.; Dong, J.-F.; Zhuang, H.-L.; Pan, Y.; Li, J.-F. Graphene network in copper sulfide leading to enhanced thermoelectric properties and thermal stability. *Nano Energy* **2018**, *49*, 267–273. [CrossRef]
170. Zhu, Z.; Zhang, Y.; Song, H.; Li, X.-J. Enhancement of thermoelectric performance of Cu_2Se by K doping. *Appl. Phys. A* **2018**, *124*, 871. [CrossRef]
171. Bulat, L.P.; Osvenskii, V.B.; Ivanov, A.A.; Sorokin, A.I.; Pshenay-Severin, D.A.; Bublik, V.T.; Tabachkova, N.Y.; Panchenko, V.P.; Lavrentev, M.G. Experimental and theoretical study of the thermoelectric properties of copper selenide. *Semiconductors* **2017**, *51*, 854–857. [CrossRef]
172. Bulat, L.P.; Ivanov, A.A.; OsvenskiiV, B.; Pshenay-Severin, D.A.; Sorokin, A.I. Thermal conductivity of Cu_2Se taking into account the influence of mobile copper ions. *Phys. Solid State* **2017**, *59*, 2097–2102. [CrossRef]
173. Kim, H.; Ballikaya, S.; Chi, H.; Ahn, J.-P.; Ahn, K.; Uher, C.; Kaviany, M. Ultralow thermal conductivity of β-Cu_2Se by atomic fluidity and structure distortion. *Acta Mater.* **2015**, *86*, 247–253. [CrossRef]
174. Tyagi, K.; Gahtori, B.; Bathula, S.; Jayasimhadri, M.; Singh, N.K.; Sharma, S.; Haranath, D.; Srivastava, A.; Dhar, A. Enhanced thermoelectric performance of spark plasma sintered copper-deficient nanostructured copper selenide. *J. Phys. Chem. Solids* **2015**, *81*, 100–105. [CrossRef]
175. Byeon, D.; Sobota, R.; Delime-Codrin, K.; Choi, S.; Hirata, K.; Adachi, M.; Kiyama, M.; Matsuura, T.; Yamamoto, Y.; Matsunami, M.; et al. Discovery of colossal Seebeck effect in metallic Cu_2Se. *Nat. Commun.* **2019**, *10*, 72. [CrossRef]
176. Ballikaya, S.; Sertkol, M.; Oner, Y.; Bailey, T.P.; Uher, C. Fracture structure and thermoelectric enhancement of Cu_2Se with substitution of nanostructured Ag_2Se. *Phys. Chem. Chem. Phys.* **2019**, *21*, 13569–13577. [CrossRef] [PubMed]
177. Zhong, B.; Zhang, Y.; Li, W.; Chen, Z.; Cui, J.; Li, W.; Xie, Y.; Hao, Q.; He, Q. High superionic conduction arising from aligned large lamellae and large figure of merit in bulk $Cu_{1.94}Al_{0.02}Se$. *Appl. Phys. Lett.* **2014**, *105*, 123902. [CrossRef]
178. Zhao, L.; Islam, S.M.K.N.; Wang, J.; Cortie, D.; Wang, X.; Cheng, Z.; Wang, J.; Ye, N.; Dou, S.X.; Shi, X.; et al. Significant enhancement of figure-of-merit in carbon-reinforced Cu_2Se nanocrystalline solids. *Nano Energy* **2017**, *41*, 164–171. [CrossRef]
179. Zhao, K.; Zhu, C.; Qiu, P.; Blichfeld, A.B.; Eikeland, E.; Ren, D.; Iversen, B.B.; Xu, F.; Shi, X.; Chen, L. High thermoelectric performance and low thermal conductivity in $Cu_{2-y}S_{1/3}Se_{1/3}Te_{1/3}$ liquid-like materials with nanoscale mosaic structures. *Nano Energy* **2017**, *42*, 43–50. [CrossRef]
180. Sirusi, A.A.; Ballikaya, S.; Chen, J.H.; Uher, C.; Ross, J.H. Band ordering and dynamics of $Cu_{2-x}Te$ and $Cu_{1.98}Ag_{0.2}Te$. *J. Phys. Chem. C* **2016**, *120*, 14549–14555. [CrossRef]
181. Park, D.; Ju, H.; Oh, T.; Kim, J. Fabrication of one-dimensional Cu_2Te/Te nanorod composites and their enhanced thermoelectric properties. *Cryst. Eng. Comm.* **2018**, *21*, 1555–1563. [CrossRef]
182. Mukherjee, S.; Parasuraman, R.; Umarji, A.M.; Rogl, G.; Rogl, P.; Chattopadhyay, K. Effect of Fe alloying on the thermoelectric performance of Cu_2Te. *J. Alloys Compd.* **2020**, *817*, 152729. [CrossRef]
183. Mukherjee, S.; Ghoshb, S.; Chattopadhyayac, K. Ultralow thermal conductivity and high thermoelectric figure of merit in Cu_2Te–Ag_2Te composites. *J. Alloys Compd.* **2020**, *848*, 156540. [CrossRef]

184. Okasaki, H. Deviation from the Einstein Relation in Average Crystals. II. Self-Diffusion of Ag$^+$ Ions in α-Ag$_2$Te. *J. Phys. Soc. Jpn.* **1977**, *43*, 213–221. [CrossRef]
185. Zhao, K.; Liu, K.; Yue, Z.; Wang, Y.; Song, Q.; Li, J.; Guan, M.; Xu, Q.; Qiu, P.; Zhu, H.; et al. Are Cu$_2$Te-Based Compounds Excellent Thermoelectric Materials? *Adv. Mater.* **2019**, *31*, 1903480. [CrossRef]
186. Gao, J.; Miao, L.; Lai, H.; Zhu, S.; Peng, Y.; Wang, X.; Koumoto, K.; Cai, H. Thermoelectric Flexible Silver Selenide Films: Compositional and Length Optimization. *IScience* **2020**, *23*, 100753. [CrossRef] [PubMed]
187. Ding, Y.; Qiu, Y.; Cai, K.; Yao, Q.; Chen, S.; Chen, L.; He, J. High performance n-type Ag$_2$Se film on nylon membrane for flexible thermoelectric power generator. *Nat. Commun.* **2019**, *10*, 841. [CrossRef]
188. Lim, K.H.; Wong, K.W.; Liu, Y.; Zhang, Y.; Cadavid, D.; Cabot, A.; Ng, K.M. Critical role of nanoinclusions in silver selenide nanocomposites as a promising room temperature thermoelectric material. *J. Mater. Chem. C* **2019**, *7*, 2646–2652. [CrossRef]
189. Perez-Taborda, J.A.; Caballero-Calero, O.; Vera-Londono, L.; Briones, F.; Martin-Gonzalez, M. High Thermoelectric zT in n-Type Silver Selenide films at Room Temperature. *Adv. Energy Mater.* **2018**, *8*, 1702024. [CrossRef]
190. Chen, J.; Sun, Q.; Bao, D.; Liu, T.; Liu, W.-D.; Liu, C.; Tang, J.; Zhou, D.; Yang, L.; Chen, Z.-G. Hierarchical Structures Advance Thermoelectric Properties of Porous n-type β-Ag$_2$Se. *ACS Appl. Mater. Interfaces* **2020**, *12*, 51523–51529. [CrossRef] [PubMed]
191. Zhou, W.-X.; Wu, D.; Xie, G.; Chen, K.-Q.; Zhang, G. α-Ag$_2$S: A Ductile Thermoelectric Material with High ZT. *ACS Omega* **2020**, *5*, 5796–5804. [CrossRef] [PubMed]
192. Tarachand, C.; Sharma, V.; Ganesan, V.; Okram, G.S. Thermoelectric properties of CuS/Ag$_2$S nanocomposites synthesed by modified polyol method. In *DAE Solid State Physics Symposium*; AIP Publishing LLC: Odisha, India, 2016; Volume 1731, p. 110024. [CrossRef]
193. Wang, T.; Chen, H.-Y.; Qiu, P.-F.; Shi, X.; Chen, L.-D. Thermoelectric properties of Ag$_2$S superionic conductor with intrinsically low lattice thermal conductivity. *Acta Phys. Sin.* **2019**, *68*, 090201. [CrossRef]
194. Kim, G.; Byeon, D.; Singh, S.; Hirata, K.; Choi, S.; Matsunami, M.; Takeuchi, T. Mixed-phase effect of a high Seebeck coefficient and low electrical resistivity in Ag$_2$S. *J. Phys. D Appl. Phys.* **2021**, *54*, 115503. [CrossRef]
195. Wu, R.; Li, Z.; Li, Y.; You, L.; Luo, P.; Yang, J.; Luo, J. Synergistic optimization of thermoelectric performance in p-type Ag$_2$Te through Cu substitution. *J. Mater.* **2019**, *5*, 489–495. [CrossRef]
196. Gao, J.; Miao, L.; Liu, C.; Wang, X.; Peng, Y.; Wei, X.; Zhou, J.; Chen, Y.; Hashimoto, R.; Asaka, T.; et al. A novel glass-fiber-aided cold-pressmethod for fabrication of n-type Ag$_2$Te nanowires thermoelectric film on flexible copy-paper substrate. *J. Mater. Chem. A* **2017**, *5*, 24740–24748. [CrossRef]
197. Zhu, H.; Luo, J.; Zhao, H.; Liang, J. Enhanced thermoelectric properties of p-type Ag$_2$Te by Cu substitution. *J. Mater. Chem. A* **2015**, *3*, 10303–10308. [CrossRef]
198. Lee, S.; Shin, H.S.; Song, J.Y.; Jung, M.-H. Thermoelectric Properties of a Single Crystalline Ag$_2$Te Nanowire. *J. Nanomater.* **2017**, *2017*, 4308968. [CrossRef]
199. Zhu, T.; Bai, H.; Zhang, J.; Tan, G.; Yan, Y.; Liu, W.; Su, X.; Wu, J.; Zhang, Q.; Tang, X. Realizing High Thermoelectric Performance in Sb-Doped Ag$_2$Te Compounds with a Low-Temperature Monoclinic Structure. *ACS Appl. Mater. Interfaces* **2020**, *12*, 39425–39433. [CrossRef]
200. Hu, Q.; Zhu, Z.; Zhang, Y.; Li, X.-J.; Song, H.; Zhang, Y. Remarkably high thermoelectric performance of Cu$_{2-x}$Li$_x$Se bulks with nanopores. *J. Mater. Chem. A* **2018**, *6*, 23417–23424. [CrossRef]
201. Ioffe, A.F. *Physics of Semiconductors*; Academic Press: New York, NY, USA, 1960.
202. Qin, Y.; Yang, L.; Wei, J.; Yang, S.; Zhang, M.; Wang, X.; Yang, F. Doping Effect on Cu$_2$Se Thermoelectric Performance: A Review. *Materials* **2020**, *13*, 5704. [CrossRef] [PubMed]
203. Qiu, P.; Zhu, Y.; Qin, Y.; Shi, X.; Chen, L. Electrical and thermal transports of binary copper sulfides Cu$_x$S with x from 1.8 to 1.96. *APL Mater.* **2016**, *4*, 104805. [CrossRef]
204. Liang, J.; Wang, T.; Qiu, P.; Yang, S.; Ming, C.; Chen, H.; Song, Q.; Zhao, K.; Wei, T.-R.; Ren, D.; et al. Flexible thermoelectrics: From silver chalcogenides to full-inorganic devices. *Energy Environ. Sci.* **2019**, *12*, 2983–2990. [CrossRef]
205. Bickulova, N.N. Crystal Structure, Lattice Dynamics and Ion Transport in Superionic Conductors Based on Copper and Silver Chalcogenides. Ph.D. Thesis, Bashkir State University, Ufa, Russia, 2005; 47p.
206. Cadoff, I.B.; Miller, E. *Thermoelectric Materials and Devices*; Reinhold: New York, NY, USA, 1960; 344p.
207. May, A.; Fleurial, J.-P.; Snyder, J. Thermoelectric performance of lanthanum telluride produced via mechanical alloying. *Phys. Rev. B* **2008**, *78*, 125205. [CrossRef]
208. Wu, Y.; Chen, Z.; Nan, P.; Xiong, F.; Lin, S.; Zhang, X.; Chen, Y.; Chen, L.; Ge, B.; Pei, Y. Lattice Strain Advances Thermoelectrics. *Joule* **2019**, *3*, 1276–1288. [CrossRef]
209. Liu, H.; Yang, J.; Shi, X.; Danilkin, S.A.; Yu, D.; Wang, C.; Zhang, W.; Chen, L. Reduction of thermal conductivity by low energy multi-Einstein optic modes. *J. Mater.* **2016**, *2*, 187–195. [CrossRef]
210. Bouyrie, Y.; Candolfi, C.; Pailhès, S.; Koza, M.; Malaman, B.; Dauscher, A.; Tobola, J.; Boisron, O.; Saviot, L.; Lenoir, B. From crystal to glass-like thermal conductivity in crystalline minerals. *Phys. Chem. Chem. Phys.* **2015**, *17*, 19751–19758. [CrossRef]
211. Danilkin, S.A.; Skomorokhov, A.N.; Hoser, A.; Fuess, H.; Rajevac, V.; Bickulova, N.N. Crystal Structure and Lattice Dynamics of Superionic Copper Selenide Cu$_{2-δ}$Se. *J. Alloys Compd.* **2004**, *35*, 57–61. [CrossRef]
212. Dutta, M.; Samanta, M.; Ghosh, T.; Voneshen, D.J.; Biswas, K. Evidence of Highly Anharmonic Soft Lattice Vibrations in a Zintl Rattler. *Angew. Chem. Int. Ed.* **2021**, *60*, 4259–4265. [CrossRef] [PubMed]

213. Matthias Agne, T.; Voorhees, P.W.; Snyder, G.J. Phase transformation contributions to heat capacity and impact on thermal diffusivity, thermal conductivity, and thermoelectric performance. *Adv. Mater.* **2019**, *31*, 1902980. [CrossRef] [PubMed]
214. Liu, H.; Yuan, X.; Lu, P.; Shi, X.; Xu, F.; He, Y.; Tang, Y.; Bai, S.; Zhang, W.; Chen, L.; et al. Ultrahigh Thermoelectric Performance by Electron and Phonon Critical Scattering in $Cu_2Se_{1-x}I_x$. *Adv. Mater.* **2013**, *25*, 6607–6612. [CrossRef] [PubMed]
215. Kang, S.D.; A Danilkin, S.; Aydemir, U.; Avdeev, M.; Studer, A.; Snyder, G.J. Apparent critical phenomena in the superionic phase transition of $Cu_{2-x}Se$. *New J. Phys.* **2016**, *18*, 013024. [CrossRef]
216. Chen, H.; Yue, Z.; Ren, D.; Zeng, H.; Wei, T.; Zhao, K.; Yang, R.; Qiu, P.; Chen, L.; Shi, X. Thermal Conductivity during Phase Transitions. *Adv. Mater.* **2018**, *31*, e1806518. [CrossRef]
217. Vasilevskiy, D.; Keshavarz, M.K.; Simard, J.-M.; Masut, R.A.; Turenne, S.; Snyder, G.J. Assessing the Thermal Conductivity of $Cu_{2-x}Se$ Alloys Undergoing a Phase Transition via the Simultaneous Measurement of Thermoelectric Parameters by a Harman-Based Setup. *J. Electron. Mater.* **2018**, *47*, 3314–3319. [CrossRef]
218. Vasilevskiy, D.; Masut, R.A.; Turenne, S. A Phenomenological Model of Unconventional Heat Transport Induced by Phase Transition in $Cu_{2-x}Se$. *J. Electron. Mater.* **2019**, *48*, 1883–1888. [CrossRef]
219. Cheng, Y.; Yang, J.; Jiang, Q.; He, D.; He, J.; Luo, Y.; Zhang, D.; Zhou, Z.; Ren, Y.; Xin, J. New insight into InSb-based thermoelectric materials: From a divorced eutectic design to a remarkably high thermoelectric performance. *J. Mater. Chem. A* **2017**, *5*, 5163–5366. [CrossRef]
220. Brown, D.R.; Day, T.; Borup, K.; Christensen, S.; Iversen, B.; Snyder, J. Phase transition enhanced thermoelectric figure-of-merit in copper chalcogenides. *APL Mater.* **2013**, *1*, 052107. [CrossRef]
221. Agne, M.T.; Hanus, R.; Snyder, G.J. Minimum thermal conductivity in the context of diffuson-mediated thermal transport. *Energy Environ. Sci.* **2018**, *11*, 609–616. [CrossRef]
222. Mansour, B.A.; Tahoon, K.H.; El-Sharkawy, A.A. Thermophysical properties and mechanism of heat transfer of non-stoichiometric Cu_2S. *Phys. Status Solidi* **1995**, *148*, 423–430. [CrossRef]
223. Grønvold, F.; Westrum, E.F. Thermodynamics of copper sulfides I. Heat capacity and thermodynamic properties of copper(I) sulfide, Cu_2S, from 5 to 950 K. *J. Chem. Thermodyn.* **1987**, *19*, 1183–1198. [CrossRef]
224. Qiu, P.; Agne, M.T.; Liu, Y.; Zhu, Y.; Chen, H.; Mao, T.; Yang, J.; Zhang, W.; Haile, S.M.; Zeier, W.G.; et al. Suppression of atom motion and metal deposition in mixed ionic electronic conductors. *Nat. Commun.* **2018**, *9*, 1–8. [CrossRef]
225. Yokota, I. On the Electrical Conductivity of Cuprous Sulfide: A Diffusion Theory. *J. Phys. Soc. Jpn.* **1953**, *8*, 595–602. [CrossRef]
226. Korte, C.; Janek, J. Nonosothermal transport properties of α-$Ag_2+\delta S$: Partial thermopowers of electrons and ions, the Soret effect and heats of transport. *J. Phys. Chem. Solids* **1997**, *58*, 623–637. [CrossRef]

Article

Superparamagnetic Fe₃O₄@CA Nanoparticles and Their Potential as Draw Solution Agents in Forward Osmosis

Irena Petrinic [1,*], Janja Stergar [1,2], Hermina Bukšek [1], Miha Drofenik [1,3], Sašo Gyergyek [1,3], Claus Hélix-Nielsen [1,4,*] and Irena Ban [1]

[1] Faculty of Chemistry and Chemical Engineering, University of Maribor, Smetanova 17, SI-2000 Maribor, Slovenia; janja.stergar@um.si (J.S.); hermina.buksek@um.si (H.B.); miha.drofenik@um.si (M.D.); saso.gyergyek@um.si (S.G.); irena.ban@um.si (I.B.)
[2] Institute of Biomedical Sciences, Faculty of Medicine, University of Maribor, Taborska Ulica 8, SI-2000 Maribor, Slovenia
[3] Department of Materials Synthesis, Jožef Stefan Institute, Jamova cesta 29, SI-1000 Ljubljana, Slovenia
[4] Department of Environmental Engineering, Technical University of Denmark, Miljøvej 113, 2800 Kgs. Lyngby, Denmark
* Correspondence: irena.petrinic@um.si (I.P.); clhe@env.dtu.dk (C.H.-N.)

Abstract: In this study, citric acid (CA)-coated magnetite Fe₃O₄ magnetic nanoparticles (Fe₃O₄@CA MNPs) for use as draw solution (DS) agents in forward osmosis (FO) were synthesized by co-precipitation and characterized by Fourier transform infrared spectroscopy (FTIR), thermogravimetric analysis (TGA), dynamic light scattering (DLS), transmission electron microscopy (TEM) and magnetic measurements. Prepared 3.7% w/w colloidal solutions of Fe₃O₄@CA MNPs exhibited an osmotic pressure of 18.7 bar after purification without aggregation and a sufficient magnetization of 44 emu/g to allow DS regeneration by an external magnetic field. Fe₃O₄@CA suspensions were used as DS in FO cross-flow filtration with deionized (DI) water as FS and with the active layer of the FO membrane facing the FS and NaCl as a reference DS. The same transmembrane bulk osmotic pressure resulted in different water fluxes for NaCl and MNPs, respectively. Thus the initial water flux with Fe₃O₄@CA was 9.2 LMH whereas for 0.45 M NaCl as DS it was 14.1 LMH. The reverse solute flux was 0.08 GMH for Fe₃O₄@CA and 2.5 GMH for NaCl. These differences are ascribed to a more pronounced internal dilutive concentration polarization with Fe₃O₄@CA as DS compared to NaCl as DS. This research demonstrated that the proposed Fe₃O₄@CA can be used as a potential low reverse solute flux DS for FO processes.

Keywords: citrate-coated magnetic nanoparticle; forward osmosis; draw solution; osmotic pressure; non-ideality analysis

1. Introduction

Magnetic nanoparticles (MNPs) have attracted attention in research and industry in the chemical, environmental and medical fields. MNPs have shown promising performance in removing contaminants or reducing toxicity [1–5] and have generated interest in research into technical applications for treatment of polluted water and water purification processes [6–10]. In recent years, MNP's have also attracted attention as materials for generating the driving force to transport water in a forward osmosis (FO) membrane separation process, one of the emerging membrane technologies that can meet the increasing global demand for water recycling and reuse [11]. FO is a technical term describing a natural phenomenon of osmosis: the transport of water molecules across a semi-permeable membrane [12]. In contrast to pressure-driven membrane processes, FO is driven by the osmotic pressure difference across the FO membrane. In ideal FO, a concentrated draw solution (DS) with high osmotic pressure extracts water from a dilute feed solution (FS) while ideally rejecting all FS solutes [13,14]. In order to function effectively as a draw agent in FO, the osmotic pressure of the DS must significantly exceed that of the FS. Simple inorganic salts, such as NaCl, remain the most widely used DS agents due to their ability to have high osmotic pressures while maintaining low solution viscosities. The strong affinity of small inorganic ions for water is reflected in their highly exothermic enthalpies of hydration [15]. Strong solvent-solute interactions provide high solution osmotic pressures while making the regeneration of DS more difficult. In resolving this problem, the development of easily removable DS agents, which allow for regeneration through exploitation of solute size, thermal sensitivity or magnetic properties, is desirable [16].

Although the FO technology is gaining popularity in niche applications including cold-concentration in the food and beverage industry, drawbacks, such as high reverse solute flux and high regeneration costs, still restrict wider application [17–19]. A reverse solute flux can potentially contaminate the FS apart from decreasing the osmotic pressure difference across the membrane. The suitability of a draw solute is defined by the ability to develop sufficient osmotic pressure in order to ensure a high water flux while maintaining low (ideally no) diffusion from the DS to the FS. The molecular size and ionic structure of the DS agent define its applicability in FO. A smaller molecular size of the DS agents reduces the internal concentration polarization and thus associated with a higher water flux however, this is generally also associated with higher reverse solute flux. In comparison, larger DS agent molecules are associated with less internal concentration polarization (and thus lower water flux) but also a lower reverse solute diffusion [20], the latter a prerequisite for use in cold-concentration.

Over the past years, a variety of compounds have been investigated as draw solutes, including sulfur dioxide [21], various inorganic salts and a variety of sugars [22,23], thermally unstable ammonium salts [24,25] polyelectrolytes [26] and MNP's [27–29]. Iron oxide MNPs such as maghemite (γ-Fe_2O_3) and magnetite (Fe_3O_4) MNPs have been widely studied due to well-defined properties such as small uniform particle size in the range of 10 to 100 nm, biocompatibility, heat generation in response to an alternating magnetic field (AMF), easy surface manipulation for targeted release, low toxicity profile, and good colloidal stability [30]. Although pristine iron oxide MNPs have desirable properties (size and stability) as DS they cannot generate significant osmotic pressure without appropriate surface modification [31,32]. To improve water solubility and surface hydrophilicity, MNP's functionalized with strong hydrophilic groups are considered as one of the feasible solutions to generate sufficient osmotic pressures, as well as allow for facile regeneration [33] as exemplified by Fe_3O_4 MNPs, which can be easily separated from water by an external magnetic field [34]. To maximize the osmotic pressure that is generated from the functional groups coated on the iron oxide MNPs, a high number of polymer hydrophilic groups must be available to interact with the solvent water molecules. Citric acid (CA) is a small molecule which is suitable for facile functionalization and confers hydrophilic properties to MNPs by virtue of three –COOH groups with one or two –COOH groups absorbed on the surface of the MNPs and at least one remaining free [32,35–37]. Recently Khazaie et al. and his co-workers synthesized Fe_3O_4 nanoparticles and covalently functionalized them with tri-sodium citrate. The DS with an optimal ratio of CA: MNPs (2:1) with concentrations of 20, 40, 60, and 80 g L^{-1} showed water fluxes of 7.1, 10.7, 14.2, and 17.1 LMH, with osmotic pressures of 114, 117, 120 and 124 bar respectively [38]. Applications with MNP coated with sodium citrate has also reported by [39–41].

Here we investigate the synthesis of Fe_3O_4 MNP's coated with CA and evaluate their feasibility as DS agents in FO. MNP's were synthesized by co-precipitation of Fe^{2+} and Fe^{3+} aqueous solutions with hydrophilic CA by adding a base in a one-pot synthesis. The research is comprised of two sections: (1): the synthesis of Fe_3O_4@CA MNP's with characterization using FTIR, TGA, DLS, TEM and magnetic measurements and (2): DS testing using an AIM™ hollow fibre FO (HFFO) module where water flux and reverse solute flux is benchmarked against NaCl as DS using deionized (DI) water as the FS.

2. Materials and Methods

2.1. Materials

The chemicals used for the synthesis of Fe_3O_4 MNP's coated with CA were iron (III) chloride hexahydrate ($FeCl_3 \times 6H_2O$, Sigma Aldrich, Darmstadt, Germany), iron (II) chloride tetrahydrate ($FeCl_2 \times 4H_2O$, Sigma Aldrich), sodium hydroxide (NaOH, Sigma Aldrich) and CA ($C_6H_8O_7$, Laiwu Taihe Biochemistry Co., Laiwu, Shadong Province, China). All the chemicals used in the experiments were of analytical grade. Inert nitrogen (N_2, Messer, Ruše, Slovenia) atmosphere was used during synthesis. Ethanol (C_2H_5OH, Sigma Aldrich) was used as washing materials. DI water is used throughout all experiments. The chemical used for FO filtrations was sodium chloride (NaCl, LabExpert, Berdyansk, UK). DI water was used for NaCl solution preparation, for cleaning the membrane module, and as FS during performing baseline measurement.

2.2. Synthesis of Fe_3O_4@CA MNPs

For the synthesis of Fe_3O_4@CA MNPs a typical co-precipitation process was applied [35]. A 25 mL solution containing 1.28 M Fe^{3+}, 0.64 M Fe^{2+} (molar ratio 2:1), and 1.28 M CA was placed in a three-necked flask at 80 °C (oil bath) with vigorous stirring (340 rpm) for 20 min under N_2 atmosphere. Then 250 mL 1 M NaOH solution was added drop by drop using a dropping funnel and the solution was continuously stirred for an additional 60 min at 80 °C under an N_2 atmosphere. The reaction scheme is as follows (Equation (1)):

$$FeCl_2 + 2\,FeCl_3 + 8\,NaOH \rightarrow Fe_3O_4 \downarrow + 8\,Na^+ + 8\,Cl^- + 4\,H_2O \tag{1}$$

In the final product, in addition to the desired Fe_3O_4 MNP's, a large amount of dissolved sodium and chloride ions is also present. To reduce their concentration, the suspension was washed several times as described here (discussed in more details in Section 3.1). Fe_3O_4@CA MNPs were uniformly dispersed in 100 mL of DI as a stock solution by 30-min sonication in an ultrasonic bath. The synthesis process is shown schematically in Figure 1.

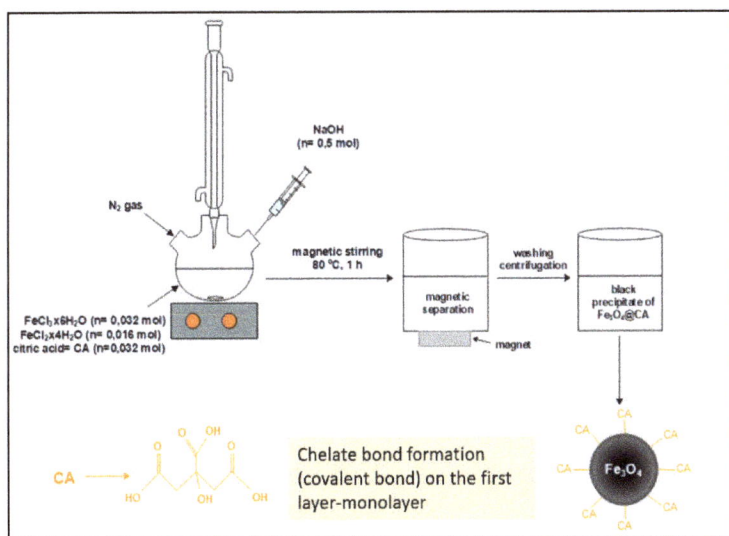

Figure 1. Schematic diagram of co-precipitation method of Fe_3O_4@CA MNP's.

2.3. Characterization of Fe$_3$O$_4$@CA MNP's

The functional groups of Fe$_3$O$_4$@CA MNP's were analysed by Fourier transform infrared spectroscopy (FTIR, mod. 5000, Perkin-Elmer Inc., Beaconsfield, UK). The FTIR spectrum was measured in the range 4000–400 cm^{-1}. The weight loss of Fe$_3$O$_4$@CA MNP's was characterized by thermogravimetric analysis (TGA, TGA/SDTA, 851^e Mettler Toledo, Greifensee, Switzerland). The measurement was performed under air from 30 to 800 °C at a heating rate of 10 K/min. The average hydrodynamic particle size, zeta potential, and isoelectric point (IEP) were determined by dynamic light scattering (DLS, Zetasizer Nano ZS, Malvern, Worcestershire, UK) using purified Fe$_3$O$_4$@CA MNP's water dispersions at 0.1 g/L with DI and pH adjusted from the current pH value to 2.0 during the measurements with an auto titrator [42]. The size and morphology of the nanoparticles were investigated by transmission electron microscopy (TEM, JEM-2010F, JEOL, Tokyo, Japan). The Fe$_3$O$_4$@CA MNP suspension was deposited on a copper-grid-supported carbon film specimen holder and left to dry at ambient conditions. The empirical size distribution of the MNPs was estimated by measuring an area of the MNP's on the TEM image. The average particle size d_{TEM} is given as a number-weighted average equivalent diameter (from Gaussian fit of the empirical distribution), the diameter of a circle having the same area as the imaged particle. The surface ligand concentration n_s (the number of ligands per particle) were estimated using Equation (2) [43]:

$$n_s = \omega \times 1/6\pi d^3 \rho_{Fe_3O_4} \times 6.023/M_{CA}\left(1 - \frac{\omega}{100}\right) \qquad (2)$$

where, ω is weight loss (%), d is particle diameter (nm), estimated from the TEM images to be 3–7 nm assuming spherical-shaped nanoparticles, ρ_{Fe3O4} is 5.18 g/cm^3 and MCA is 192.14 g/mol.

The room temperature magnetization curve of the Fe$_3$O$_4$@CA MNP's as a dry powder was measured using a vibrating sample magnetometer (VSM; model 7307, Lake Shore Cryotronics, Westerville, OH, USA). The saturation magnetization Ms of the sample is given as the average of the magnetiza-tions measured at magnetic field strength H of −10 kOe and 10 kOe.

The osmolality of the DS prepared from MNP's was measured with an osmometer (Gonotec-Osmomat 030, Berlin, Germany) and the osmotic pressure, π of Fe$_3$O$_4$@CA MNP's solutions were calculated using Equation (3):

$$\pi = \left(\sum \varphi \times n \times c\right) \times R \times T \qquad (3)$$

where R is gas constant (8.31447 J/mol K), T is the absolute temperature (K), n is the amount of substance (mol), c is the molarity of a solution (mol/L), and ϕ is an osmotic coefficient (-). The characterisation methods in detailed is presented in a previous paper [29].

2.4. FO Filtrations

The FO filtration experiments were performed using the AIM™ HFFO module. The specification of the AIM™ HFFO module and experimental set-up used in the study was presented in a previous paper [44]. The cross-flow velocity used for experiments was 120 mL/min in co-current mode. Characteristics of used DS and FS are presented in Table 1.

Table 1. Experimental overview.

	DS	FS	V_{DS} (mL)	V_{FS} (mL)
1	MNP (3.7% w/w)	DI water	100	250
2	0.45 M NaCl	DI water	100	250

Water flux, J_w (LMH) across the membrane was calculated using Equation (4) [45]:

$$J_w = \frac{\Delta V}{A \, \Delta t} \qquad (4)$$

where: ΔV total volume change of permeate water (L), A effective membrane area (m²) and Δt is the time (h).

The reverse salt flux, J_s (GMH) was determined using Equation (5) [45]:

$$J_s = \frac{\gamma_t V_t - \gamma_0 V_0}{A \, \Delta t} \qquad (5)$$

where: γ_0 = initial concentration of the FS (g/L), V_0 = initial volume of the FS (L), γ_t = solute concentration at time t (g/L), V_t = volume of the FS measured at time t (L), A = effective membrane area (m²) and Δt = time (h).

The recovery fraction (R) indicates the amount of feed recovered as permeate using Equation (6) [46]:

$$R \, (\%) = \frac{V_p}{V_f} \, 100 \qquad (6)$$

where: V_P = the permeate volume of water (L), and V_F = the volume of water in the FS (L).

3. Results and Discussion

The results are presented as follows: First, MNP's were synthesized by co-precipitation of Fe^{2+} and Fe^{3+} aqueous solutions with hydrophilic CA by adding a base in a one-pot synthesis, see Figure 1. Second, the as-prepared Fe_3O_4@CA MNP's were fully characterized using TEM, FTIR, TGA, DLS, zeta potential and magnetic measurements (Section 3.1). Third, the non-ideality of the solutions of un-grafted and grafted CA molecules onto MNP's was investigated (Section 3.2) and the osmotic properties and FO performance described (Section 3.3). Finally, FTIR and TGA analyses were done of Fe_3O_4@CA MNP's after the FO process (Section 3.4).

3.1. Magnetic Particle Preparation

After synthesis, the Fe_3O_4@CA MNPs solution was cooled to room temperature (21 °C), and the black product was separated by a permanent magnet (0.8 T). Fe_3O_4@CA MNPs were rinsed out four times with DI water, once with ethanol and finally with DI again. In Table 2 conductivity, pH and osmotic pressures of Fe_3O_4@CA MNPs after synthesis and washing cycles are presented.

Table 2. Conductivity, pH and osmotic pressure values of Fe_3O_4@CA MNPs after synthesis and washing cycles.

	Conductivity (mS/cm)	pH	Osmotic Pressure (bar)
After synthesis	69.4	12.5	33.2
Drained water	70.8	12.9	32.2
1. Cleaning with DI water	28.8	12.4	10.3
Drained water	27.7	12.5	10.2
2. Cleaning with DI water	16.1	12.1	5.4
Drained water	15.5	12.2	5.4
3. Cleaning with DI water	7.6	11.9	2.5
Drained water	7.3	11.9	2.6
4. Cleaning with DI water	3.3	11.6	0.9
Drained water	2.7	11.4	0.9
5. Cleaning with ethanol	0.38	11.9	89.6
Drained ethanol	/	/	/
6. Cleaning with DI water	0.62	11.3	/
Drained water	0.5	11.3	/
Prepared DS	0.77	10.3	13.2

From Table 2, the changes in conductivity, pH and osmotic pressure after synthesis and multiple washing with ethanol and water can be followed (explained in more details in the Results (Section 3.4.2.)).

3.2. Characterization of Fe_3O_4@CA MNPs

The coordination of the CA (carboxylic groups) to the nanoparticle surface is confirmed by FTIR spectroscopy (Figure 2). In the FTIR spectrum of bare MNPs (black), the band at 580 cm^{-1} corresponds to the vibration of the Fe-O bonds in the crystal lattice of Fe_3O_4 (magnetite) [47,48]. A broad band at 3400 cm^{-1} can be assigned to the structural OH groups of water molecules [36]. The FTIR spectrum of pure CA (red) dispersive and broadened O-H stretching vibration at 3400 cm^{-1} and C=O stretching vibration of (R-COOH) group at 1739 cm^{-1} are present. Asymmetrical and symmetrical stretching vibrations of the carboxylate group were observed at 1547 cm^{-1} and 1386 cm^{-1}, respectively.

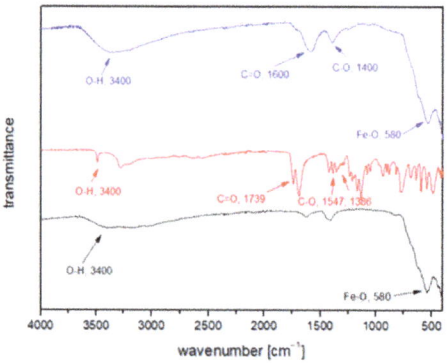

Figure 2. FTIR spectra of magnetite MNPs (black), CA (red), and Fe_3O_4@CA MNP's (blue).

Compared to bare MNPs, two additional peaks were observed for Fe_3O_4@CA MNP's at the wavelengths 1600 cm^{-1} and 1400 cm^{-1} (blue), representing the symmetrical and asymmetrical stretching of C=O and CO vibration of the three carboxyl groups (-COOH) as well as the large and intense band from 3200 to 3400 cm^{-1} corresponding to the OH group confirms the presence of non-dissociated OH groups of the CA and water traces [36,49,50]. Thus, our FTIR measurements confirmed that the CA binds chemically to the magnetite surface by carboxylate chemisorption and citrate ions are formed.

A thermogravimetric analysis (TGA) was performed to measure the organic content of Fe_3O_4@CA MNP's. The results of the TGA are shown in Figure 3 which shows the weight loss of Fe_3O_4@CA. Here, the correction (3%) due to the oxidation of magnetite in hematite was taking into the account and the mass loss of the acetic acid was therefore accordingly corrected.

The weight loss curve observed in a range of 100 to 800 °C is attributed to the evaporation of water, the decomposition of the bound citrate molecules on the Fe_3O_4 MNPs' surfaces 3% [41,51] and to oxygen during magnetite oxidation 3.3%. The initial weight loss below 160 °C due to the evaporation of physically adsorbed water was 5% and the organic content of coated Fe_3O_4@CA was determined to be 6.3%. This is the mass percentage of all surface coated organic groups on the MNPs corresponding to desorption of CA molecules on the magnetite particles. The number of COOH groups bound to the surface of the nanoparticles can be calculated according to Equation (2). Assuming a spherical shape of the MNPs with an average diameter of five nm, then the estimated total number of COOH groups is 71 per particle.

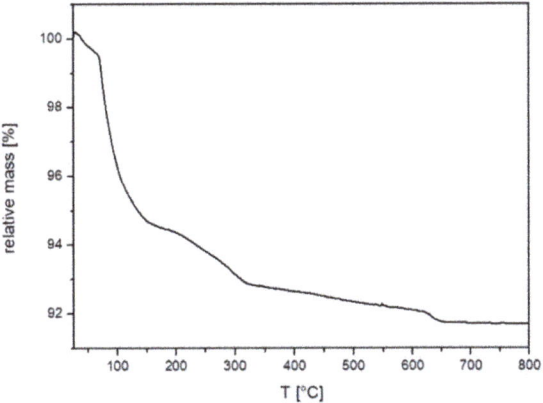

Figure 3. TGA curve of Fe_3O_4@CA MNPs.

The surface charges of the bare Fe_3O_4 MNPs and Fe_3O_4@CA MNPs were characterized by zeta potential measurements as a function of pH from 7 to 2 and z-average hydrodynamic diameter changes as a function of pH. Figure 4 presents the results for Fe_3O_4@CA MNPs.

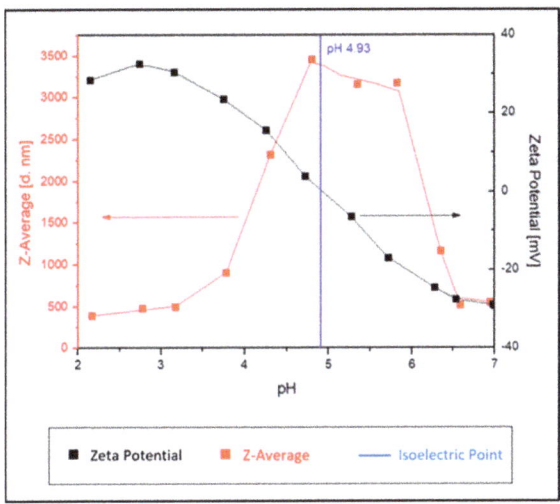

Figure 4. Change of zeta potential for Fe_3O_4@CA MNPs with pH values.

As reported in the literature, the isoelectric point (IEP) for neat Fe_3O_4 is ~6.8 [41,52]. Fe_3O_4@CA MNP's have a negative zeta potential of −30 mV at pH = 7 as shown in Figure 4. This observed phenomenon is likely caused by the adsorption of CA molecules on the bare MNPs' surface, where the surface charges were influenced by the introduction of carboxylate groups [41,53]. The zeta potential value becomes more negative with increasing pH due to the increase of OH^- ions in the dissociated solution and deprotonation of the carboxyl groups of CA. This confirmed the presence of negatively charged carboxylate groups' on the MNPs surface. The ensuing electrostatic repulsion ensures their colloidal stability in aqueous suspension where some of the carboxylate groups from CA are adsorbed/coordinated on the surface of MNPs while uncoordinated species protrude into the water medium [54]. Thus, the effective charge at pH 7 provides stabilization of Fe_3O_4@CA

MNPs. The effective hydrodynamic diameter of Fe$_3$O$_4$@ CA MNPs, measured by DLS at 25 °C, changes from 500 to 3500 nm with varying pH. Since the pH was adjusted with HCl, the effective particle size increased when the IEP 4.93 was reached and then decreased back to the original size when the pH dropped below the IEP 4.93.

The osmotic pressure of the Fe$_3$O$_4$@CA solution was determined by the freezing point depression method. Table 3 shows concentration, osmotic pressure, size, number of molecules (n_s), IEP, and magnetic saturation of Fe$_3$O$_4$@CA MNPs.

Table 3. Concentration, pH, osmotic pressure, size, surface ligand concentration, IEP, and magnetic saturation of Fe$_3$O$_4$@CA MNP's.

Sample Name	Concentration [%]	pH	Osmotic Pressure [bar]	d_x [nm]	n_S [molecules/nm^2]	IEP	M_s [emu/g]
Fe$_3$O$_4$@CA	3.7	7	18.7	3–7	0.904	4.93	44

Morphology and size of the nanoparticles Fe$_3$O$_4$@CA were observed using TEM. Upon drying on the TEM specimen support the nanoparticles formed agglomerates (Figure 5a,b). However, at the edges of the agglomerates, the deposit of nanoparticles is relatively thin and individual nanoparticles can be resolved (Figure 5b). The nanoparticles are approximately spherical and their estimated average diameter d_{TEM} is 5.2 nm ± 0.9 nm (Figure 5c).

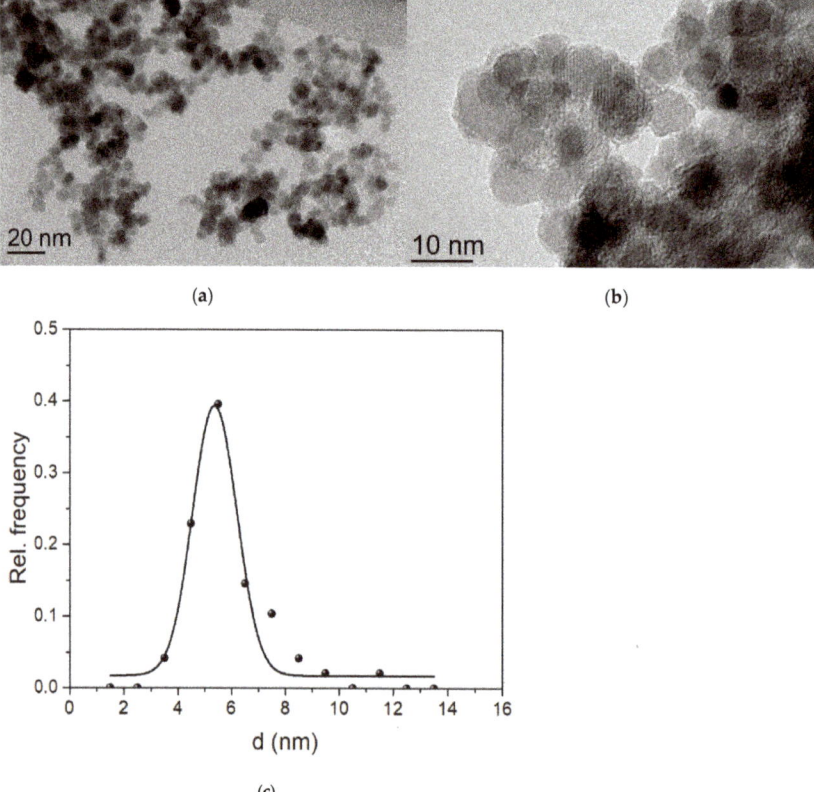

Figure 5. (a,b) TEM images of Fe$_3$O$_4$@CA MNPs; (c) corresponding NP size distribution from the TEM.

The room-temperature magnetization curve of the nanoparticles Fe_3O_4@CA suggests a superparamagnetic state of the nanoparticles, due to the lack of measured coercivity and remanence (Figure 6). The saturation magnetization of the Fe_3O_4@CA M_s of 45 emu/g is significantly smaller than reported values for bulk Fe_3O_4, which are in the range between 92 emu/g and 100 emu/g [54]. In the Fe_3O_4@CA magnetic nanoparticles are "diluted" with diamagnetic physically adsorbed water and CA (Figure 4) which reduces the measured M_s. The TGA analysis suggests that the diamagnetic mater contributes approx. 8 wt.% of the simple Fe_3O_4@CA, therefore estimated value of M_s of the Fe_3O_4 is 49 emu/g. The value is still smaller than value characteristic for bulk material and is consistent with small size of the Fe_3O_4 MNPs [55]. Because of the surface effects, namely the magnetically distorted surface layer small nanoparticles in general exhibit lower values of saturation magnetization than bulk material of the same composition [56].

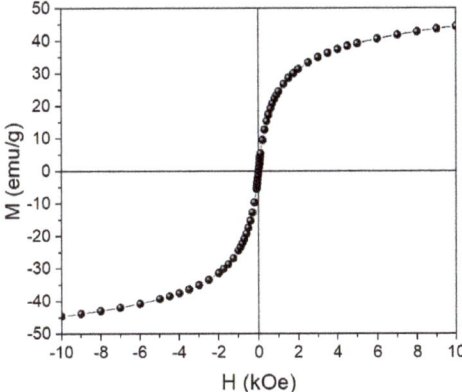

Figure 6. The room-temperature magnetization curve of the Fe_3O_4@CA MNPs (the line serves as a guide to the eye).

3.3. Properties of Fe_3O_4@CA

The advantage of coated nanoparticles is that they may have higher osmotic pressures than solutions consisting of the grafting agents alone at the same concentration [16]. The increase in osmotic pressure values can be attributed to the increased solvent-accessible surface area and thus improved hydration. The non-ideality parameters I and S were determined for Fe_3O_4@CA, see Figure 7 and M^*_{CA} calculated using the semi-empirical model Equations (7) and (8):

$$\frac{m_w}{m_s} = S \times \frac{1}{\pi} + I \qquad (7)$$

$$S = \frac{R \times T \times \rho}{M^*} \qquad (8)$$

where m_w and m_s is the mass of water and solute respectively; S and I are the non-ideality parameters, π the osmotic pressure, ρ is the density of water at temperature T, M is the molecular weight of the solute and R, T gas constant and temperature.

Table 4. The fitting parameters of the non-ideality analysis Fe_3O_4@CA.

	S	I
MNP@CA	434	3.92
CA	142	−0.17

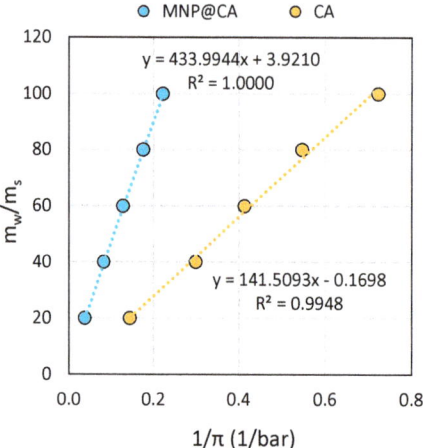

Figure 7. Non-ideality analysis of neat CA (orange) and MNP@CA nanocomposites (blue). m_W and m_S: mass of water and mass of solute, π: osmotic pressure. Dashed lines represent fits to Equation (8) (r > 0.995) and fitting parameters are given in Table 4.

The results show that the osmotically derived molecular weights are M^*_{CA} = 176.12 g/mol for pure CA and M^*_{CA} = 57.9 g/mol for MNP@CA. The value M^*_{CA} = 57.9 g/mol is significantly lower than the nominal standard value of 192.12 g/mol, indicating that the MNP-grafted CA molecule are almost four-fold stronger as osmotic agent as CA alone.

3.4. Forward Osmosis Process Evaluation

3.4.1. Determination of J_w and J_s

Fe_3O_4 MNPs at concentrations of CA of 3.7% were used to prepare Fe_3O_4@CA as DS in a lab-scale cross-flow FO filtration setup. In Figure 8 water flux is presented where DI was used as FS and Fe_3O_4@CA as well as the 0.45 M NaCl was used as DS, respectively, in order to make a comparison between Fe_3O_4@CA solution and standard DS (NaCl solution). For this purpose, we wanted to imitate the FO process when using Fe_3O_4@CA solution as DS, including measuring conditions (e.g., starting volumes of FS and DS, initial osmotic power of DS). Imitation was performed using NaCl solution as DS, therefore, 0.45 M concentration of NaCl was used to give initial osmotic pressure comparable to the solution of Fe_3O_4@CA. The filtrations with Fe_3O_4@CA were stopped after 2 h to ensure sufficient data to have an insight into the filtration dynamics. The filtration with 0.45 M NaCl as DS was finished spontaneously in 1 h (75.2 g of FS left) because the water flux was higher in comparison when Fe_3O_4@CA was used as DS agent despite the similar initial osmotic transmembrane gradient.

The initial water flux of 0.45 M NaCl solution as reference is higher (14.1 LMH) when comparing to the Fe_3O_4@CA (9.3 LMH). A uniform water flux decline was observed for all solutions as a function of time. The overall water flux decreased 30% and 70% for 0.45 M NaCl and Fe_3O_4@CA, respectively, due to dilution of DS and reduction in the osmotic pressure difference between FS and DS.

When Fe_3O_4@CA was used as DS, the diffusion of water molecules decreased due to the higher concentration polarization of the Fe_3O_4@CA from the draw side (96.1 mL), presented in Table 5. Therefore, the dilutive ICP in FO mode is stronger for Fe_3O_4@CA and the effective osmotic gradient across the active layer is lower, see Figure 9.

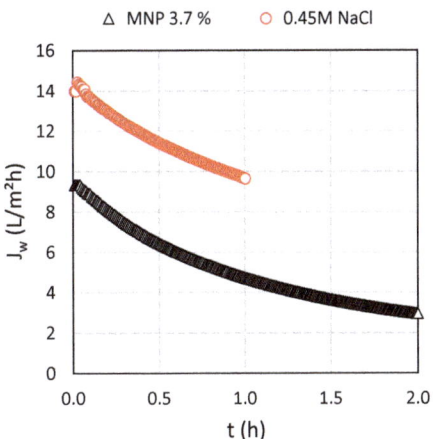

Figure 8. Water fluxes versus filtration time. Higher initial flux was reached using 0.45 M NaCl solution (14.1 LMH) when compared to Fe$_3$O$_4$@CA (9.3 LMH), respectively.

Table 5. The initial and final volumes of the DS and FS.

	DS	FS	V_{DS} (mL) Initial	V_{FS} (mL) Initial	V_{DS} (mL) at 1 h	V_{DS} (mL) at 2 h	V_{FS} (mL) at 1 h	V_{FS} (mL) at 2 h
1	MNP	DI water	100	250	185.1	206.2	163.6	142.5
2	0.45 M NaCl	DI water	100	250	273.7	/	75.2	/

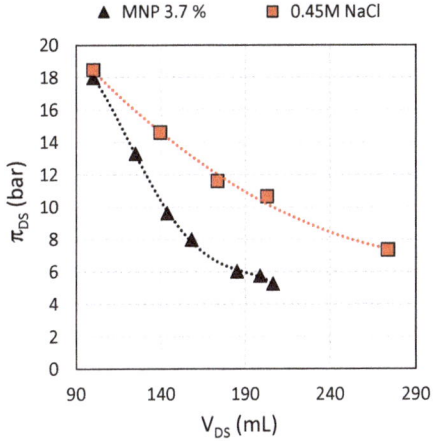

Figure 9. DS osmotic pressures with increased volume, for 0.45 M NaCl and Fe$_3$O$_4$@CA as DS and DI water as FS, respectively.

The decrease in water flux with an increase in the conductivity indicates that the main driving force across the FO membrane is the osmotic pressure difference between the DS and FS. The conductivity was monitored (Figure 10a) and reverse solute flux, J_s (Figure 10b) was calculated using Equation (5) where concentrations in FS were obtained out of measured conductivity data in the FS. Although the 0.45 M NaCl solution gives rise to higher water flux (29.1% higher) compared to Fe$_3$O$_4$@CA, the reverse solute flux is also larger (2.5 versus 0.08 GMH). The conductivity increased (as well as J_s) for the 0.45 M NaCl solution during one hour.

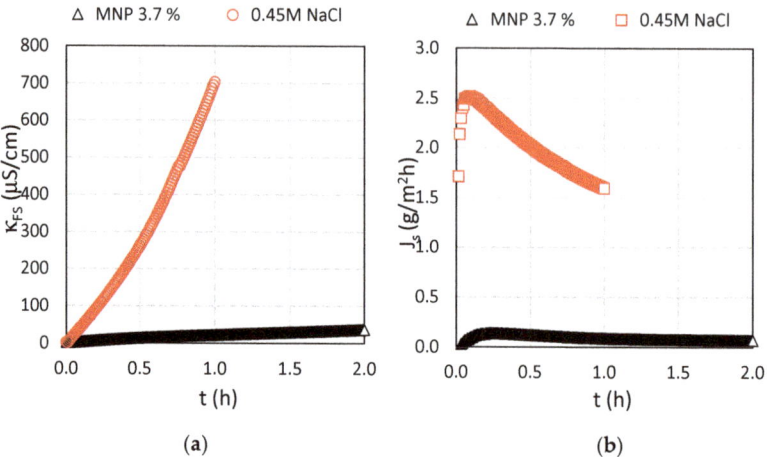

Figure 10. (a) Conductivity of FS and (b) revere salt fluxes during the time for 0.45 M NaCl and Fe$_3$O$_4$@CA as DS and DI water as FS, respectively.

The suitability of a draw solute is defined by its ability to develop osmotic pressure, which leads to higher water flux, and lower diffusion to the feed side. From the molecular view, the size and ionic structure of the draw solute defines its applicability in the FO process. The lower molecular size increases the diffusivity of the DS and reduces the internal concentration polarization that leads to the higher water flux, while the reverse salt flux is also enhanced. The larger molecules are designed as DS to prevent reverse salt diffusion.

In this study, small molecules such as NaCl that were used for comparison generated high osmotic pressure at low solution viscosities and mitigate ICP due to their high reverse solute flux [20].

The synthesized Fe$_3$O$_4$@CA particles are relatively large, and the reverse diffusion was low, however, the potency of DS to build up a reasonable osmotic pressure and water flux was enhanced by increasing particle hydrophilicity.

The used AIM™ HFFO module was cleaned using only DI water for 30 min and then a new baseline experiment was performed under the same conditions. The water flux of the cleaned AIM™ HFFO module (restored to about 97%) was almost the same as that of the pristine module. This indicates that simple physical flushing can effectively rejuvenate the used module membrane and the average water flux can be restored to about 93.2% even after 2 cycles.

3.4.2. Development of the FS and DS Osmotic Pressures

The osmotic pressure of the Fe$_3$O$_4$ functionalized with CA is relatively high due to the higher hydrophilicity because the CA molecule possesses two terminal COOHs, one central COOH, and one C–OH group, which are all active in bonding with iron oxide via the Fe–OH groups [35,36]. For a given concentration and temperature, the osmotic pressure depends on the solubility and molecular weight of the DS moieties. From the CA (H$_3$C$_6$H$_5$O$_7$) speciation, CA will be mostly deprotonated as HC$_6$H$_5$O$_7^{2-}$ and C$_6$H$_5$O$_7^{3-}$ over a broad pH range (5.0–9.0). Thus, at pH = 7 there will be mostly HC$_6$H$_5$O$_7^{2-}$ and C$_6$H$_5$O$_7^{3-}$ functional groups of CA [57]. The deprotonated CA made MNP@CA surfaces highly negatively charged [32]. However, both MNP@CA (and CA alone) will have counter ions closely attached and will thus not contribute to bulk solution conductivity as such.

When comparing the development in FS conductivity (Figure 10a) with the FS osmotic pressure development (Figure 11a) it is observed that the osmotic pressure of the FS with 0.45 M NaCl as DS reached 0.5 bar in one hour while at the same time, the osmotic pressure for the FS with Fe$_3$O$_4$@CA as DS reached 4 bar, respectively. This suggests that DS solutes diffuse across the membranes into the FS. The DS contains four components: Fe$_3$O$_4$@CA MNP, unconjugated CA, ethanol and NaOH. As the conductance increase in the FS is negligible (compared to NaCl) the reverse solute flux could in principle be due to any of the three first components or a combination thereof (since NaOH would be dissociated and thus give rise to an increase in solution conductivity). However, a 4 bar FS osmotic pressure at the end of the experiment (see Figure 12a) generated solely by Fe$_3$O$_4$@CA MNPs would correspond to a 0.8% w/w solution which is opaque. Since the FS is transparent it is unlikely that Fe$_3$O$_4$@CA MNP's are crossing the membrane. This leaves only ethanol and/or unconjugated CA as potential reverse flux solute candidates. The FS ethanol concentration was measured to be 0.07% which corresponds to an osmotic pressure of 2.8 bar. Thus, ethanol diffusion can explain some of the FS osmotic pressure increase during the experiment. With regards to unconjugated CA is crossing the membrane, it is noted that CA has a molecular weight of 192 Da which is relatively large as compared to ionic salts. However, it has previously been shown that peptides with molecular weights of 375 and 692 Da can cross a thin film composite Aquaporin Inside™ Membrane (AIM) FO membrane and the transport mechanism was determined to be diffusion-based [58]. Thus, it is likely that unconjugated CA (with Na$^+$ as associated counter ion) also contributes to the increase in FS osmotic pressure.

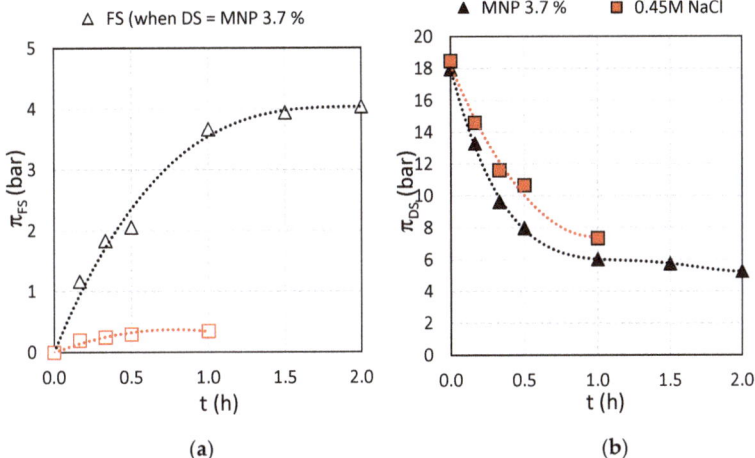

Figure 11. Osmotic pressures as function of time for (**a**) FS and (**b**) DS for 0.45 M NaCl and Fe$_3$O$_4$@CA as DS and DI water as FS, respectively.

When comparing with other studies where magnetic nanoparticles were used a DS in FO, several studies have focused on poly-sodium acrylate (PSA) coated MNPs as DS. Thus, a water flux of 5.3 LMH and a DS osmotic pressure of 11.4 bar has been reported using PSA-MNPs at 0.078%, wt% (~1.3 g/L) DS concentration ([57]. Ge and co-workers [59] reported an osmotic pressure of 11–12 bar at a concentration of ~15%, wt% PSA (Mn 1800) as DS. Similar osmotic pressures were obtained with 24–48%, wt% PSA DS (in the form of free polyelectrolyte osmotic agent) [31]. The significant difference in osmotic pressure between PSA-MNP solutions and free PSA solutions reflect the non-ideality of PSA-MNP solutions similar to what we observe in this study. However, the robustness of PSA-MNP based DSs (i.e., their ability to maintain their osmotic pressure after repeated concentrations) is still an issue [29].

In our previous study [29] a 7% solution of robust PSA-coated MNPs yielded an initial J_w of 4.2 LMH and an osmotic pressure of 9 bar and a reverse solute flux J_s of 0.05 GMH. Thus, the results presented here where Fe$_3$O$_4$@CA MNPs a 3.7% solution yielded approximately a two-fold higher initial J_w (Figure 8) and osmotic pressure (Figure 9) implies that the Fe_3O_4@CA MNPs DS is about four-fold more potent as DS compared to a DS based on PSA-coated MNPs. However, the nominal reverse solute flux for Fe_3O_4@CA of 0.08 GMH (Figure 10b) is slightly higher than for PSA-coated MNP based DS.

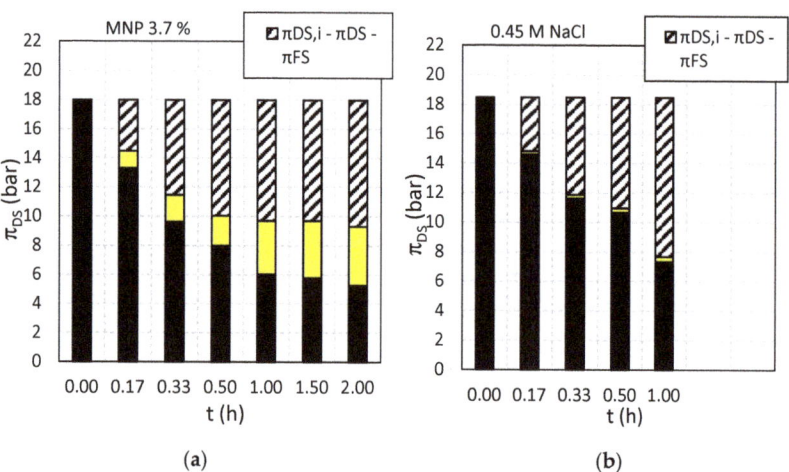

Figure 12. Osmotic pressures when DS is (**a**) Fe_3O_4@CA and (**b**) 0.45 M NaCl, during the time. Black columns present the osmotic pressure in the DS during filtration, which is decreasing because of two factors: (i) outcoming salts from DS to FS (yellow part of columns) and (ii) incoming water from the FS that is diluting DS (hatched part of columns).

3.5. Characterization after FO Process

The functional groups of chemical bonds between Fe_3O_4 and CA after the FO process were characterized by FTIR (Figure 13a). It is confirmed that the Fe_3O_4@CA spectral features are preserved, so the CA coating on Fe_3O_4 MNPs was maintained. The TGA results in Figure 13b show the weight loss of Fe_3O_4@CA before (black curve) and after the FO process (red curve). The overall weight loss of Fe_3O_4@CA before the FO process was determined 8.3% and after the FO process 6.6%. The total estimated number of COOH groups is still 85 molecules per particle after FO process.

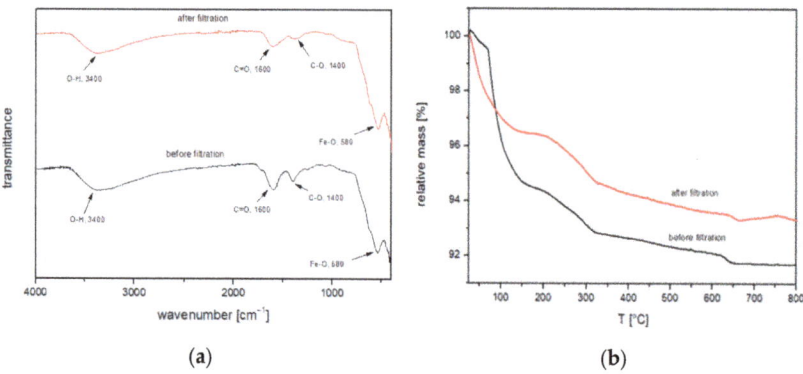

Figure 13. (**a**) FTIR spectrum of Fe_3O_4@CA MNPs before (red) and after (black) FO process and (**b**) TGA analysis curve of Fe_3O_4@CA after FO process.

4. Conclusions

The CA-coated Fe_3O_4 nanoparticles (Fe_3O_4@CA MNPs) have been synthesized for use as draw solution (DS) agents in forward osmosis (FO) and systematically analyzed. Non-ideality of the solutions of ungrafted and grafted CA molecules onto MNPs indicates that the MNP-grafted CA molecule are almost four-fold stronger as osmotic agent as CA alone. Carboxyl groups of CA have an affinity for MNPs, thus preventing aggregation. The osmotic pressure of as-prepared MNP@CA was high enough to be used as a DS in FO. Although the 0.45 M NaCl solution gives rise to higher water flux (for 29.1%) compared to Fe_3O_4@CA, the reverse solute flux is also larger (2.5 to 1.8 GMH). These differences are ascribed to a more pronounced internal concentration polarization associated with Fe_3O_4@CA as DS as compared to NaCl as DS. The results showed that the conductivity of the feed solution stays the same low level before and after FO filtration, and that the reverse solute flux of Fe_3O_4@CA MNPs is negligible. Still the FS osmotic pressure is increasing during the filtration, which is ascribed to diffusion of DS ethanol and unconjugated CA. Thus, an effective purification of Fe_3O_4@CA MNPs is a prerequisite in order to be able to benefit fully from the low inherent reverse solute flux. Therefore, FO can be considered as a potential candidate for a broad range of concentration applications where current technologies still suffer from critical limitations

Author Contributions: Conceptualization I.B., I.P. and C.H.-N.; experimental investigation J.S., H.B., M.D. and S.G.; writing—original draft preparation, review and editing, I.B., I.P., H.B., J.S., M.D., S.G. and C.H.-N. All authors have read and agreed to the published version of the manuscript.

Funding: The authors are grateful for support from the Novo Nordisk Foundation (NNF18OC0034918).

Institutional Review Board Statement: Not applicable.

Informed Consent Statement: Not applicable.

Data Availability Statement: Data available on request.

Acknowledgments: This research was funded by Slovenian National Agency, grant numbers P3-0036, P2-0006, P2-0089, J1-9169, and J3-1762, L4-1843, and NEPWAT funded by the Novo Nordisk Foundation.

Conflicts of Interest: The authors declare no conflict of interest.

References

1. Farmany, A.; Mortazavi, S.S.; Mahdavi, H. Ultrasond-assisted synthesis of Fe_3O_4/SiO_2 core/shell with enhanced adsorption capacity for diazinon removal. *J. Magn. Magn. Mater.* **2016**, *416*, 75–80. [CrossRef]
2. Lu, X.; Deng, S.; Wang, B.; Huang, J.; Wang, Y.; Yu, G. Adsorption behavior and mechanism of perfluorooctane sulfonate on nanosized inorganic oxides. *J. Colloid Interface Sci.* **2016**, *474*, 199–205. [CrossRef] [PubMed]
3. Roto, R.; Yusran, Y.; Kuncaka, A. Magnetic adsorbent of Fe_3O_4@SiO_2 core-shell nanoparticles modified with thiol group for chloroauric ion adsorption. *Appl. Surf. Sci.* **2016**, *377*, 30–36. [CrossRef]
4. Turcu, R.; Socoliuc, V.; Craciunescu, I.; Petran, A.; Paulus, A.; Franzreb, M.; Vasile, E.; Vekas, L. Magnetic microgels, a promising candidate for enhanced magnetic adsorbent particles in bioseparation: Synthesis, physicochemical characterization, and separation performance. *Soft Matter* **2015**, *11*, 1008–1018. [CrossRef]
5. Tratnyek, P.G.; Johnson, R.L. Nanotechnologies for environmental cleanup. *Nano Today* **2006**, *1*, 44–48. [CrossRef]
6. Al-Hobaib, A.; Al-Sheetan, K.; El Mir, L. Effect of iron oxide nanoparticles on the performance of polyamide membrane for ground water purification. *Mater. Sci. Semicond. Process.* **2016**, *42*, 107–110. [CrossRef]
7. Ghaemi, N.; Madaeni, S.S.; Daraei, P.; Rajabi, H.; Zinadini, S.; Alizadeh, A.; Heydari, R.; Beygzadeh, M.; Ghouzivand, S. Polyethersulfone membrane enhanced with iron oxide nanoparticles for copper removal from water: Application of new functionalized Fe_3O_4 nanoparticles. *Chem. Eng. J.* **2015**, *263*, 101–112. [CrossRef]
8. Sabbatini, P.; Yrazu, F.; Rossi, F.; Thern, G.; Marajofsky, A.; de Cortalezzi, M.F. Fabrication and characterization of iron oxide ceramic membranes for arsenic removal. *Water Res.* **2010**, *44*, 5702–5712. [CrossRef]
9. Shen, J.F.; Tang, J.; Nie, Z.H.; Wang, Y.D.; Ren, Y.; Zuo, L. Tailoring size and structural distortion of Fe_3O_4 nanoparticles for the purification of contaminated water. *Bioresour. Technol.* **2009**, *100*, 4139–4146. [CrossRef] [PubMed]
10. Yantasee, W.; Warner, C.L.; Sangvanich, T.; Addleman, R.S.; Carter, T.G.; Wiacek, R.J.; Fryxell, G.E.; Timchalk, C.; Warner, M.G. Removal of Heavy Metals from Aqueous Systems with Thiol Functionalized Superparamagnetic Nanoparticles. *Environ. Sci. Technol.* **2007**, *41*, 5114–5119. [CrossRef] [PubMed]

11. Zhao, S.; Zou, L.; Tang, C.Y.; Mulcahy, D. Recent developments in forward osmosis: Opportunities and challenges. *J. Membr. Sci.* **2012**, *396*, 1–21. [CrossRef]
12. Wei, R.; Zhang, S.; Cui, Y.; Ong, R.C.; Chung, N.T.-S.; Helmer, B.J.; de Wit, J.S. Highly permeable forward osmosis (FO) membranes for high osmotic pressure but viscous draw solutes. *J. Membr. Sci.* **2015**, *496*, 132–141. [CrossRef]
13. Salehi, H.; Rastgar, M.; Shakeri, A. Anti-fouling and high water permeable forward osmosis membrane fabricated via layer by layer assembly of chitosan/graphene oxide. *Appl. Surf. Sci.* **2017**, *413*, 99–108. [CrossRef]
14. Shakeri, A.; Salehi, H.; Rastgar, M. Chitosan-based thin active layer membrane for forward osmosis desalination. *Carbohydr. Polym.* **2017**, *174*, 658–668. [CrossRef]
15. Yamada, S.; Tanaka, M. Softness of some metal ions. *J. Inorg. Nucl. Chem.* **1975**, *37*, 587–589. [CrossRef]
16. Roach, J.D.; Bondaruk, M.M.; Burney, Z. Nonideal Solution Behavior in Forward Osmosis Processes Using Magnetic Nanoparticles. In *Osmotically Driven Membrane Processes—Approach, Development and Current Status*; IntechOpen: London, UK, 2018.
17. Cui, Y.; Liu, X.-Y.; Chung, T.-S.; Weber, M.; Staudt, C.; Maletzko, C. Removal of organic micro-pollutants (phenol, aniline and nitrobenzene) via forward osmosis (FO) process: Evaluation of FO as an alternative method to reverse osmosis (RO). *Water Res.* **2016**, *91*, 104–114. [CrossRef] [PubMed]
18. Xue, W.; Yamamoto, K.; Tobino, T. Membrane fouling and long-term performance of seawater-driven forward osmosis for enrichment of nutrients in treated municipal wastewater. *J. Membr. Sci.* **2016**, *499*, 555–562. [CrossRef]
19. Han, G.; Liang, C.-Z.; Chung, T.-S.; Weber, M.; Staudt, C.; Maletzko, C. Combination of forward osmosis (FO) process with coagulation/flocculation (CF) for potential treatment of textile wastewater. *Water Res.* **2016**, *91*, 361–370. [CrossRef] [PubMed]
20. Qasim, M.; Darwish, N.A.; Sarp, S.; Hilal, N. Water desalination by forward (direct) osmosis phenomenon: A comprehensive review. *Desalination* **2015**, *374*, 47–69. [CrossRef]
21. Ling, M.M.; Chung, T.-S. Desalination process using super hydrophilic nanoparticles via forward osmosis integrated with ultrafiltration regeneration. *Desalination* **2011**, *278*, 194–202. [CrossRef]
22. Zhao, D.; Chen, S.; Guo, C.; Zhao, Q.; Lu, X. Multi-functional forward osmosis draw solutes for seawater desalination. *Chin. J. Chem. Eng.* **2016**, *24*, 23–30. [CrossRef]
23. Kessler, J.; Moody, C. Drinking water from sea water by forward osmosis. *Desalination* **1976**, *18*, 297–306. [CrossRef]
24. McCutcheon, J.R.; McGinnis, R.L.; Elimelech, M. A novel ammonia—carbon dioxide forward (direct) osmosis desalination process. *Desalination* **2005**, *174*, 1–11. [CrossRef]
25. McGinnis, R.L.; Elimelech, M. Energy requirements of ammonia–carbon dioxide forward osmosis desalination. *Desalination* **2007**, *207*, 370–382. [CrossRef]
26. Ge, Q.; Su, J.; Amy, G.L.; Chung, N.T.-S. Exploration of polyelectrolytes as draw solutes in forward osmosis processes. *Water Res.* **2012**, *46*, 1318–1326. [CrossRef]
27. Ling, M.M.; Wang, K.Y.; Chung, N.T.-S. Highly Water-Soluble Magnetic Nanoparticles as Novel Draw Solutes in Forward Osmosis for Water Reuse. *Ind. Eng. Chem. Res.* **2010**, *49*, 5869–5876. [CrossRef]
28. Guo, C.X.; Huang, S.; Lu, X. A solventless thermolysis route to large-scale production of ultra-small hydrophilic and biocompatible magnetic ferrite nanocrystals and their application for efficient protein enrichment. *Green Chem.* **2014**, *16*, 2571–2579. [CrossRef]
29. Ban, I.; Markuš, M.; Gyergyek, S.; Drofenik, M.; Korenak, J.; Helix-Nielsen, C.; Petrinić, I. Synthesis of Poly-Sodium-Acrylate (PSA)-Coated Magnetic Nanoparticles for Use in Forward Osmosis Draw Solutions. *Nanomaterials* **2019**, *9*, 1238. [CrossRef]
30. Wu, W.; Wu, Z.; Yu, T.; Jiang, C.; Kim, W.-S. Recent progress on magnetic iron oxide nanoparticles: Synthesis, surface functional strategies and biomedical applications. *Sci. Technol. Adv. Mater.* **2015**, *16*, 023501. [CrossRef]
31. Zufia-Rivas, J.; Morales, P.; Veintemillas-Verdaguer, S. Effect of the Sodium Polyacrylate on the Magnetite Nanoparticles Produced by Green Chemistry Routes: Applicability in Forward Osmosis. *Nanomaterials* **2018**, *8*, 470. [CrossRef]
32. Liu, J.; Dai, C.; Hu, Y. Aqueous aggregation behavior of citric acid coated magnetite nanoparticles: Effects of pH, cations, anions, and humic acid. *Environ. Res.* **2018**, *161*, 49–60. [CrossRef] [PubMed]
33. Chung, N.T.-S.; Zhang, S.; Wang, K.Y.; Su, J.; Ling, M.M. Forward osmosis processes: Yesterday, today and tomorrow. *Desalination* **2012**, *287*, 78–81. [CrossRef]
34. Hong, R.; Feng, B.; Chen, L.; Liu, G.; Li, H.; Zheng, Y.; Wei, D. Synthesis, characterization and MRI application of dextran-coated Fe3O4 magnetic nanoparticles. *Biochem. Eng. J.* **2008**, *42*, 290–300. [CrossRef]
35. Manal-Mounir, R.; Eglal, H.; Gadallah, A.; Ali Azab, H.M. Comparative study for the preparation of superparamagnetic-citric coated magnetic nanoparticle and fodesalination application. *ARPN J. Appl. Sci. Eng.* **2018**, *13*, 1150–1162.
36. Ge, Q.; Yang, L.; Cai, J.; Xu, W.; Chen, Q.; Liu, M. Hydroacid magnetic nanoparticles in forward osmosis for seawater desalination and efficient regeneration via integrated magnetic and membrane separations. *J. Membr. Sci.* **2016**, *520*, 550–559. [CrossRef]
37. Srivastava, S.; Awasthi, R.; Gajbhiye, N.S.; Agarwal, V.; Singh, A.; Yadav, A.; Gupta, R.K. Innovative synthesis of citrate-coated superparamagnetic Fe_3O_4 nanoparticles and its preliminary applications. *J. Colloid Interface Sci.* **2011**, *359*, 104–111. [CrossRef]
38. Khazaie, F.; Sheshmani, S.; Shokrollahzadeh, S.; Shahvelayati, A.S. Desalination of Saline Water via Forward Osmosis Using Magnetic Nanoparticles Covalently Functionalized with Citrate Ions as Osmotic Agent. *Environ. Technol.* **2020**, 1–26. [CrossRef] [PubMed]
39. Guizani, M.; Saito, M.; Ito, R.; Funamizu, N. Engineering of size-controlled magnetic nanoparticles for use as a draw solution in a forward osmosis process. *Desalin. Water Treat.* **2019**, *154*, 21–29. [CrossRef]

40. Kadhim, R.M.; Al-Abodi, E.E.; Al-Alawy, A.F. Citrate-coated magnetite nanoparticles as osmotic agent in a forward osmosis process. *Desalin. Water Treat.* **2018**, *115*, 45–52. [CrossRef]
41. Na, Y.; Yang, S.; Lee, S. Evaluation of citrate-coated magnetic nanoparticles as draw solute for forward osmosis. *Desalination* **2014**, *347*, 34–42. [CrossRef]
42. Kaszuba, M.; Corbett, J.; Watson, F.M.; Jones, A. High-concentration zeta potential measurements using light-scattering techniques. *Philos. Trans. R. Soc. A Math. Phys. Eng. Sci.* **2010**, *368*, 4439–4451. [CrossRef] [PubMed]
43. Park, S.Y.; Ahn, H.-W.; Chung, J.W.; Kwak, S.-Y. Magnetic core-hydrophilic shell nanosphere as stability-enhanced draw solute for forward osmosis (FO) application. *Desalination* **2016**, *397*, 22–29. [CrossRef]
44. Petrinic, I.; Bukšek, H.; Galambos, I.; Gerencsér-Berta, R.; Sheldon, M.S.; Helix-Nielsen, C. Removal of naproxen and diclofenac using an aquaporin hollow fibre forward osmosis module. *Desalin. Water Treat.* **2020**, *192*, 415–423. [CrossRef]
45. Oh, S.-H.; Im, S.-J.; Jeong, S.; Jang, A. Nanoparticle charge affects water and reverse salt fluxes in forward osmosis process. *Desalination* **2018**, *438*, 10–18. [CrossRef]
46. McCutcheon, J.R.; McGinnis, R.L.; Elimelech, M. Desalination by ammonia–carbon dioxide forward osmosis: Influence of draw and feed solution concentrations on process performance. *J. Membr. Sci.* **2006**, *278*, 114–123. [CrossRef]
47. Zhang, B.; Tu, Z.; Zhao, F.; Wang, J. Superparamagnetic iron oxide nanoparticles prepared by using an improved polyol method. *Appl. Surf. Sci.* **2013**, *266*, 375–379. [CrossRef]
48. Gaihre, B.; Khil, M.S.; Lee, D.R.; Kim, H.Y. Gelatin-coated magnetic iron oxide nanoparticles as carrier system: Drug loading and in vitro drug release study. *Int. J. Pharm.* **2009**, *365*, 180–189. [CrossRef] [PubMed]
49. Singh, D.; Gautam, R.K.; Kumar, R.; Shukla, B.K.; Shankar, V.; Krishna, V. Citric acid coated magnetic nanoparticles: Synthesis, characterization and application in removal of Cd (II) ions from aqueous solution. *J. Water Process. Eng.* **2014**, *4*, 233–241. [CrossRef]
50. Răcuciu, M. Synthesis protocol influence on aqueous magnetic fluid properties. *Curr. Appl. Phys.* **2009**, *9*, 1062–1066. [CrossRef]
51. Cheng, C.; Wen, Y.; Xu, X.; Gu, H. Tunable synthesis of carboxyl-functionalized magnetite nanocrystal clusters with uniform size. *J. Mater. Chem.* **2009**, *19*, 8782–8788. [CrossRef]
52. Campelj, S.; Makovec, D.; Drofenik, M. Preparation and properties of water-based magnetic fluids. *J. Phys. Condens. Matter* **2008**, *20*, 204101. [CrossRef] [PubMed]
53. Imran, M.; Riaz, S.; Sanaullah, I.; Khan, U.; Sabri, A.N.; Naseem, S. Microwave assisted synthesis and antimicrobial activity of Fe_3O_4-doped ZrO_2 nanoparticles. *Ceram. Int.* **2019**, *45*, 10106–10113. [CrossRef]
54. Cornell, R.M.; Schwertmann, U.U. *The Iron Oxides: Structure Properties, Reactions, Occurrences and Uses*, 2nd ed.; Completely Revised and Extended Edition; Wiley-VCH: Weinheim, Germany, 2003.
55. Kodama, R. Magnetic nanoparticles. *J. Magn. Magn. Mater.* **1999**, *200*, 359–372. [CrossRef]
56. Elbagerma, M.A.; Edwards, H.G.M.; Azimi, G.; Scowen, I.J. Raman spectroscopic determination of the acidity constants of salicylaldoxime in aqueous solution. *J. Raman Spectrosc.* **2011**, *42*, 505–511. [CrossRef]
57. Dey, P.; Izake, E.L. Magnetic nanoparticles boosting the osmotic efficiency of a polymeric FO draw agent: Effect of polymer conformation. *Desalination* **2015**, *373*, 79–85. [CrossRef]
58. Bajraktari, N.; Madsen, H.T.; Gruber, M.F.; Truelsen, S.F.; Jensen, E.L.; Jensen, H.; Hélix-Nielsen, C. Separation of Peptides with Forward Osmosis Biomimetic Membranes. *Membranes* **2016**, *6*, 46. [CrossRef] [PubMed]
59. Ge, Q.; Su, J.; Chung, T.-S.; Amy, G. Hydrophilic Superparamagnetic Nanoparticles: Synthesis, Characterization, and Performance in Forward Osmosis Processes. *Ind. Eng. Chem. Res.* **2010**, *50*, 382–388. [CrossRef]

Article

High-Temperature Nanoindentation of an Advanced Nano-Crystalline W/Cu Composite

Michael Burtscher [1,*,†], Mingyue Zhao [1,†], Johann Kappacher [2], Alexander Leitner [2,‡], Michael Wurmshuber [1], Manuel Pfeifenberger [3,§], Verena Maier-Kiener [2] and Daniel Kiener [1]

[1] Department of Materials Science, Chair of Materials Physics, University of Leoben, Jahnstraße 12, 8700 Leoben, Austria; zhaomingyue510@126.com (M.Z.); michael.wurmshuber@unileoben.ac.at (M.W.); daniel.kiener@unileoben.ac.at (D.K.)
[2] Department of Materials Science, Chair of Physical Metallurgy and Metallic Materials, University of Leoben, Roseggerstraße 12, 8700 Leoben, Austria; johann.kappacher@unileoben.ac.at (J.K.); alexander.leitner@posteo.at (A.L.); verena.maier-kiener@unileoben.ac.at (V.M.-K.)
[3] Erich Schmid Institute of Materials Science, Austrian Academy of Sciences, University of Leoben, Jahnstaße 12, 8700 Leoben, Austria; m.pfeifenberger@posteo.de
* Correspondence: michael.burtscher@unileoben.ac.at
† These authors contributed equally.
‡ Currently working at RHI Magnesita N.V., Magnesitstraße 2, 8700 Leoben, Austria.
§ Currently working at Anton Paar GmbH, Anton Paar Strasse 20, 8054 Graz, Austria.

Abstract: The applicability of nano-crystalline W/Cu composites is governed by their mechanical properties and microstructural stability at high temperatures. Therefore, mechanical and structural investigations of a high-pressure torsion deformed W/Cu nanocomposite were performed up to a temperature of 600 °C. Furthermore, the material was annealed at several temperatures for 1 h within a high-vacuum furnace to determine microstructural changes and surface effects. No significant increase of grain size, but distinct evaporation of the Cu phase accompanied by Cu pool and faceted Cu particle formation could be identified on the specimen's surface. Additionally, high-temperature nanoindentation and strain rate jump tests were performed to investigate the materials mechanical response at elevated temperatures. Hardness and Young's modulus decrease were noteworthy due to temperature-induced effects and slight grain growth. The strain rate sensitivity in dependent of the temperature remained constant for the investigated W/Cu composite material. Also, the activation volume of the nano-crystalline composite increased with temperature and behaved similar to coarse-grained W. The current study extends the understanding of the high-temperature behavior of nano-crystalline W/Cu composites within vacuum environments such as future fusion reactors.

Keywords: W/Cu composite; nanocrystalline; high-pressure torsion; microstructure; nanoindentation

1. Introduction

W/Cu composites are typically used as heat sinks in power electronics and construction material for high-power switches, but are also intended as a shielding material in environments exposed to radiation [1–3]. Thereby, the material is subjected to high local and cyclic loads due to temperature changes as well as additional irradiation or ion bombardment [2,4]. These composite materials exhibit high thermal stability concerning their microstructure combined with predictable thermal and mechanical properties. Depending on their composition, strength or ductility is limited at room temperature and, therefore, the vulnerability for cyclic damage or fatal crack propagation due to overload events is prevalent. To further increase the applicability and counteract the brittle behavior, strength, but also ductility, of the composite material must be increased [5,6]. One possibility to do so, and concurrently increase both material parameters, is to constitute a proper grain refinement [7–9]. This strategy was successfully applied by performing a high-pressure torsion (HPT) process on a coarse-grained (cg) W/Cu alloy [10]. This results in a homogenous nano-crystalline (nc) W/Cu composite with enhanced mechanical properties at room temperature (RT) [10].

Studies regarding nano-lamellar W/Cu multilayer coatings revealed a decomposition of the W and Cu layers to globular W grains within a Cu matrix starting at 700 °C [11]. Additionally, the formation of voids and pores on the surface of W/Cu specimens when annealed within a vacuum furnace at this temperature is reported [11]. This circumstance was further exploited to produce a nc W foam on the surface of a nc W/Cu alloy [12]. During in situ deformation experiments on Cu/TiN, Cu/W or Cu/Cr layered micro-pillars at elevated temperatures, the formation of differently shaped Cu particles along their surface were identified [13–15]. Their presence is explained by a shear-assisted diffusion process along with these interfaces. The Cu-multilayer composite materials exhibit stable mechanical properties with increasing temperature under compression during micro-pillar experiments [15]. Furthermore, the investigation of nc W foam yielded promising results concerning hardness and Young's modulus [12]. Situated on the surface of shielding components, the nc W/Cu composite material is typically exposed to He^+ implantation causing the formation of He bubbles accompanied by crack propagation constituting common irradiation damage. This may be avoided due to the introduction of open channels and pores in a foam [16–19]. In addition, this composite material provides distinct self-healing potential, as the Cu evaporates on the exposed surface and restores the protecting W foam. Ductility may be improved by adjustment of the grain size, but also through modified alloying concepts to enhance the brittle to ductile temperature or to increase the grain boundary strength [10,20]. According to density-functional theory studies on W regarding the cohesion strength between grains, the presence of Cu was found to have no significant effect on the extent of embrittlement [20]. However, this class of nc W/Cu composite material offers several application possibilities if the mechanical properties are increased and the microstructural stability is enhanced. Therefore, the present study aims to investigate the mechanical properties and microstructural stability of the nc W/Cu composite concerning exposure to elevated temperatures.

2. Materials and Methods

The material examined in this study constitutes a nc W-Cu composite fabricated via the HPT process. A cg W-33 wt% Cu composite was provided by Plansee SE, Reutte, Austria as a cuboid, cut from a plate with a thickness of 10 mm. This cuboid was wire-eroded to an 8 mm diameter rod, and subsequently cut into discs with a thickness of 0.9 mm. The HPT process was performed on the cg W/Cu composite discs at room temperature (RT), applying a pressure of 7.5 GPa and a rotational speed of 1.2 rpm. Sixty rotations were applied to obtain a homogeneous and saturated nc microstructure at radii ranging from 0.5 to 4 mm of the discs. At a distance of 3.65 and 3.8 mm from the center of the disc, an applied strain of about 950 ± 60 was reached [10,21]. A detailed introduction to the HPT procedure of cg W/Cu composite can be found in reference [10]. Subsequently, a heat-treatment at 300 °C for 1 h within a vacuum furnace (X.Tube, Xerion Berlin Laboratories®GmbH, Berlin, Germany) was performed at about 3.0×10^{-4} Pa. This allows for reduction of the amount of forced mechanical intermixing between W and Cu during HPT process without resulting in significant grain coarsening or recrystallization [10,22]. Hence, the mean grain size of W and Cu grains was determined to 14.4 ± 6.3 nm in diameter [1]. This ranges in the same regime as the as-deformed specimen, exhibiting a grain size of 5–15 nm [10]. Therefore, only a minor influence of the heat treatment and, thus, testing at temperatures up to 300 °C, are suggested. Representative images showing the nanostructure of the investigated W/Cu composite can be found in ref. [12].

Subsequently, heat-treatments at 400, 500, 600, 700 and 900 °C for 1 h within a high-vacuum furnace were conducted on separate specimens to determine the temperature stability of the nc W/Cu microstructure. The specimens were previously mirror-polished to ensure an appropriate surface condition for nanoindentation tests, as well as to disclose the effects on the surface. Subsequently, a scanning electron microscope (SEM, LEO1525, Carl Zeiss GmbH, Oberkochen, Germany) was used to investigate the microstructure and surface conditions of the heat-treated specimens. Here, an acceleration voltage of 15 kV, typical magnifications ranging from 2000 to 30,000 times, working distances between 3.5 and 5.5 mm and an image resolution of 1024 × 768 pixel were used as imaging parameters. To achieve a more detailed understanding of surface effects, cross-sections were excavated by use of a Zeiss Auriga Laser FIB system containing a focused ion beam column (Orsay Physics Ga^+ ion FIB) and a scanning electron column (Gemini Schottky field emission, Carl Zeiss GmbH, Oberkochen, Germany). To realize extended cross-sections, a femtosecond laser from type Origami 10 XP (Onefive GmbH, Zürich, Switzerland) and a pulse duration of about 500 fs was used [23]. Further polishing was performed by the use of the attached FIB device using ion cutting currents from 2 nA to 200 pA.

The RT and nanoindentation experiments up to 400 °C were performed on a G200 platform (KLA Corporation, Milpitas, CA, USA). This setup, including a continuous stiffness measurement (CSM) unit, enables the determination of hardness and Young's modulus continuously throughout the indentation depth. A diamond Berkovich tip (Synton MDP, Nidau, Switzerland) exhibiting a tip radius of 250 nm was calibrated following the Oliver-Pharr procedure and the indentation strain rate ($\dot{P}/P$) was kept constant at 0.05 s^{-1} during the measurement [24,25]. A maximum load of 700 mN at an indentation depth of 2500 nm was applied during the measurements. The resultant hardness values were determined and averaged within a range of 300 to 425 nm as well as 1200 and 1400 nm. Furthermore, strain rate jump tests were performed at indentation depths of 500 and 1500 nm with respective constant indentation strain rates of 0.01 and 0.005 s^{-1} up to a total indentation depth of 2500 nm [26]. To avoid oxidation during the experiments, a protective gas atmosphere was applied including a constant gas flux of 0.4 to 0.6 L/min.

In the case of HT nanoindentation tests from 400 to 600 °C, an InSEM-HT device (Nanomechanics Inc. KLA, Oak Ridge, TN, USA) including a CSM module was utilized. To ensure stable temperature conditions and guarantee an isothermal contact between tip and specimen, both are heated and controlled separately [27]. To avoid specimen oxidation or degradation of the indenter tip at high testing temperatures, the system is mounted within a Tescan Vega3 SEM (Tescan, Brno, Czech Republic) and is operated under high-vacuum conditions [28]. The used Berkovich type indenter tip consists of silicon carbide (Synton-MDP, Nidau, Switzerland). The machine compliance and area function calibration of the tip were determined by indentation tests on fused silica at RT [25]. Possible degradation of the tip was monitored by regular indentations on fused silica. A maximum indentation depth of 600 nm and maximum force of 50 mN were utilized. Strain rate jump tests were performed with a constant indentation strain rate of 0.4 s^{-1} to a displacement of 300 nm, followed by a 0.04 s^{-1} jump for 125 nm, and a subsequent return to 0.4 s^{-1}. Following Maier et al. [26], the strain rate sensitivity (m) and the apparent activation volume (v^*) were determined.

Notably, high-temperature nanoindentation tests were performed from RT up to 600 °C on the same HPT deformed and subsequently polished specimen with both testing setups. Here, similar distances from the center point were selected to ensure comparable deformation ratios and consequently, the corresponding identic microstructure. The exact positions are visible in Figure S1a and are located directly next to each other.

3. Results

3.1. Microstructural Development

Secondary electron (SE) SEM images of the HPT deformed specimens' surface after the heat-treatments are displayed in Figure 1a–f. The specimens were heated to 300, 400, 500, 600, 700 and 900 °C for 1 h and subsequently cooled to RT within a vacuum furnace. The W grain size is slightly increased after annealing at 400 °C compared to 300 °C, as can be seen in Figure 1a,b. Upon further increasing the temperature, Cu droplets and pools are emerging on the surface of the specimen within a temperature range of 600 to 700 °C.

Their number and size increase with rising temperature Figure 1d,e. Specimens annealed at 900 °C show again a homogenous surface structure without any Cu particles or pools visible in Figure 1f. However, the formation of dimples and pores on the surface indicates the evaporation of Cu from the surface. Due to the rough surface, a higher contrast between the nc W grains and evolving pores is evoked in SE imaging mode.

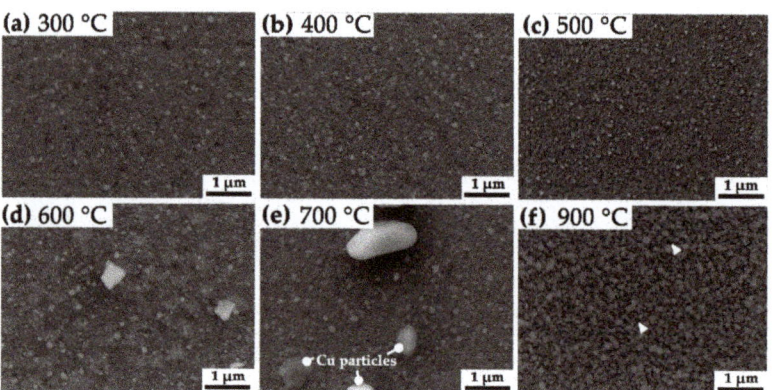

Figure 1. SE SEM images of the HPT processed W/Cu specimens after 1 h annealing at temperatures of (**a**) 300, (**b**) 400, (**c**) 500, (**d**) 600, (**e**) 700 and (**f**) 900 °C. Within (**d**,**e**) faceted Cu particles become visible on the polished surfaces. In (**f**) the formation of pores along the surface is visible and marked by white arrows.

To further investigate the effect of Cu pools located on the surface of the respective specimen, a cross-section was realized using FIB. In Figure 2a several Cu pools are visible on the surface of the specimen heat-treated at 700 °C. Additionally, several faceted Cu particles are situated next to and atop a selected Cu pool in Figure 2b. A more detailed SE image of one of the faceted Cu particles is displayed in Figure 2c. The exposed edge of a Cu pool including two different regions below the surface is visible in Figure 2e. The cross-section reveals a shallow appearance of the Cu pools with no noticeable in-depth effects on the microstructure.

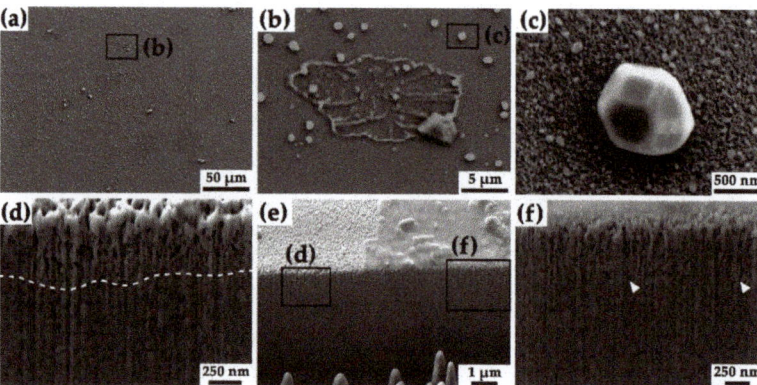

Figure 2. SE SEM images of a W/Cu composite annealed at 700 °C: (**a**) overview of the polished surface, (**b**) representative Cu pool, (**c**) faceted Cu particle, (**d**) sub-surface region not covered by a Cu pool, (**e**) overview of the cross-section containing different areas and sub-surface structures and (**f**) sub-surface region below the Cu pool. In (**d**,**f**) the dotted line restricts the porous zone and white arrows indicate single pores, respectively.

However, below the pool a lower porosity was determined (see Figure 2f) compared to the uncovered material in Figure 2d. Here, the mean depth of an open porosity was determined to be 330 nm. The Cu pool covered area solely contains some loose regions including visible pores. The mean diameter of about 50 Cu pool determined from SEM images at 700 °C was 18.3 ± 6.7 µm. The vertical contrast visible along the cross-sections in Figure 2d,f can be attributed to curtaining effects originating from FIB preparation. Due to the re-deposition of the sputtered material on the specimen's surface, the top layer near the edge visible in Figure 2d,f appears to exhibit a coarser appearance of the surface grains.

3.2. Nanoindentation and HT Nanoindentation Tests

Selected nanoindentation results of the G200 setup at RT, 100, 200, 300 and 400 °C are displayed in Figure 3. Representative load-indentation depth curves including strain rate jumps are displayed in Figure 3a. The corresponding hardness values as a function of the indentation depth are displayed in Figure 3b. Considering RT data, the force limit was hit; therefore, the aimed maximum indentation depth of 2500 nm was not reached. To compare measurements from different temperatures and later the InSEM-HT testing setup, values were determined in a depth of 300 to 450 nm, where the hardness remains almost constant.

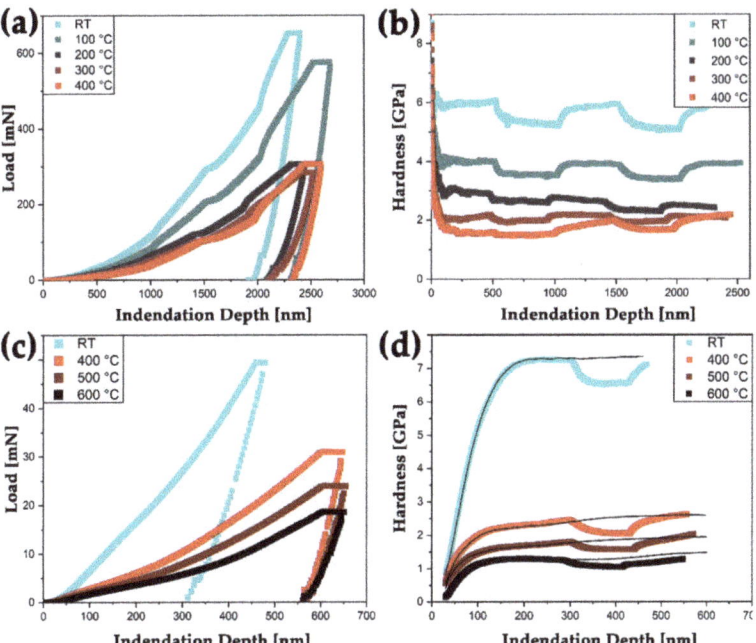

Figure 3. Representative indentation data of nc W/Cu specimen from indentation experiments gathered by the G200 setup (a,b) and the InSEM-HT nanoindentation testing setup(c,d). In (a) typical load and in (c) hardness vs. indentation depth curves at RT, 100, 200, 300 and 400 °C are displayed. In (c) load and (d) hardness values are displayed over indentation depth for testing temperatures at RT, 400, 500, and 600 °C, respectively. Within an indentation depth of 300 to 425 nm the constant strain rate was decreased from 0.4 to 0.04 s^{-1} for strain rate jump tests at the same temperatures. Additional indentation experiments with a constant strain rate were performed for comparison and are displayed as black lines in (d). For a colored representation, the reader is referred to the web version of this article.

Furthermore, representative load-indentation depth curves from the InSEM-HT setup at RT, 400, 500 and 600 °C up to either a maximal indentation depth of 600 nm or a load of 50 mN are shown in Figure 3c Their corresponding hardness values are displayed in Figure 3d and were determined with a $\dot{P}/P$ of 0.4 s^{-1}. Continuous mechanical data is available due to the used CSM testing setup [26]. Additionally, hardness profiles determined from strain rate jump tests are displayed in Figure 3d. To compare different data points at varying temperatures and also to the G200 setup, mean values from 300 to 450 nm were used. Also, the associated Young's modulus was determined within this indentation depth [29]. Here, initial surface and contact effects can be excluded, resulting in almost constant values. As expected for nc materials, no indentation size-effect was identified with increasing indentation depth [30].

For comparison, the hardness and Young's modulus values determined from different testing setups and temperatures are summarized in Figure 4. As the material is strain rate sensitive and two different constant indentation strain rates were used, the evaluated hardness from the InSEM-HT device is higher compared to the values from the G200 setup. Thus, to obtain comparable hardness values, a correction of the constant strain rate of the respective test setup was performed: values from the InSEM-HT device and G200 setup were corrected to a $\dot{P}/P$ of 0.05 and 0.4 s^{-1}, which correspond to the open symbols in Figure 4a. However, at RT and 400 °C, mean deviations of 0.58 and 0.78 GPa between the InSEM-HT and corrected G200 values (open circles) were determined. Hence, hardness values from the InSEM-HT indenter remain slightly elevated compared to data gathered using the G200 setup.

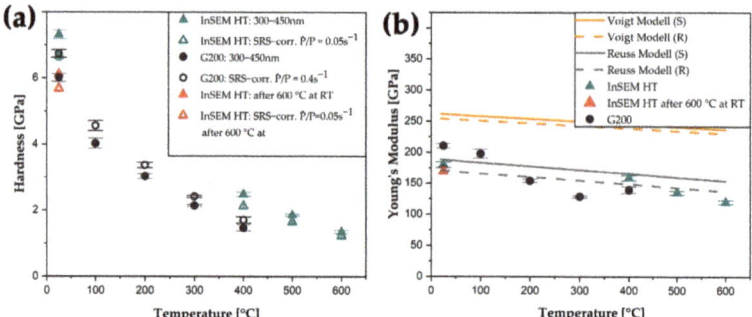

Figure 4. In (a) the evolution of the hardness as a function of the temperature of the nc W/Cu composite material is displayed. The data include a standard deviation calculated from at least five indents and was corrected according to the given constant indentation strain rate. The Young's modulus as a function of the temperature is shown in (b). Guidelines based on literature data were included to guide the reader's eyes [31,32]. Therefore, the anisotropic Young's modulus of Cu was calculated after Shtrikman (S) and Reuss (R) bounds. The open symbols constitute strain-rate-corrected data points, and the red triangles display data points recorded subsequently to the high-temperature experiments at RT. Here, only a minor deviation of 16 and 5% from the hardness and Young's modulus was determined, respectively. For a colored version of this article, the reader is referred to the online edition.

The measured Young's moduli including some trend lines calculated from literature data are shown in Figure 4b. Hence, Young's moduli of Cu and W from Chang et al. [31] and Lowrie et al. [32] with a phase ratio of 49/51 were used. With increasing temperature, the measured Young's modulus decreases and shows good accordance with the trend lines calculated after the Reuss model (R). Here, RT Young's moduli from InSEM-HT setup are slightly under- and in the case of 400 °C measurements overestimate values from G200 setup, but in general, nicely follow literature predictions [31,32].

4. Discussion

The microstructural investigation (Figure 1) confirmed that the surface structure of heat-treated nc W/Cu composite material witnesses a distinct change of surface characteristics at varying temperatures. With increasing temperature, the nc Cu matrix phase tends to sublimate into the high-vacuum oven atmosphere. Furthermore, Cu pool-like structures are observed on the surface starting at temperatures of 600 °C. Their number and size rise with increasing temperature but vanish at 900 °C. As the temperatures increase, an accumulation of Cu pool could be determined with increasing distance from the center of the HPT deformed specimen. This indicates an influence of the applied deformation strain, as this effect amplifies with increasing distance from the center. By raising the deformation ratio, the grain size, mechanical intermixing, homogeneity but also internal stresses are enhanced and ultimately reach a saturation level [10]. Mechanical intermixing and stresses were reduced during the preliminary heat-treatment at 300 °C for 1 h within a vacuum furnace. In this condition, the mean grain size of W and Cu grains was determined to 14.4 ± 6.3 nm. In the following, no visible changes on the surface or microstructure could be determined during SE SEM investigations. Post-indentation images of the specimen's surface indicate an additional accumulation of Cu pool in and around the area where the indentation tests were performed (see supplementary SP-1). This substantiates a stress-assisted diffusion of Cu along different interfaces, as previously observed during micropillar experiments on heat-treated Cu/TiN multilayers by R. Raghavan et al. [13]. The authors identified faceted Cu particles on the surface of tested pillars and Cu infiltrated pores along with deformation-related shear bands [13]. In our study, faceted Cu particles were identified on the specimen's surface by SEM investigations starting at temperatures of 600 °C. Similar to the Cu pool, their number and size increase with increasing temperature and they vanish during a heat-treatment of 900 °C for 1 h within a vacuum furnace. Studies of W/Cu multilayer specimens indicate a correlation of cracks along W grain boundaries and the presence of faceted Cu particles on the surface [11,33]. Therefore, a coupled mechanism of preferred diffusion pathways and stress-induced diffusion of Cu to the surface was postulated [11,13]. As the presence of faceted Cu particles is randomly distributed on the respective surfaces visible in Figure 1d,e, a site unspecific availability of diffusion pathways within the nc microstructure is assumed. The presence and morphology of the Cu pool were further investigated from a cross-section of the specimen heat-treated at 700 °C (Figure 2). Here, the area below the surface, which is not covered by a Cu pool, exhibits a pronounced porosity up to a mean depth of 330 nm, assuming an effective loss of 70% of Cu up to this depth under consideration of the prevalent volume share of 49 vol.% Cu, theoretically 0.62 faceted Cu particles with a mean diameter of 716 nm should be present per μm^2 on the surface of the specimen. This fraction would be equivalent to a mean coverage of 25.0% of the available surface area by Cu particles. The mean particle diameter of 716 ± 153 nm was determined from high-resolution SEM images analysis. However, the determination of SE SEM images of this specimen state yields a mean coverage from 4.3 to 5.2% of faceted Cu particles besides the Cu pools. This observation indicates a strong loss of Cu into the vacuum due to sublimation from the specimen's surface. Similar effects were observed in the case of ZnCu alloys, as the Zn starts to evaporate at temperatures of 450 °C, and the remaining Cu forms a microporous foam [34]. This effect was described as physical vacuum dealloying, as the vapor pressure of the volatile phase is by far higher compared to the Cu phase. In the case of the nc W/Cu composite, the vapor pressure of Cu at 700 °C is more than 27 magnitudes higher compared to that of the W phase [35]. Hence, at this temperature, the vacuum pressure of Cu amounts to 5.0×10^{-7} Pa and ranges, therefore, below typical high-vacuum pressures of 10^{-4} to 10^{-6} Pa. This enables the deposition of Cu on the surface as Cu pools and later as faceted Cu particles due to the stress-induced diffusion mechanism. At a temperature of 900 °C, the Cu vacuum pressure increases to 5.8×10^{-4} Pa, resulting in the sublimation of Cu from the surface [35]. Here, the limiting factor is the replenishment of Cu via diffusional processes along or through the depleted nano-porous W-rich zone.

Underneath the Cu pool in the cross-section view (see Figure 2f) by themselves, a few agglomerated pores are visible. This indicates an easier diffusion pathway or stronger driving forces to relocate Cu from deeper regions of the specimen. Therefore, the presence of microcracks is assumed to promote the formation of Cu pools on the surface. Furthermore, faceted Cu particles are also located on these Cu pool, indicating no or at least a minor influence on the distribution and size of the pools.

The faceted morphology of the Cu particles is suggested to rely on the fact that the material is striving to achieve an optimal shape to reduce its total surface energy [36]. A visualization of this, named the Wulff plot, resembles the shape of faceted Cu nanoparticles for face-centered cubic (fcc) specimens [37]. Their shape was proven to change as the formation of specific Miller planes is favorable. under different atmospheric conditions [38–40]. However, the morphology of the nc W grains on the surface of the heat-treated specimens was not noticeably changed, due to the annealing process at different temperatures, which was due to the low diffusivity of W in this temperature regime.

The results from the RT and high-temperature nanoindentation experiments must be regarded in the light of developing surface porosity and thus, changing microstructure. As the Cu phase sublimates to a larger extent with increasing temperature and time, an increasing number of open channels and pores are prevalent in the near-surface layer of the nc W/Cu composite material. As the indentation experiments were performed up to an indentation depth of 600 or 2500 nm in the case of InSEM-HT or G200 testing setup, respectively, and the respective hardness and Young's modulus were determined within a range of 300 to 450 nm, only minor influences of the evolving porosity and other microstructural changes are assumed up to a temperature of 400 °C. This is further supported by the hardness determination of the G200 experiments within a depth of 1200 to 1400 nm, resulting in comparable values within the given standard deviation. Also, hardness and modulus are deduced from the plastic and elastic deformation regions, which reach to a depth of around 5–15 times deeper than the actual indentation depth. Hence, the hardness is almost constant for indentation depths exceeding 200 nm at any temperature. Furthermore, after strain rate jump tests and returning to the initial strain rate, similar hardness values were reached with both used test setups (see Figure 4a). In a study on a nc W foam with a similar initial grain size but higher deformation applied using a severe plastic deformation process, lower hardness values of the pure foam of 2.8 GPa were determined at RT [12]. Furthermore, the minor influence of pores on the surface of nc W/Cu composite is substantiated by elevated hardness values determined from the InSEM-HT test setup at RT after the performed G200 experiments (see Figure 4a). Here, the strain rate corrected hardness from the InSEM-HT device exceeds the G200 RT hardness by 0.66 GPa and amounts to 6.67 ± 0.13 GPa. At the given temperature, this value is similar to hardness values of cg W specimens investigated by Kappacher et al. [41]. Furthermore, there is a good consistency of hardness and Young's modulus values with deposited and heat-treated nc W-Cu thin films [22,29,42]. With increasing temperature, the hardness of the nc W/Cu composite decreases steadily. This behavior may be attributed to the drastic drop in flow-stress of body-centered cubic (bcc) materials with increasing temperature [41,43]. Hence, the thermal activation of kink-pairs along the $\frac{1}{2}$ <111> screw dislocation lines at a temperature of about 0.2 T_m enables plastic deformation under applied shear stresses [44]. Above this temperature, the athermal regime is governed by both, edge and screw, dislocation processes which are independent of thermal activation [41,44]. Since the drop in hardness with increasing temperature in the here investigated nc W/Cu is even more pronounced compared to the cg W, a major influence of the Cu phase is suggested. Moreover, the critical temperature of 0.2 T_m of W was not reached during the in situ nanoindentation experiments. Different studies are proposing a hardness between 9 and 11 GPa for nc Cu at RT, with a strong influence of the grain size and a Hall-Petch breakdown below a grain size of 7 to 10 nm [45–47]. However, the specimen's mean grain size after the heat-treatment at 300 °C for 1 h amounts to 14.4 ± 6.3 nm for the W and Cu grains. This is in the same range as the as-deformed HPT processed W/Cu [10]. Therefore, a minor effect of the heat-treatment and testing temperatures up to 300 °C on the initial microstructure is assumed.

Young's modulus also decreases with increasing temperature. To guide the reader's eyes, different trendlines were included in Figure 4b. They were calculated including their phase composition according to the Voigt and Reuss model [31,32,48,49]. In the case of the strong anisotropic Young's modulus of Cu, methods after Shtrikman (S) and Reuss (R) were implemented [31,50,51]. Hence, the best accordance was achieved with the Reuss model (R), visible in Figure 4b, representing the lowest bound of Cu, which was adapted to the temperature [50]. Under consideration of the InSEM-HT data points (green triangles in Figure 4b), a deviation of the linear behavior at temperatures of 500 and 600 °C of the two different models can be evidenced. This drop in Young's modulus amounts to about 17 or 26 GPa at temperatures of 500 and 600 °C, respectively. After the performed G200 and InSEM-HT measurements, a reference measurement was conducted at RT, depicted as the red triangle in Figure 4b. Here, a Young's modulus reduction of 10 GPa indicates a minor change of testing conditions due to the formation of pores along the surface, a distinct evaporation of Cu or minor grain growth after heating to a temperature of 600 °C.

The strain rate sensitivity (m) and activation volume (v^*) were calculated using strain rate jump tests during indentation experiments according to the method of Maier et al. [26]. Typically, m is determined as a function of grain size—leading to a decreased m value with decreasing grain size for bcc materials at RT [52–55]. In the case of fcc crystals, this trend is reversed, as the governing mechanism turns from cutting of forest dislocations in coarse grains to a dislocation-grain boundary interaction along nc grains [53,56,57].

As demonstrated for ultra-fine grained Cr, m increases above the critical temperature of 0.2 T_m. This fact is attributed to a mechanism change when the critical temperature is reached and fcc-like behavior governs the plastic processes along with grain boundary contributions [44]. In more detail, thermally activated grain boundary-dislocation interactions become dominant against the contribution of screw dislocation movement by kink-pair interactions [44,54]. This was also determined with un- and deformed W, where m increases in the deformed state with increasing temperature [55,58]. In general, fcc materials exhibit lower m values compared to bcc materials, as Peierl's stress is lower and, therefore, the effect of temperature and further time-dependent influences are not as pronounced. However, especially in ufg or nc conditions, an increasing m with increasing temperature is reported [59].

In the case of nc W/Cu composites, differently pronounced and mutually overlapping effects are affecting the temperature-dependent condition of m. Since the critical temperature of W is typically reached at temperatures above 600 °C, an increase of m due to the thermally activated grain boundary-dislocation interaction mechanism can be neglected [58]. Furthermore, a constant influence of nc W grains on m is suspected within the tested temperature range [28]. The literature unveils a strong increase of m with increasing temperature within nc Cu, but is solely available up to a temperature of 300 °C [59,60]. The determined strain rate sensitivity of the nc W/Cu composite increases slightly with increasing temperature up to 400 °C (see Figure 5a). However, no clear trend can be evaluated, as all values range within the measured standard deviation. At temperatures from 400 to 600 °C, a decrease of m was determined from the InSEM-HT data. This may be attributed to the decreasing influence of the Cu phase, since a distinct amount of Cu was evaporated. To verify the measurement, a second RT test was performed and a strain rate sensitivity decrease of about 0.01 was determined. This deviation represents the drop in m from 500 to 600 °C, substantiating the effect of pores along with the surface layer.

As can be seen in Figure 5b, the calculated v^* increases from ~3.6 b^3 at RT to about 57 b^3 at 600 °C. A similar trend was reported by Kappacher et al. regarding the activation volume of cg W [41]. However, measured RT values in the present study are below nc or *ufg* representatives from the literature, wherein, typically, values around 10 b^3 were determined for both, *fcc* and *bcc* materials [28,61,62]. This suggests an even lower grain size for the *nc* W/Cu composite. Contrary to this, with increasing temperature, a steady rise of v^* was determined. This agrees well with the literature, where stabilisation of v^* above $0.2\ T_m$ could be determined after rising values for ufg W and Cr [28,62]. In the present work, the regarded unit cell was defined as the W cell, since the W grain size stays almost constant and no drastic evaporation effects could be determined for this element. In addition to the increasing temperature, the decreasing hardness further results in higher v^* values at elevated temperatures. Very limited literature data is available for the mechanical properties of *nc* or ufg materials at temperatures as high as 600 °C. Also, the influence of the 300 °C annealing heat-treatment may influence the number of preexisting dislocations, thus, affecting the mechanical behavior of the material up to the respective annealing temperature. However, the increasing activation volume indicates a strong influence of the W phase on the deformation behavior. Another aspect may be the coarsening of Cu grains, forming a continuous matrix phase around the nc W grains. This, also, would consequently result in an increased v^*, as longer pinned dislocation segments become possible.

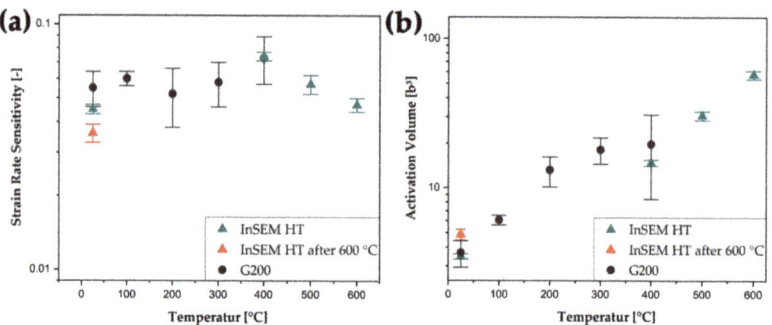

Figure 5. In (**a**) the strain rate sensitivity and (**b**) the calculated activation volume of the nc W/Cu composite are plotted as a function of testing temperature. The activation volume was normalized to the cubed Burgers vector (b^3) to enable comparison between different material systems [26,28]. The round markers indicate G200 and the triangular ones the InSEM-HT data points, respectively. Red triangles depict measurements at RT, which were performed after heating and testing up to a temperature of 600 °C. For a colored version of this article, the reader is referred to the online version.

5. Conclusions

Within this study, a nano-crystalline W/Cu composite processed via the HPT process was investigated regarding the microstructure and mechanical properties at elevated temperatures. The material was heat-treated in a vacuum furnace at various temperatures to achieve information about the stability of each phase as well as the grain size during annealing. High-temperature nanoindentation allowed for the quantification of hardness and Young's modulus of the material in situ between room temperature and 600 °C. Furthermore, strain rate jump tests were performed to calculate m and the related v^* and assess the underlying deformation mechanisms.

Scanning electron microscope investigations on the heat-treated specimens unveiled only minor changes of the microstructure up to 600 °C. At 900 °C, a distinct coarsening of W and Cu grains was evident. On the polished surface of these specimens, the formation of pores, faceted Cu particles and Cu pools was determined. Owing to the higher vapor pressure, Cu is more prone to evaporate into the high vacuum chamber of the furnace or nanoindentation device compared to W. However, with increasing temperature, the number and size of Cu pools, as well as faceted Cu particles, increases. At a temperature of 700 °C, the porous layer exhibits a thickness of about 330 nm. Below the Cu pools, fewer pores were observed since the Cu supply is enabled by deformation-induced diffusion along open Cu channels from deeper regions.

With increasing temperature, the hardness and Young's modulus decrease continuously. This behavior is reasoned by the temperature-dependent motion of dislocation;

therefore, easier deformability is achieved at higher temperatures. Concerning the strain rate sensitivity with increasing temperature, overlapping effects lead to a constant value within the measurement uncertainty. At the same time, the activation volume increases with temperature and thus behaves similar to the high-temperature behavior of coarse-grained W.

Based on the results of this investigation, a deeper understanding of the high-vacuum behavior and temperature dependence of nano-crystalline W/Cu composites could be gathered. With complementary nanoindentation investigations, the mechanical response at elevated temperatures was determined in regard of the present microstructure.

Supplementary Materials: The Figure S1 is available online at https://www.mdpi.com/article/10.3390/nano11112951/s1, Figure S1: Cu pools.

Author Contributions: Conceptualization, D.K., V.M.-K. and M.Z.; methodology, M.Z., M.P., A.L., J.K. and M.W.; software, J.K. and M.B.; validation, D.K., V.M.-K., M.B., A.L. and J.K.; formal analysis, D.K., V.M.-K. and M.B.; investigation, M.Z., M.W., J.K., A.L. and M.P.; resources, D.K and V.M.-K.; data curation, J.K. and M.B.; writing—original draft preparation, M.B. and M.Z.; writing—review and editing, J.K., M.P., M.W., D.K., A.L. and M.B.; visualization, J.K., M.Z. and M.B.; supervision, D.K. and V.M.-K.; project administration, D.K.; funding acquisition, D.K. All authors have read and agreed to the published version of the manuscript.

Funding: The authors acknowledge financial support from the European Research Council (ERC) under the European Union's Horizon 2020 research and innovation programme (Grant No. 771146 TOUGHIT).

Data Availability Statement: The data presented in this study are available on reasonable request from the corresponding author.

Conflicts of Interest: The authors declare no conflict of interest.

References

1. Zhao, M.; Issa, I.; Pfeifenberger, M.J.; Wurmshuber, M.; Kiener, D. Tailoring ultra-strong nanocrystalline tungsten nanofoams by reverse phase dissolution. *Acta Mater.* **2020**, *182*, 215–225. [CrossRef]
2. Rieth, M.; Dudarev, S.L.; Gonzalez de Vicente, S.M.; Aktaa, J.; Ahlgren, T.; Antusch, S.; Armstrong, D.; Balden, M.; Baluc, N.; Barthe, M.-F.; et al. Recent progress in research on tungsten materials for nuclear fusion applications in Europe. *J. Nucl. Mater.* **2013**, *432*, 482–500. [CrossRef]
3. Chen, S.; Bourham, M.; Rabiei, A. Neutrons attenuation on composite metal foams and hybrid open-cell Al foam. *Radiat. Phys. Chem.* **2015**, *109*, 27–39. [CrossRef]
4. El-Atwani, O.; Hinks, J.A.; Greaves, G.; Gonderman, S.; Qiu, T.; Efe, M.; Allain, J.P. In-situ TEM observation of the response of ultrafine-and nanocrystalline-grained tungsten to extreme irradiation environments. *Sci. Rep.* **2014**, *4*, 4716. [CrossRef]
5. Koch, C.C.; Morris, D.G.; Lu, K.; Inoue, A. Ductility of Nanostructured Materials. *MRS Bull.* **1999**, *24*, 54–58. [CrossRef]
6. Hohenwarter, A.; Pippan, R. Fracture and fracture toughness of nanopolycrystalline metals produced by severe plastic deformation. *Philos. Trans. Ser. A Math. Phys. Eng. Sci.* **2015**, *373*. [CrossRef]
7. Hall, E.O. The Deformation and Ageing of Mild Steel: III Discussion of Results. *Proc. Phys. Soc. B* **1951**, *64*, 747–753. [CrossRef]
8. Petch, N.J. The cleavage strength of polycrystals. *J. Iron Steel Inst.* **1953**, *174*, 25–28.
9. Armstrong, R.W. 60 Years of Hall-Petch: Past to Present Nano-Scale Connections. *Mater. Trans.* **2014**, *55*, 2–12. [CrossRef]
10. Kormout, K.S.; Pippan, R.; Bachmaier, A. Deformation-Induced Supersaturation in Immiscible Material Systems during High-Pressure Torsion. *Adv. Eng. Mater.* **2017**, *19*, 1600675. [CrossRef]
11. Moszner, F.; Cancellieri, C.; Chiodi, M.; Yoon, S.; Ariosa, D.; Janczak-Rusch, J.; Jeurgens, L. Thermal stability of Cu/W nano-multilayers. *Acta Mater.* **2016**, *107*, 345–353. [CrossRef]
12. Zhao, M.; Schlueter, K.; Wurmshuber, M.; Reitgruber, M.; Kiener, D. Open-cell tungsten nanofoams: Scaling behavior and structural disorder dependence of Young's modulus and flow strength. *Mater. Design* **2021**, *197*, 109187. [CrossRef]
13. Raghavan, R.; Wheeler, J.M.; Esqué-de los Ojos, D.; Thomas, K.; Almandoz, E.; Fuentes, G.G.; Michler, J. Mechanical behavior of Cu/TiN multilayers at ambient and elevated temperatures: Stress-assisted diffusion of Cu. *Mater. Sci. Eng. A* **2015**, *620*, 375–382. [CrossRef]
14. Raghavan, R.; Wheeler, J.M.; Harzer, T.P.; Chawla, V.; Djaziri, S.; Thomas, K.; Philippi, B.; Kirchlechner, C.; Jaya, B.N.; Wehrs, J.; et al. Transition from shear to stress-assisted diffusion of copper–chromium nanolayered thin films at elevated temperatures. *Acta Mater.* **2015**, *100*, 73–80. [CrossRef]
15. Wheeler, J.M.; Raghavan, R.; Chawla, V.; Zechner, J.; Utke, I.; Michler, J. Failure mechanisms in metal–metal nanolaminates at elevated temperatures: Microcompression of Cu–W multilayers. *Scr. Mater.* **2015**, *98*, 28–31. [CrossRef]

16. Zhao, M.; Pfeifenberger, M.J.; Kiener, D. Open-cell tungsten nanofoams: Chloride ion induced structure modification and mechanical behavior. *Results Phys.* **2020**, *17*, 103062. [CrossRef]
17. Allen, F.I.; Hosemann, P.; Balooch, M. Key mechanistic features of swelling and blistering of helium-ion-irradiated tungsten. *Scr. Mater.* **2020**, *178*, 256–260. [CrossRef]
18. El-Atwani, O.; Suslova, A.; Novakowski, T.J.; Hattar, K.; Efe, M.; Harilal, S.S.; Hassanein, A. In-situ TEM/heavy ion irradiation on ultrafine-and nanocrystalline-grained tungsten: Effect of 3 MeV Si, Cu and W ions. *Mater. Charact.* **2015**, *99*, 68–76. [CrossRef]
19. Juarez, T.; Biener, J.; Weissmüller, J.; Hodge, A.M. Nanoporous Metals with Structural Hierarchy: A Review. *Adv. Eng. Mater.* **2017**, *19*, 1700389. [CrossRef]
20. Scheiber, D.; Pippan, R.; Puschnig, P.; Ruban, A.; Romaner, L. Ab-initio search for cohesion-enhancing solute elements at grain boundaries in molybdenum and tungsten. *Int. J. Refract. Met. Hard Mater.* **2016**, *60*, 75–81. [CrossRef]
21. Hohenwarter, A.; Bachmaier, A.; Gludovatz, B.; Scheriau, S.; Pippan, R. Technical parameters affecting grain refinement by high pressure torsion. *Int. J. Mater. Res.* **2009**, *100*, 1653–1661. [CrossRef]
22. Vüllers, F.; Spolenak, R. From solid solutions to fully phase separated interpenetrating networks in sputter deposited "immiscible" W–Cu thin films. *Acta Mater.* **2015**, *99*, 213–227. [CrossRef]
23. Pfeifenberger, M.J.; Mangang, M.; Wurster, S.; Reiser, J.; Hohenwarter, A.; Pfleging, W.; Kiener, D.; Pippan, R. The use of femtosecond laser ablation as a novel tool for rapid micro-mechanical sample preparation. *Mater. Design* **2017**, *121*, 109–118. [CrossRef]
24. Lucas, B.N.; Oliver, W.C. Indentation power-law creep of high-purity indium. *Met. Mater. Trans. A* **1999**, *30*, 601–610. [CrossRef]
25. Oliver, W.C.; Pharr, G.M. An improved technique for determining hardness and elastic modulus using load and displacement sensing indentation experiments. *J. Mater. Res.* **1992**, *7*, 1564–1583. [CrossRef]
26. Maier, V.; Durst, K.; Mueller, J.; Backes, B.; Höppel, H.W.; Göken, M. Nanoindentation strain-rate jump tests for determining the local strain-rate sensitivity in nanocrystalline Ni and ultrafine-grained Al. *J. Mater. Res.* **2011**, *26*, 1421–1430. [CrossRef]
27. Korte, S.; Stearn, R.J.; Wheeler, J.M.; Clegg, W.J. High temperature microcompression and nanoindentation in vacuum. *J. Mater. Res.* **2012**, *27*, 167–176. [CrossRef]
28. Kappacher, J.; Renk, O.; Kiener, D.; Clemens, H.; Maier-Kiener, V. Controlling the high temperature deformation behavior and thermal stability of ultra-fine-grained W by re alloying. *J. Mater. Res.* **2021**, *36*.
29. Liang, L.; Li, M.; Qin, F.; Wei, Y. Temperature effect on elastic modulus of thin films and nanocrystals. *Philos. Mag.* **2013**, *93*, 574–583. [CrossRef]
30. Milman, Y.; Golubenko, A.; Dub, S.N. Indentation size effect in nanohardness. *Acta Mater.* **2011**, *59*, 7480–7487. [CrossRef]
31. Chang, Y.A.; Himmel, L. Temperature Dependence of the Elastic Constants of Cu, Ag, and Au above Room Temperature. *J. Appl. Phys.* **1966**, *37*, 3567–3572. [CrossRef]
32. Lowrie, R.; Gonas, A.M. Dynamic Elastic Properties of Polycrystalline Tungsten, 24°–1800 °C. *J. Appl. Phys.* **1965**, *36*, 2189–2192. [CrossRef]
33. Auciello, O.; Chevacharoenkul, S.; Ameen, M.S.; Duarte, J. Controlled ion beam sputter deposition of W/Cu/W layered films for microelectronic applications. *J. Vac. Sci. Technol. A Vac. Surf. Film.* **1991**, *9*, 625–631. [CrossRef]
34. Sun, Y.; Ren, Y.; Yang, K. New preparation method of micron porous copper through physical vacuum dealloying of Cu–Zn alloys. *Mater. Lett.* **2016**, *165*, 1–4. [CrossRef]
35. Alcock, C.B.; Itkin, V.P.; Horrigan, M.K. Vapour Pressure Equations for the Metallic Elements: 298–2500 K. *Can. Metall. Q.* **1984**, *23*, 309–313. [CrossRef]
36. Willard Gibbs, J. *On the Equilibrium of Heterogeneous Substances*; Nr. 1; Transactions of the Connecticut Academy of Arts and Sciences: New Haven, CT, USA, 1874; Volume 3.
37. Wulff, G. Zur Frage der Geschwindigkeit des Wachsthums und der Auflösung der Krystallflächen. *Z. Kristallogr. Cryst. Mater.* **1901**, *34*. [CrossRef]
38. Barmparis, G.D.; Lodziana, Z.; Lopez, N.; Remediakis, I.N. Nanoparticle shapes by using Wulff constructions and first-principles calculations. *Beilstein J. Nanotechnol.* **2015**, *6*, 361–368. [CrossRef]
39. Wang, Y.; Wang, Z.; Dinh, C.-T.; Li, J.; Ozden, A.; Golam Kibria, M.; Seifitokaldani, A.; Tan, C.-S.; Gabardo, C.M.; Luo, M.; et al. Catalyst synthesis under CO_2 electroreduction favours faceting and promotes renewable fuels electrosynthesis. *Nat. Catal.* **2020**, *3*, 98–106. [CrossRef]
40. Hansen, P.L.; Wagner, J.B.; Helveg, S.; Rostrup-Nielsen, J.R.; Clausen, B.S.; Topsøe, H. Atom-resolved imaging of dynamic shape changes in supported copper nanocrystals. *Science* **2002**, *295*, 2053–2055. [CrossRef]
41. Kappacher, J.; Leitner, A.; Kiener, D.; Clemens, H.; Maier-Kiener, V. Thermally activated deformation mechanisms and solid solution softening in W-Re alloys investigated via high temperature nanoindentation. *Mater. Design* **2020**, *189*, 108499. [CrossRef]
42. Monclús, M.A.; Karlik, M.; Callisti, M.; Frutos, E.; Llorca, J.; Polcar, T.; Molina-Aldareguía, J.M. Microstructure and mechanical properties of physical vapor deposited Cu/W nanoscale multilayers: Influence of layer thickness and temperature. *Thin Solid Film.* **2014**, *571*, 275–282. [CrossRef]
43. Gröger, R.; Vitek, V. Temperature and strain rate dependent flow criterion for bcc transition metals based on atomistic analysis of dislocation glide. *Int. J. Mater. Res.* **2009**, *100*, 315–321. [CrossRef]
44. Maier, V.; Hohenwarter, A.; Pippan, R.; Kiener, D. Thermally activated deformation processes in body-centered cubic Cr—How microstructure influences strain-rate sensitivity. *Scr. Mater.* **2015**, *106*, 42–45. [CrossRef]

45. Li, J.; Lu, B.; Zhang, Y.; Zhou, H.; Hu, G.; Xia, R. Nanoindentation response of nanocrystalline copper via molecular dynamics: Grain-size effect. *Mater. Chem. Phys.* **2020**, *241*, 122391. [CrossRef]
46. Huang, C.-C.; Chiang, T.-C.; Fang, T.-H. Grain size effect on indentation of nanocrystalline copper. *Appl. Surf. Sci.* **2015**, *353*, 494–498. [CrossRef]
47. Schiøtz, J.; Jacobsen, K.W. A maximum in the strength of nanocrystalline copper. *Science* **2003**, *301*, 1357–1359. [CrossRef]
48. Kamaya, M. A procedure for estimating Young's modulus of textured polycrystalline materials. *Int. J. Solids Struct.* **2009**, *46*, 2642–2649. [CrossRef]
49. Kim, H.S.; Hong, S.I.; Kim, S.J. On the rule of mixtures for predicting the mechanical properties of composites with homogeneously distributed soft and hard particles. *J. Mater. Process. Technol.* **2001**, *112*, 109–113. [CrossRef]
50. Simmons, G.; Wang, H. *Single Crystal Elastic Constants and Calculated Aggregate Properties: A Handbook*, 2nd ed.; M.I.T. Press: Cambridge, MA, USA, 1971.
51. Shen, T.D.; Koch, C.C.; Tsui, T.Y.; Pharr, G.M. On the elastic moduli of nanocrystalline Fe, Cu, Ni, and Cu–Ni alloys prepared by mechanical milling/alloying. *J. Mater. Res.* **1995**, *10*, 2892–2896. [CrossRef]
52. Wei, Q.; Jiao, T.; Ramesh, K.; Ma, E.; Kecskes, L.; Magness, L.; Doeding, R.; Kazykhanov, V.; Valiev, R. Mechanical behavior and dynamic failure of high-strength ultrafine grained tungsten under uniaxial compression. *Acta Mater.* **2006**, *54*. [CrossRef]
53. Wei, Q.; Cheng, S.; Ramesh, K.; Ma, E. Effect of nanocrystalline and ultrafine grain sizes on the strain rate sensitivity and activation volume: Fcc versus bcc metals. *Mater. Sci. Eng. A* **2004**, *381*, 71–79. [CrossRef]
54. Maier, V.; Schunk, C.; Göken, M.; Durst, K. Microstructure-dependent deformation behaviour of bcc-metals—Indentation size effect and strain rate sensitivity. *Philos. Mag.* **2015**, *95*, 1766–1779. [CrossRef]
55. Fukuda, M.; Tabata, T.; Hasegawa, A.; Nogami, S.; Muroga, T. Strain rate dependence of tensile properties of tungsten alloys for plasma-facing components in fusion reactors. *Fus. Eng. Des.* **2016**, *109–111*, 1674–1677. [CrossRef]
56. Chen, J.; Lu, L.; Lu, K. Hardness and strain rate sensitivity of nanocrystalline Cu. *Scr. Mater.* **2006**, *54*, 1913–1918. [CrossRef]
57. Zhu, T.; Li, J.; Samanta, A.; Kim, H.G.; Suresh, S. Interfacial plasticity governs strain rate sensitivity and ductility in nanostructured metals. *Proc. Natl. Acad. Sci. USA* **2007**, *104*, 3031–3036. [CrossRef]
58. Kiener, D.; Fritz, R.; Alfreider, M.; Leitner, A.; Pippan, R.; Maier-Kiener, V. Rate limiting deformation mechanisms of bcc metals in confined volumes. *Acta Mater.* **2019**, *166*, 687–701. [CrossRef]
59. Wang, Y.; HAMZA, A.; Ma, E. Temperature-dependent strain rate sensitivity and activation volume of nanocrystalline Ni. *Acta Mater.* **2006**, *54*, 2715–2726. [CrossRef]
60. Suo, T.; Li, Y.; Xie, K.; Zhao, F.; Zhang, K.-S.; Deng, Q. Experimental investigation on strain rate sensitivity of ultra-fine grained copper at elevated temperatures. *Mech. Mater.* **2011**, *43*, 111–118. [CrossRef]
61. Asaro, R.J.; Suresh, S. Mechanistic models for the activation volume and rate sensitivity in metals with nanocrystalline grains and nano-scale twins. *Acta Mater.* **2005**, *53*, 3369–3382. [CrossRef]
62. Fritz, R.; Wimler, D.; Leitner, A.; Maier-Kiener, V.; Kiener, D. Dominating deformation mechanisms in ultrafine-grained chromium across length scales and temperatures. *Acta Mater.* **2017**, *140*, 176–187. [CrossRef]

Article

Physical Surface Modification of Carbon-Nanotube/Polydimethylsiloxane Composite Electrodes for High-Sensitivity DNA Detection

Junga Moon [1], Huaide Jiang [1] and Eun-Cheol Lee [1,2,*]

[1] Department of Nano Science and Technology, Graduate School, Gachon University, Seongnam-si 13120, Gyeonggi-do, Korea; mka3202@naver.com (J.M.); huaide20@gmail.com (H.J.)
[2] Department of Physics, Gachon University, Seongnam-si 13120, Gyeonggi-do, Korea
* Correspondence: ecleel@gachon.ac.kr; Tel.: +82-31-750-8752

Citation: Moon, J.; Jiang, H.; Lee, E.-C. Physical Surface Modification of Carbon-Nanotube/Polydimethylsiloxane Composite Electrodes for High-Sensitivity DNA Detection. *Nanomaterials* **2021**, *11*, 2661. https://doi.org/10.3390/nano11102661

Academic Editor: Daniela Iannazzo

Received: 9 September 2021
Accepted: 7 October 2021
Published: 10 October 2021

Publisher's Note: MDPI stays neutral with regard to jurisdictional claims in published maps and institutional affiliations.

Copyright: © 2021 by the authors. Licensee MDPI, Basel, Switzerland. This article is an open access article distributed under the terms and conditions of the Creative Commons Attribution (CC BY) license (https://creativecommons.org/licenses/by/4.0/).

Abstract: The chemical modification of electrode surfaces has attracted significant attention for lowering the limit of detection or for improving the recognition of biomolecules; however, the chemical processes are complex, dangerous, and difficult to control. Therefore, instead of the chemical process, we physically modified the surface of carbon-nanotube/polydimethylsiloxane composite electrodes by dip coating them with functionalized multi-walled carbon nanotubes (F-MWCNTs). These electrodes are used as working electrodes in electrochemistry, where they act as a recognition layer for sequence-specific DNA sensing through π–π interactions. The F-MWCNT-modified electrodes showed a limit of detection of 19.9 fM, which was 1250 times lower than that of pristine carbon/polydimethylsiloxane electrodes in a previous study, with a broad linear range of 1–1000 pM. The physically modified electrode was very stable during the electrode regeneration process after DNA detection. Our method paves the way for utilizing physical modification to significantly lower the limit of detection of a biosensor system as an alternative to chemical processes.

Keywords: biosensor; physical surface modification; dip coating; functionalized carbon nanotube; electrochemical impedance spectroscopy

1. Introduction

Highly sensitive and selective detection of specific DNA sequences is crucial for biotechnological applications such as clinical diagnosis [1,2] and environmental [3,4] and food monitoring [5]. Therefore, extensive efforts have been devoted to developing techniques for detecting highly sensitive sequence-specific DNA sensors based on optical [6,7], electrical [8,9] and electrochemical [1,3,10,11] methods. Among these techniques, electrochemical methods have received considerable attention because they can be used to fabricate fast, simple, highly sensitive, and miniature DNA sensors [12,13]. Electrochemical impedance spectroscopy (EIS) is a very accurate electrochemical method, which is sensitive to changes on the electrode surface and is suitable for label-free DNA detection [14,15]. In several studies, EIS-based sensors have been fabricated using a gold electrode as the working electrode [16–18]. However, gold is very expensive and a laborious and inefficient process for immobilizing the probe is usually required for gold electrodes, which is unfavorable for commercialization.

Additionally, as an alternative to glassy carbon electrodes (GCEs), Au electrodes are extensively used as working electrodes in electrochemical DNA detection. However, because GCEs themselves do not recognize DNA molecules, chemical processes for immobilizing probes, which are more complicated than those for Au electrodes, are required when GCEs are used for electrochemical DNA detection [19–21]. Moreover, to further reduce the limit of detection (LOD), chemical surface modification can involve other nanomaterials, such as carbon nanotubes [22] and graphene [23]. In these cases, the chemical surface modifications consist of multiple steps that link probe DNA and nanomaterials to the electrode surfaces [24–26] and require several reagents. To avoid the complexity of chemical processes, in previous studies, multi-walled carbon nanotube (MWCNT)/polydimethylsiloxane (PDMS) composite electrodes were fabricated for DNA detection, and MWCNTs were used as the main material for the electrode [27,28]. In contrast to GCE, the electrode can behave as a recognition layer for DNA sensors without chemical modification; in addition, it is flexible and easy to fabricate. However, the LOD for DNA is 25 pM, which is higher (32–350 fM) than those of some chemically treated GCE-based sensors [29,30] and lower (38–275 pM) than those of other chemically treated GCE-based sensors [31,32].

In this study, to further decrease the LOD of a MWCNT/PDMS electrode, we modified the physical surface of MWCNT/PDMS composite electrodes using functionalized MWCNTs (henceforth called F-MWCNTs) using a dip-coating process. This electrode was used as the working electrode for the electrochemical DNA sensor. Compared to previous sensors based on pristine MWCNT/PDMS electrodes [27,28], the LOD decreased by approximately 1250 times, from 25 pM to 20 fM. The electrode was stably regenerated by ethanol and water cleaning, which indicated that the physical modification method was highly stable. Our results suggest that physical modification using F-MWCNT and dip-coating process could be a good alternative to chemical surface modifications for attaching nanomaterials to the electrode surfaces and that the F-MWCNT-modified electrode is a good alternative to expensive Au electrodes.

2. Materials and Methods

2.1. Materials and Reagents

The MWCNTs (diameter: 10–15 nm; length: 30–40 μm) were obtained from Hanwha Chemical (Daejeon, Korea) and the DNA samples, including probe DNA (P), complementary target DNA (T1), one base non-complementary target DNA (T2), and non-complementary target DNA (T3), were obtained from Bionics (Seoul, Korea) (Table 1). Nitric acid (HNO_3, purity 69%) was obtained from Avantor (Radnor, PA, USA) and sulfuric acid (H_2SO_4, purity 98%) was obtained from Daejung (Siheung, Korea). The polydimethylsiloxane (PDMS) and potassium ferrocyanide ($K_4Fe(CN)_6$) were obtained from Sigma-Aldrich (St. Louis, MO, USA). The potassium ferricyanide ($K_3Fe(CN)_6$) was obtained from Junsei Chemical Co., Ltd. (Tokyo, Japan), membrane filters were obtained from Sigma-Aldrich (St. Louis, MO, USA) and 1 × phosphate-buffered saline (1 × PBS) solution (pH = 7.4) was obtained from Tech and Innovation Corporation (Chuncheon, Korea). Absolute ethanol was obtained from Fisher Scientific Inc. (Hampton, NH, USA) and isopropyl alcohol (IPA) solution was obtained from Sigma-Aldrich (St. Louis, MO, USA). The resistivity of deionized water (DI water) used throughout this study was 18.2 MΩ·cm. The Ag/AgCl reference electrode, with a potential of 0.197 V vs. SHE, and platinum wire, which was used as a counter electrode, was obtained from Princeton Applied Research (Oak Ridge, TN, USA). All other chemicals were of analytical reagent grade. To prepare a sample solution for the DNA detection experiments, target DNA with a specific concentration was added to the PBS solution containing 1000 nM probe DNA and 4 mM potassium ferrocyanide/potassium ferricyanide.

Table 1. Probe and target nucleotide sequences. The bases leading to the mismatches are underlined.

Type	Name	Sequence
Probe DNA	P	5′-GTG TTG TCT CCT AGG TTG GCT CTG-3′
Complementary target DNA	T1	5′-CAG AGC CAA CCT AGG AGA CAA CAC-3′
One base-non-complementary target DNA	T2	5′-CAG AGC CAA CCT CGG AGA CAA CAC-3′
Non-complementary target DNA	T3	5′-ATA TCG ACC TTG GCC GAG ACG GTG-3′

2.2. Instruments

All electrochemical methods, including EIS, cyclic voltammetry (CV), and differential pulse voltammetry (DPV) were measured using a CHI622D (CH Instruments, Inc., Austin, TX, USA). A tip sonicator (HD 2070, Bandelin sonopuls, Berlin, Germany) was utilized to disperse the MWCNT solvent. A dip coater (PTL-UMB, MTI Co., Richmond, CA, USA) was used to modify the MWCNT/PDMS electrode surface with the F-MWCNT solution and a UV-visible spectrophotometer (Cary 50, Varian, Mulgrave, Australia) was used to measure ultraviolet (UV) absorption spectroscopy. A scanning electron microscope (SEM; S4800, Hitachi, Tokyo, Japan) was used to obtain cross-sectional images and surface morphology of the F-MWCNT/MWCNT/PDMS electrodes and Fourier-transform infrared spectroscopy (FTIR) was used to obtain an infrared spectrum of transmittance of the F-MWCNT solution (L160000A, PerkinElmer, Waltham, MI, USA).

2.3. Procedures

2.3.1. Preparation of Functionalized CNTs

The preparation of the functionalized CNTs is summarized in Figure S1a. First, the MWCNTs (0.2 g) were sonicated in a 3:1 (v/v) mixture of nitric acid and sulfuric acid solution [33]. Subsequently, the MWCNTs were filtered and rinsed with DI water using vacuum filtration until the pH of the filtrate was neutral. Finally, the filtrate was dried in an oven (Figure S1b) and sonicated for 5 min with 60 mL of DI water to make the F-MWCNT solution. As shown in Figure S1c, our F-MWCNTs were uniformly dispersed in DI water because they have better dispersion than MWCNTs in water [34]. The FTIR spectra (Figure 1), showed that there were O-H, C=O, and C-O vibrations at 3372.11, 1642.21, and 1226.28 cm^{-1}, respectively, which clearly indicated that carboxylic and hydroxyl groups had been successfully attached to the surface of the MWCNTs.

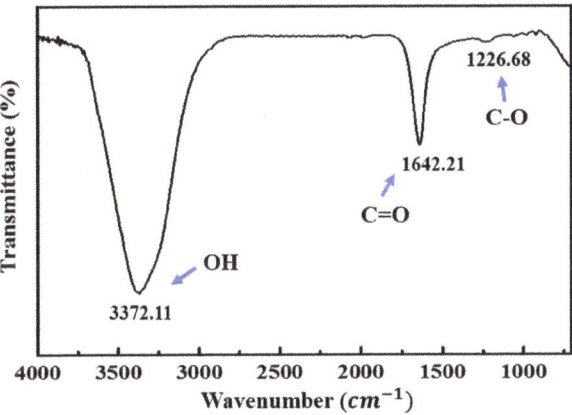

Figure 1. FTIR spectra of MWCNT modified with COOH.

2.3.2. Process Method of the F-MWCNT/MWCNT/PDMS Electrode

A schematic representation of the technique used to fabricate the F-MWCNT/MWCNT/PDMS composite electrode is shown in Figure 2. First, the MWCNT composite electrode layer was fabricated according to a previous study's methodology [28], which is explained in detail in Section 1 of the Supplementary Materials. Next, we coated the MWCNT/PDMS film with F-MWCNT solution according to the preparation method described above. Finally, the film was dip-coated at a dipping speed of 500 μm/s. In our experiments, the optimum repetition number for the same dip coating was found to be 60.

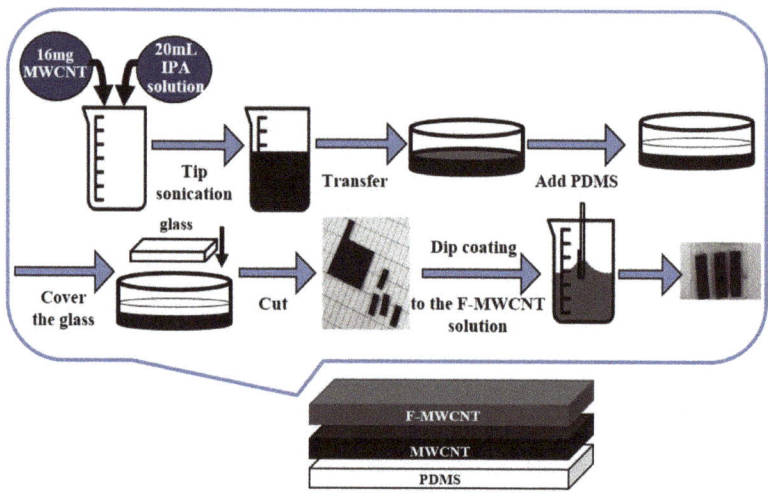

Figure 2. Fabrication process for the F-MWCNT/MWCNT/PDMS composite electrode.

2.3.3. Measurement Using EIS

Before the DNA detection experiment using EIS measurement, a sample solution was prepared by including one of the DNA targets (T1, T2, and T3) with a specific concentration of 1 nM DNA probe (P) and 4 mM $[Fe(CN)_6]^{3-/4-}$ in 1mL of PBS solution. The sample solvent was maintained at 25 °C for 7 min because the UV absorption of the sample solution was saturated after 7 min (Figure S2), which indicated that the DNA reaction in the solution had stabilized. After starting the EIS measurements, 9 min and 0.26 V were required to obtain the stabilized value of the charge transfer resistance (R_{ct}) (Figure S3). Thus, a preparation time of 16 min was needed for EIS-based DNA detection in our experiments. Our three-electrode system, which consisted of an Ag/AgCl reference electrode, platinum counter electrode, and 5 mm × 5 mm F-MWCNT/MWCNT/PDMS working electrode, was placed in sample solutions, as shown in Figure S4. The EIS measurements were performed in the frequency range of 0.1 Hz to 10000 Hz, with an AC (alternating current) amplitude of 5mV and a DC (direct current) bias of 0.26 V for DNA detection.

3. Results and Discussion

3.1. SEM Measurments of the Composite Electrode

The cross-section and surface morphology of the F-MWCNT-modified and pristine MWCNT/PDMS electrodes were characterized using SEM. For the pristine MWCNT/PDMS electrode, most MWCNTs were buried in PDMS and only a few tubular structures of MWCNTs were visible near the electrode surface, as shown in Figure 3a. After 60 dip-coating cycles with the F-MWCNT solution, many F-MWCNTs were observed on the surface of the F-MWCNT-modified MWCNT/PDMS electrode (see Figure 3b). These may act as active centers for DNA interactions. Figure 3c shows the cross-section of the F-MWCNT/MWCNT composite electrode. The results indicate that the mean thickness of the PDMS layers was approximately 430 µm. In addition, the mean thickness of the recognition layer was 15 µm, which increased by approximately 5 nm after the deposition of F-MWCNTs on the surface.

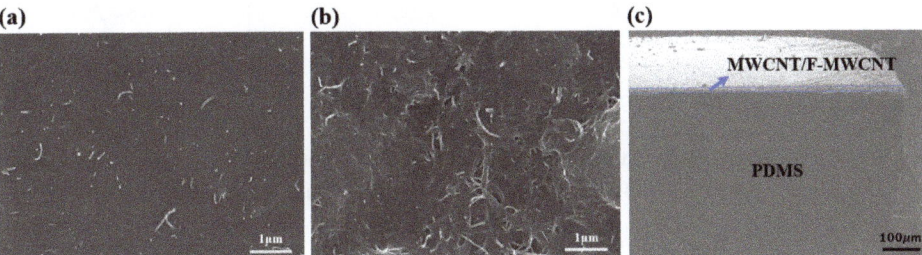

Figure 3. SEM image of the surface of MWCNT before (**a**) and after (**b**) the dip coating process with F-MWCNT solution. (**c**) Cross section of the F-MWCNT/MWCNT/PDMS layer.

3.2. Characterization of the F-MWCNT-Modified Electrodes

Figure 4a showed how the EIS characteristic of the electrode was changed for different numbers of dip coating. R_{ct} for 0, 20, 40, and 60 dip coating cycles (Figure 4b) were obtained by fitting the Nyquist plots using F-MWCNT-modified electrodes in PBS solvent containing 4 mM $[Fe(CN)_6]^{3-/4-}$, with the equivalent circuit shown in Figure 4a. The circuit model is widely used for describing processes at the electrochemical interfaces [35,36], consisting of the active electrolyte resistance (R_Ω), double-layer capacitance (C_d), charge transfer resistance (R_{ct}), and Warburg impedance (Z_w). R_{ct} and Z_w describe the electron transfer and the mass transport of the electroactive species near the solution–electrode interface, respectively [37]. The R_{ct} of the film decreased when increasing the number of dip coating cycles, as shown in Figure 4b; as compared to R_{ct} for no dip coating (5500 Ω), that for 60 dip coatings (200 Ω) was about 27.5 times smaller. When the number of dip coatings was over 60, we found that the F-MWCNT/MWCNT layers were easily peeled off from the electrode during the electrochemical analysis. Thus, the number of dip coating cycles in the standard fabrication process was set to 60.

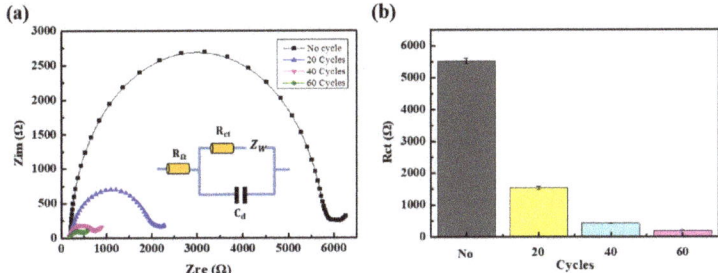

Figure 4. (**a**) Nyquist plots. (**b**) R_{ct} values before using dip coating and after dip coating forv20, 40, and 60 cycles.

We think these results were obtained for two reasons. First, the charge transfer resistance might be associated with the number of MWCNTs on the electrode surface. In the MWCNT electrode, only a small number of MWCNTs and active area were exposed because some of the MWCNTs were buried in the PDMS. However, after depositing F-MWCNTs, the number of MWCNTs and the active area of the electrode were increased. Second, we think the energy barrier of the redox species reaching the electrode was lowered due to Coulomb or steric interactions [38]. We speculate that the energy barrier of $[Fe(CN)_6]^{3-/4-}$ approaching the electrode surface is lowered by hydrogen bonding affinity to carboxylic and hydroxyl groups in F-MWCNTs.

Cyclic voltammetry (CV) is one voltammetry technique for measuring the current response of a redox active solution to a linearly cycled potential sweep. To investigate whether the reactions near the electrodes are diffusion-controlled, we performed CV measurements with

F-MWCNT/MWCNT/PDMS electrodes for 4mM [Fe(CN)$_6$]$^{3-/4-}$ in the PBS solution, varying the scan rate from 0.05 to 0.6 V/s (see Figure 5a). Figure 5b indicates that the anodic and cathodic peak currents (I_{pa} and I_{pc}, respectively) increased linearly with the square root of the scan rate (v). The linear equations can be described as; $I_{pa}(A) = (8.598 \times 10^{-4}) \times v^{\frac{1}{2}} + 1.311 \times 10^{-4}$ with an $R^2 = 0.9953$ and $I_{pc}(A) = -0.001 \times v^{\frac{1}{2}} - 1.006 \times 10^{-4}$ with an $R^2 = 0.9989$. These results indicate that the oxidation–reduction reactions were diffusion controlled, indicating that the composite electrode is appropriate for quantitative electrochemical analysis [39,40].

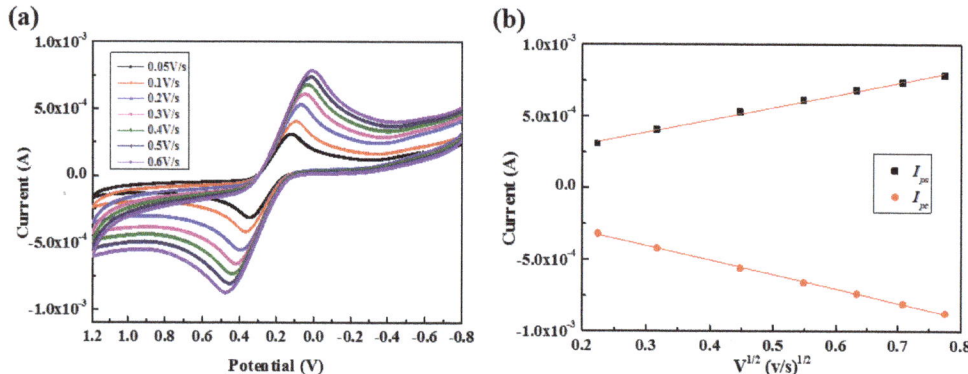

Figure 5. (a) Cyclic voltammograms of 4mM [Fe(CN)$_6$]$^{3-/4-}$ for a F-MWCNT/MWCNT/PDMS electrode with scan rates from 0.05 to 0.6 V/s and (b) anodic and cathodic peak currents (I_{pa} and I_{pc}, respectively) as a function of square root of the scan rates (v).

3.3. DNA Detection Using F-MWCNT-Modified Electrodes

As discussed above, a F-MWCNT/MWCNT/PDMS electrode could be used as a DNA sensor. The schematic of the DNA detection mechanism is described in Figure 6. It is well known that single-stranded DNA (s-DNA) is adsorbed on the surfaces of F-MWCNT or MWCNT through π–π interactions, whereas double-stranded DNA (d-DNA) is not [4,41–45]. For target T3 and probe DNA, the sequences of the probe and target are all mismatched, so that the single-stranded target and probe DNA might be adsorbed onto the recognition layer (F-MWCNT/MWCNT) of the electrode through π–π interactions [4,43,44]. The adsorbed s-DNA can act as the barrier of charge transfer between the oxidation–reduction couple ([Fe(CN)$_6$]$^{3-/4-}$) and the MWCNTs, increasing R_{ct}, whereas for target T1 and probe DNA, d-DNA can form through hybridization between T1 and probe DNA because their sequences match completely. The vast majority of d-DNA remains far from the electrode rather than being adsorbed onto the recognition layer [45]. Thus, because d-DNA does not behave as a charge transfer barrier, R_{ct} is expected to be lower than that for the former case with s-DNA adsorbed on the electrode. In the case of T2, the probe and target DNA have a one-base mismatch. Because the level of sequence mismatch is between those for T1 and T3, resistance might also be between those of T1 and T3.

We performed EIS measurements for the sample solutions described in the experimental section, which include probe DNA (P) and one target DNA (T1, T2, or T3), with the target as shown in Figure 7a. R_{ct} was calculated from the fitting of the Nyquist plots in Figure 7a, the values of which were about 940, 1190, and 2000 Ω for the T1, T2, and T3 targets, respectively, as shown in Figure 7b. The highest R_{ct} values were obtained for probe and T3, whereas the lowest were for probe and T1, as expected due to the mechanism discussed above. We defined the DNA detection sensitivity, r, by the equation r= $\Delta R/R°$, where $R°$ is the R_{ct} when hybridized with T3 and ΔR is the difference between the R_{ct} for the present target and $R°$. According to this equation, the sensitivities for perfectly matched and one-base mismatched targets (T1 and T2) were 52.5% and 40.4%, respectively;

the sensitivity differs by 12.1% even for a one-base mismatch, indicating the good sequence selectivity of our system.

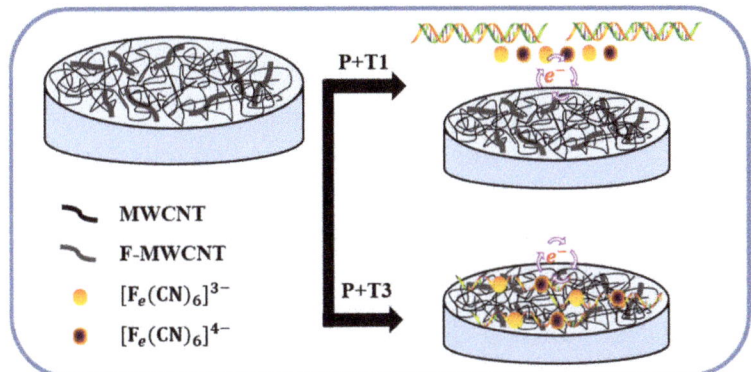

Figure 6. Schematic diagram of DNA detection mechanism.

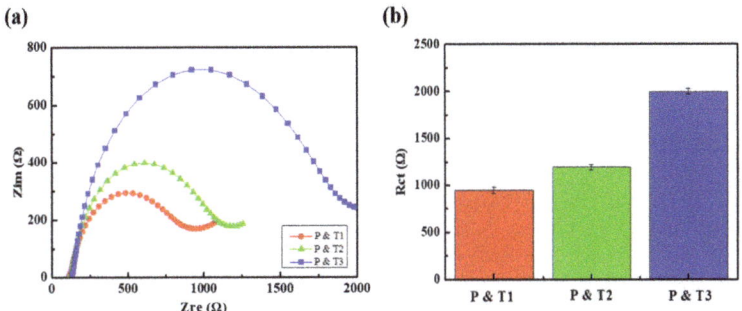

Figure 7. (a) Nyquist plots and (b) R_{ct} values for 1 nM probe DNA and 1 nM target DNA (T1, T2 and T3) in 4 mM $[Fe(CN)_6]^{3-/4-}$ solution with 1× PBS buffer (pH = 7.4).

3.4. Detection Limit of the DNA Sensor

As shown in Figure 8b, R_{ct} in the EIS analysis with target T1 was linearly dependent on the logarithm of the T1 concentration in the range of 1–1000 pM in the sample solutions. In Figure 8a, a decrease in the R_{ct} is clearly observed alongside an increase in the concentration of T1. The reason might be that as the concentration of T1 increases, the amount of DNA attached to the active site increases, which increases the current and reduces R_{ct}. The linear regression equation was determined to be $R_{ct} = -2.17 \times 10^2 \cdot logC_1 + 1.58 \times 10^3$ with an $R^2 = 0.992$, where C_1 is the concentration of T1 ranging from 1 to 1000 pM, as shown in Figure 8b. The LOD was extrapolated to be 19.9 fM using a signal-to-noise ratio of 3:1. Our LOD decreased by 1250 times compared to that obtained in a previous study, wherein a MWCNT electrode was used as a working electrode for DNA detection in EIS [27]. The 27.5 times reduction in R_{ct} by F-MWCNT modification to the MWCNT/PDMS electrode surface might be an important reason for the drastic decrease in LOD. The change in the adsorption/desorption characteristics of the s-DNA and d-DNA on the electrode surface by F-MWCNT modification may also have contributed to lowering the LOD. Furthermore, F-MWCNTs can interact with DNA through hydrogen bonding and electrostatic interactions, as well as π–π stacking interactions, and these interactions are more complicated than the interactions in MWCNTs. Although a comprehensive study of DNA adsorption characteristics on F-MWCNTs was beyond the scope of this study, it is known that graphene oxide, which has a chemical structure similar to that of F-MWCNT, absorbs s-DNA well, but does not efficiently absorb d-DNA [46]. Graphene-oxide-based

DNA sensing can be performed in less than 30 min, whereas it takes several hours to achieve a similar detection rate with carbon nanotubes, which do not have functional groups [47]. Based on these results, we speculate that F-MWCNTs are more efficient for discriminating s-DNA and d-DNA than MWCNTs owing to their additional interactions, such as hydrogen bonding.

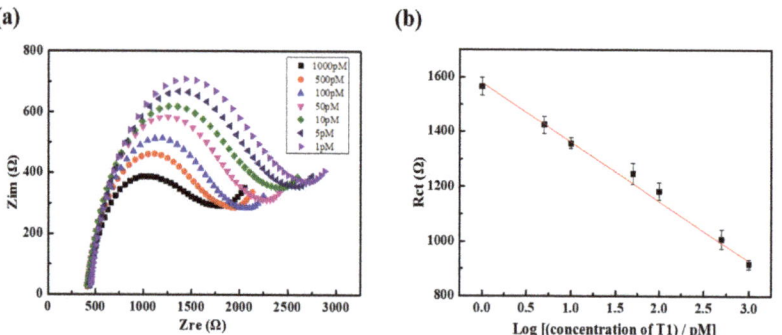

Figure 8. (a) Nyquist plots for the 1 nM probe DNA and various target (T1) concentrations and (b) plot of R_{ct} values against the logarithm of the various target (T1) concentrations.

3.5. Stability of the Modified Electrode

The stability of the fabricated DNA sensor over 7 days was investigated by measuring the average R_{ct} daily (Figure 9 and Figure S5). For the experiment, 2 mL of sample solvent containing 1 nM P and T1 DNA was poured into the modified electrode. It was first stored at room temperature for 1 week and then examined via its EIS response (Figure 9a) after hybridization in 4 mM $[Fe(CN)_6]^{3-/4-}$ solution with 1× PBS buffer. Between every experiment, the electrode was cleaned by ultrasonication in absolute ethanol for 3 min and DI water for 3 min. Figure 9b shows the charge transfer resistance (R_{ct}) and the relative standard deviation of R_{ct} is 2.31%, indicating the good stability of the fabricated electrode. The CV curves of the fabricated electrode are shown in Figure S5 (one cycle and 100 cycles). Even after 100 cycles, the results show that there was no significant change. This indicates that our physical surface modification using F-MWCNTs is very stable during the regeneration process. It is important to note that our electrode method is not a chemical but a physical process; therefore, it is easy to control, thereby reducing LOD, and is applicable to other carbon sensors.

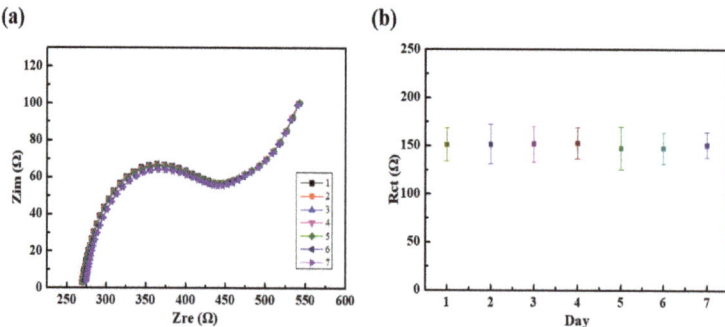

Figure 9. (a) Nyquist plots and (b) R_{ct} values for 1 nM probe DNA and target 1 DNA measured every 24 h for 7 days in 4 mM $[Fe(CN)_6]^{3-/4-}$ solution with 1× PBS buffer (pH = 7.4).

4. Summary

We investigated the physical surface modification of multi-walled carbon-nanotube/polymer electrodes treated using F-MWCNTs and a dip-coating process. The F-MWCNTs could be successfully deposited using solution-based physical processes, such as dip coating, because of their high dispersibility in water. The electrode surface can be used as a recognition layer, enabling the simple fabrication of DNA sensors. The vital properties of our structure are that by modifying it with F-MWCNTs, the active area of the electrode was increased, which resulted in a low charge transfer resistance. Compared to EIS-based DNA sensors that use electrodes chemically modified with nanomaterials, our electrode has a simpler and more controllable fabrication process with a very low LOD of 19.9 fM. The physically modified electrodes were stable during the regeneration process. Our results indicate that the physical surface modification of electrodes is a promising alternative approach to chemical surface.

Supplementary Materials: The following are available online at https://www.mdpi.com/article/10.3390/nano11102661/s1, Figure S1: (a) Fabrication process for the functionalized MWCNT (MWCNT-COOH). (b) Membrane filter remaining after processing by vacuum filtration and (c) after sonicating with DI water.; Figure S2: Determination of hybridization time. (a) Sixteen UV absorption curves measured every 30 s for 480 s. (b) Absorbance at a wavelength of 260 nm as a function of time; Figure S3: (a) Nyquist plot for the different accumulation times at 0.26 V from 1 min to 15 min. (b) R_{ct} value as a function of time. The sample PBS solutions contained 1 nM probe DNA and 4 mM $[Fe(CN)_6]^{3-/4-}$.; Figure S4: (a) Schematic and (b) image showing the integration of the three electrodes platform.; Figure S5: Cyclic voltammetry responses of the F-MWCNT/MWCNT/PDMS electrode after 1 cycle and after 100 cycles. The electrode was immersed in PBS solution containing 4 mM $[Fe(CN)_6]^{3-/4-}$. Table S1: Fitted values of the equivalent circuit parameters from dark Nyquist plots of devices without and with 5% Cs.

Author Contributions: Conceptualization, J.M. and E.-C.L.; methodology, H.J.; investigation, J.M.; supervision, E.-C.L.; writing—original draft, J.M., H.J. and E.-C.L.; writing—review and editing, E.-C.L.; All authors have read and agreed to the published version of the manuscript.

Funding: This work was supported by the National Research Foundation of Korea (NRF), funded by the Ministry of Science and ICT (Grant Nos. NRF-2016R1A2B2015389 and NRF-2021R1F1A1051089).

Institutional Review Board Statement: Not applicable.

Informed Consent Statement: Not applicable.

Data Availability Statement: Data sharing is not applicable to this article.

Conflicts of Interest: The authors declare no conflict of interest.

References

1. Abbaspour, A.; Norouz-Sarvestani, F.; Noori, A.; Soltani, N. Aptamer-conjugated silver nanoparticles for electrochemical dual-aptamer-based sandwich detection of staphylococcus aureus. *Biosens. Bioelectron.* **2015**, *68*, 149–155. [CrossRef]
2. Feng, L.; Chen, Y.; Ren, J.; Qu, X. A graphene functionalized electrochemical aptasensor for selective label-free detection of cancer cells. *Biomaterials* **2011**, *32*, 2930–2937. [CrossRef] [PubMed]
3. Ezoji, H.; Rahimnejad, M. Electrochemical behavior of the endocrine disruptor bisphenol A and in situ investigation of its interaction with DNA. *Sens. Actuators B Chem.* **2018**, *274*, 370–380. [CrossRef]
4. Lian, Y.; He, F.; Wang, H.; Tong, F. A new aptamer/graphene interdigitated gold electrode piezoelectric sensor for rapid and specific detection of Staphylococcus aureus. *Biosens. Bioelectron.* **2015**, *65*, 314–319. [CrossRef] [PubMed]
5. Somerson, J.; Plaxco, K.W. Electrochemical aptamer-based sensors for rapid point-of-use monitoring of the mycotoxin ochratoxin a directly in a food stream. *Molecules* **2018**, *23*, 912. [CrossRef] [PubMed]
6. Gaylord, B.S.; Heeger, A.J.; Bazan, G.C. DNA detection using water-soluble conjugated polymers and peptide nucleic acid probes. *Proc. Natl. Acad. Sci. USA* **2002**, *99*, 10954–10957. [CrossRef] [PubMed]
7. Pavlov, V.; Xiao, Y.; Shlyahovsky, B.; Willner, I. Aptamer-functionalized Au nanoparticles for the amplified optical detection of thrombin. *J. Am. Chem. Soc.* **2004**, *126*, 11768–11769. [CrossRef]
8. Ohno, Y.; Maehashi, K.; Matsumoto, K. Label-Free Biosensors Based on Aptamer-Modified Graphene Field-Effect. *J. Am. Chem. Soc.* **2010**, *132*, 18012–18013. [CrossRef]
9. Shin, J.-K.; Lee, J.-H.; Kim, D.-S.; Park, H.-J.; Choi, P.; Jeong, Y.-T.; Lim, G. An FET-type charge sensor for highly sensitive detection of DNA sequence. *Biosens. Bioelectron.* **2004**, *20*, 69–74. [CrossRef]
10. Cai, H.; Xu, C.; He, P.; Fang, Y. Colloid Au-enhanced DNA immobilization for the electrochemical detection of sequence-specific DNA. *J. Electroanal. Chem.* **2001**, *510*, 78–85. [CrossRef]

11. Wang, J.; Rivas, G.; Cai, X. Screen-Printed Electrochemical Hybridization Biosensor for the Detection of DNA Sequences from the Escherichia Coli Pathogen. *Electroanalysis* **1997**, *9*, 395–398. [CrossRef]
12. Luong, J.H.T.; Sheu, F.-S.; Al-Rubeaan, K.; Zheng, D.; Vashist, S.K. Advances in carbon nanotube based electrochemical sensors for bioanalytical applications. *Biotechnol. Adv.* **2010**, *29*, 169–188. [CrossRef]
13. Zhang, F.T.; Cai, L.Y.; Zhou, Y.L.; Zhang, X.X. Immobilization-free DNA-based homogeneous electrochemical biosensors. *TrAC Trends Anal. Chem.* **2016**, *85*, 17–32. [CrossRef]
14. Ahmed, R.; Reifsnider, K. Study of influence of electrode geometry on impedance spectroscopy. *Int. J. Electrochem. Sci.* **2011**, *6*, 1159–1174.
15. Han, L.; Liu, P.; Petrenko, V.A.; Liu, A.H. A label-free electrochemical impedance cytosensor based on specific peptide-fused phage selected from landscape phage library. *Sci. Rep.* **2016**, *6*, 22199. [CrossRef] [PubMed]
16. Castillo, G.; Lamberti, I.; Mosiello, L.; Hianik, T. Impedimetric DNA Aptasensor for Sensitive Detection of Ochratoxin A in Food. *Electroanalysis* **2012**, *24*, 512–520. [CrossRef]
17. Li, A.; Yang, F.; Ma, Y.; Yang, X. Electrochemical impedance detection of DNA hybridization based on dendrimer modified electrode. *Biosens. Bioelectron.* **2007**, *22*, 1716–1722. [CrossRef] [PubMed]
18. Pan, C.; Guo, M.; Nie, Z.; Xiao, X.; Yao, S. Aptamer-based electrochemical sensor for label-free recognition and detection of cancer cells. *Electroanalysis* **2009**, *21*, 1321–1326. [CrossRef]
19. Jin, H.; Zhao, C.; Gui, R.; Gao, X.; Wang, Z. Reduced graphene oxide/nile blue/gold nanoparticles complex-modified glassy carbon electrode used as a sensitive and label-free aptasensor for ratiometric electrochemical sensing of dopamine. *Anal. Chim. Acta* **2018**, *1025*, 154–162. [CrossRef]
20. Kang, X.; Mai, Z.; Zou, X.; Cai, P.; Mo, J. A sensitive nonenzymatic glucose sensor in alkaline media with a copper nanocluster/multiwall carbon nanotube-modified glassy carbon electrode. *Anal. Biochem.* **2007**, *363*, 143–150. [CrossRef]
21. Liu, Q.; Zhu, X.; Huo, Z.; He, X.; Liang, Y.; Xu, M. Electrochemical detection of dopamine in the presence of ascorbic acid using PVP/graphene modified electrodes. *Talanta* **2012**, *97*, 557–562. [CrossRef] [PubMed]
22. Cai, H.; Cao, X.; Jiang, Y.; He, P.; Fang, Y. Carbon nanotube-enhanced electrochemical DNA biosensor for DNA hybridization detection. *Anal. Bioanal. Chem.* **2003**, *375*, 287–293. [CrossRef] [PubMed]
23. Huang, K.-J.; Niu, D.-J.; Sun, J.-Y.; Han, C.-H.; Wu, Z.-W.; Li, Y.-L.; Xiong, X.-Q. Novel electrochemical sensor based on functionalized graphene for simultaneous determination of adenine and guanine in DNA. *Colloids Surf. B Biointerfaces* **2011**, *82*, 543–549. [CrossRef] [PubMed]
24. Amouzadeh Tabrizi, M.; Shamsipur, M. A label-free electrochemical DNA biosensor based on covalent immobilization of salmonella DNA sequences on the nanoporous glassy carbon electrode. *Biosens. Bioelectron.* **2015**, *69*, 100–105. [CrossRef]
25. Benvidi, A.; Tezerjani, M.D.; Jahanbani, S.; Mazloum Ardakani, M.; Moshtaghioun, S.M. Comparison of impedimetric detection of DNA hybridization on the various biosensors based on modified glassy carbon electrodes with PANHS and nanomaterials of RGO and MWCNTs. *Talanta* **2016**, *147*, 621–627. [CrossRef] [PubMed]
26. Gupta, V.K.; Yola, M.L.; Qureshi, M.S.; Solak, A.O.; Atar, N.; Üstündağ, Z. A novel impedimetric biosensor based on graphene oxide/gold nanoplatform for detection of DNA arrays. *Sens. Actuators B Chem.* **2013**, *188*, 1201–1211. [CrossRef]
27. Jiang, H.; Lee, E.C. Highly selective, reusable electrochemical impedimetric DNA sensors based on carbon nanotube/polymer composite electrode without surface modification. *Biosens. Bioelectron.* **2018**, *118*, 16–22. [CrossRef]
28. Li, J.; Lee, E.C. Carbon nanotube/polymer composite electrodes for flexible, attachable electrochemical DNA sensors. *Biosens. Bioelectron.* **2015**, *71*, 414–419. [CrossRef] [PubMed]
29. Du, M.; Yang, T.; Li, X.; Jiao, K. Fabrication of DNA/graphene/polyaniline nanocomplex for label-free voltammetric detection of DNA hybridization. *Talanta* **2012**, *88*, 439–444. [CrossRef] [PubMed]
30. Zhang, Y.; Wang, J.; Xu, M. A sensitive DNA biosensor fabricated with gold nanoparticles/ploy (p-aminobenzoic acid)/carbon nanotubes modified electrode. *Colloids Surf. B Biointerfaces* **2010**, *75*, 179–185. [CrossRef]
31. Niu, S.; Zhao, M.; Hu, L.; Zhang, S. Carbon nanotube-enhanced DNA biosensor for DNA hybridization detection using rutin-Mn as electrochemical indicator. *Sens. Actuators B Chem.* **2008**, *135*, 200–205. [CrossRef]
32. Miao, X.; Guo, X.; Xiao, Z.; Ling, L. Electrochemical molecular beacon biosensor for sequence-specific recognition of double-stranded DNA. *Biosens. Bioelectron.* **2014**, *59*, 54–57. [CrossRef]
33. Liu, J.; Rinzler, A.G.; Dai, H.; Hafner, J.H.; Bradley, R.K.; Boul, P.J.; Lu, A.; Iverson, T.; Shelimov, K.; Huffman, C.B.; et al. Fullerene Pipes. *Science* **1998**, *280*, 1253–1256. [CrossRef]
34. Jagadish, K.; Srikantaswamy, S.; Byrappa, K.; Shruthi, L.; Abhilash, M.R. Dispersion of multiwall carbon nanotubes in organic solvents through hydrothermal supercritical condition. *J. Nanomater.* **2015**, *2015*. [CrossRef]
35. Huang, Y.; Bell, M.C.; Suni, I.I. Impedance biosensor for peanut protein Ara h 1. *Anal. Chem.* **2008**, *80*, 9157–9161. [CrossRef]
36. Rodriguez, M.C.; Kawde, A.-N.; Wang, J. Aptamer biosensor for label-free impedance spectroscopy detection of proteins based on recognition-induced switching of the surface charge. *Chem. Commun.* **2005**, 4267–4269. [CrossRef] [PubMed]
37. Wang, Z.; Murphy, A.; O'Riordan, A.; O'Connell, I. Equivalent Impedance Models for Electrochemical Nanosensor-Based Integrated System Design. *Sensors* **2021**, *21*, 3259. [CrossRef] [PubMed]
38. Daniels, J.S.; Pourmand, N. Label-free impedance biosensors: Opportunities and challenges. *Electroanal. An Int. J. Devoted to Fundam. Pract. Asp. Electroanal.* **2007**, *19*, 1239–1257. [CrossRef] [PubMed]

39. Ghoreishi, S.M.; Behpour, M.; Hajisadeghian, E.; Golestaneh, M. Electrochemical determination of acetaminophen at the surface of a glassy carbon electrode modified with multi-walled carbon nanotube. *J. Chil. Chem. Soc.* **2013**, *58*, 1513–1516. [CrossRef]
40. Peng, Y.; Qi, B.-P.; Wang, B.-S.; Bao, L.; Hu, H.; Pang, D.-W.; Zhang, Z.-L.; Tang, B. An efficient edge-functionalization method to tune the photoluminescence of graphene quantum dots. *Nanoscale* **2015**, *7*, 5969–5973. [CrossRef]
41. Shankar, A.; Mittal, J.; Jagota, A. Binding between DNA and carbon nanotubes strongly depends upon sequence and chirality. *Langmuir* **2014**, *30*, 3176–3183. [CrossRef]
42. Shapter, J.G.; Sibley, A.J.; Shearer, C.J.; Andersson, G.G.; Ellis, A.V.; Gibson, C.T.; Yu, L.; Quinton, J.S.; Fenati, R. Adsorption and Desorption of Single-Stranded DNA from Single-Walled Carbon Nanotubes. *Chem. An Asian J.* **2017**, *12*, 1625–1634. [CrossRef]
43. Star, A.; Joiner, C.S.; Gabriel, J.-C.P.; Valcke, C.; Tu, E.; Niemann, J. Label-free detection of DNA hybridization using carbon nanotube network field-effect transistors. *Proc. Natl. Acad. Sci. USA* **2006**, *103*, 921–926. [CrossRef] [PubMed]
44. Zheng, M.; Jagota, A.; Semke, E.D.; Diner, B.A.; McLean, R.S.; Lustig, S.R.; Richardson, R.E.; Tassi, N.G. DNA-assisted dispersion and separation of carbon nanotubes. *Nat. Mater.* **2003**, *2*, 338–342. [CrossRef] [PubMed]
45. Zhao, X.; Johnson, J.K. Simulation of adsorption of DNA on carbon nanotubes. *J. Am. Chem. Soc.* **2007**, *129*, 10438–10445. [CrossRef] [PubMed]
46. He, S.; Song, B.; Li, D.; Zhu, C.; Qi, W.; Wen, Y.; Wang, L.; Song, S.; Fang, H.; Fan, C. A graphene nanoprobe for rapid, sensitive, and multicolor fluorescent DNA analysis. *Adv. Funct. Mater.* **2010**, *20*, 453–459. [CrossRef]
47. Liu, B.; Salgado, S.; Maheshwari, V.; Liu, J. DNA adsorbed on graphene and graphene oxide: Fundamental interactions, de-sorption and applications. *Curr. Opin. Colloid Interface Sci.* **2016**, *26*, 41–49. [CrossRef]

Article

Sintering Bonding of SiC Particulate Reinforced Aluminum Metal Matrix Composites by Using Cu Nanoparticles and Liquid Ga in Air

Zeng Gao [1], Congxin Yin [1], Dongfeng Cheng [1,*], Jianguang Feng [1], Peng He [2], Jitai Niu [2] and Josip Brnic [3]

[1] School of Materials Science and Engineering, Henan Polytechnic University, Jiaozuo 454003, China; gaozeng@hpu.edu.cn (Z.G.); Ycx7469@163.com (C.Y.); jayfeng_cool@126.com (J.F.)
[2] State Key Laboratory of Advanced Welding and Joining, Harbin Institute of Technology, Harbin 150001, China; hepeng@hit.edu.cn (P.H.); niujitai@163.com (J.N.)
[3] Faculty of Engineering, University of Rijeka, 51000 Rijeka, Croatia; brnic@riteh.hr
* Correspondence: cdf_alex@hpu.edu.cn; Tel.: +86-391-396-6901

Abstract: SiC particulate reinforced aluminum metal matrix composites (SiC$_p$/Al MMCs) are characterized by controllable thermal expansion, high thermal conductivity and lightness. These properties, in fact, define the new promotional material in areas and industries such as the aerospace, automotive and electrocommunication industries. However, the poor weldability of this material becomes its key problem for large-scale applications. Sintering bonding technology was developed to join SiC$_p$/Al MMCs. Cu nanoparticles and liquid Ga were employed as self-fluxing filler metal in air under joining temperatures ranging from 400 °C to 500 °C, with soaking time of 2 h and pressure of 3 MPa. The mechanical properties, microstructure and gas tightness of the joint were investigated. The microstructure analysis demonstrated that the joint was achieved by metallurgical bonding at contact interface, and the sintered layer was composed of polycrystals. The distribution of Ga was quite homogenous in both of sintered layer and joint area. The maximum level of joint shear strength of 56.2 MPa has been obtained at bonding temperature of 450 °C. The specimens sintering bonded in temperature range of 440 °C to 460 °C had qualified gas tightness during the service, which can remain 10^{-10} Pa·m^3/s.

Keywords: sintering bonding; Cu nanoparticles; liquid Ga; aluminum metal matrix composites; gas tightness

1. Introduction

Certain properties of SiC particulate reinforced aluminum metal matrix composites (SiC$_p$/Al MMCs), like controllable thermal expansion, high thermal conductivity, etc., promote this material as very acceptable in various applications [1–4]. However, bonding of SiC$_p$/Al MMCs is still one of the unsettled problems which limit widespread use of considered materials in structural and functional applications [5]. A benefit from vacuum stirring casting technology, 15 vol.% SiC$_p$/Al MMCs can be mass produced with lower production costs compared to the other production technology, such as powder metallurgic method and liquid impregnation technology. Following the previous bonding technology of aluminum alloy, most researches have been conducted on fusion welding, using arc, laser and electron beam as heating sources to join SiC$_p$/Al MMCs [6–9]. However, attempts to weld this type create some practical problems [10]. The first limitation is the chemical reactions in the matrix/reinforcement interfaces initiated by high temperature. In this particular case, this SiC reinforcement will react with the aluminum in the welding following Equation (1):

$$3SiC + 4Al = 3Si + Al_4C_3, \quad (1)$$

The chemical product Al$_4$C$_3$ degrades the mechanical properties of joints due to its brittleness and inhomogeneous distribution [11]. Another limitation is the distribution of reinforcement particles. During fusion welding, there is a tendency to disturb the distribution of SiC particles in the molten pool and may disappear completely, even MMCs are used as filler metal [12]. In addition to this, the poor fluidity of the molten pool may cause the welding defects, such as pore, slag inclusions and incomplete backfill, etc. [6,13]. In addition to fusion welding, some other welding techniques can be applied to SiC$_p$/Al MMCs bonding, such as solid-state diffusion bonding, brazing, friction stir welding, etc. However, these welding methods involve several difficulties to join SiC$_p$/Al MMCs in practical applications. For example, solid-state diffusion bonding of such composites results in excessive plastic deformation under high applied pressure [14]. Given the brazing technique, it is difficult to develop the filler metal that can very well wet both the aluminum matrix and SiC particles due to the difference in physical and chemical properties of base metal and ceramic reinforcement [15]. Friction stir welding of SiC$_p$/Al MMCs causes serious tool wear due to the existence of wear-resistant SiC particles in aluminum. In addition, it is easy to produce tunnel defects inside of the weld during the friction stir welding process [16].

In the present study, a preliminary research was conducted to investigate the sintering bonding performance of SiC$_p$/Al MMCs using Cu nanoparticles (Cu NPs) and liquid Ga in an atmospheric environment. Ga can eliminate the aluminum oxide and form a low-temperature eutectic liquid phase with aluminum at temperature of 26.6 °C [17,18]. Meanwhile, the peritectic reaction between Ga and Cu occurs at temperature of 254 °C, as known from Cu-Ga binary phase diagram. Cu nanoparticles are of great interest due to their potential applications in the fields of electronic packaging [19]. Reducing the size of Cu particles to nanoscale, their characteristics can vary greatly from those of bulk state. Typically, the decrease of melting point and increase of diffusion coefficient are remarkable compared with that of bulk state [20–23]. Based on the character of liquid Ga and Cu nanoparticles, it would be promising to use liquid Ga and Cu nanoparticle as filler metal to join SiC$_p$/Al MMCs without flux [24]. Furthermore, the understanding of sintering bonding, interaction between liquid Ga and Cu nanoparticles is still limited. In this work, different temperatures were considered to evaluate the influence of nanoparticles on joint microstructure, mechanical properties and gas tightness.

2. Materials and Methods

In this research, the material to be bonded was 6063 aluminum matrix composites reinforced with 15 vol.% SiC particles (15 vol.% SiC$_p$/6063 Al MMCs), which was commercially manufactured by vacuum stirring casting method. SiC$_p$/6063 Al MMCs is widely used in aerospace industry, electronic packaging, automotive industry, etc. The density of fabricated 15 vol.% SiC$_p$/6063 Al MMCs was measured to be 2.82 g/cm^3. The microstructure of as-received 15 vol.% SiC$_p$/6063 Al MMCs mainly consists of α-Al and SiC particles with sizes between 10 and 25 μm, as shown in Figure 1. Due to the characteristics of solidification dynamics of liquid metal, SiC particles are mainly distributed on α-Al grain boundaries during the manufacturing process. The properties of this material will be improved because of the effect of fine grain strengthening and dislocation strengthening. The solid-liquidus temperature of 15 vol.% SiC$_p$/6063 Al MMCs is in the range of 580–639 °C. Cu nanoparticles (Cu NPs) were purchased from the commercial company. The size of Cu NPs is around 50 nm. Ga is a low melting point metal, with a melting point of 29.7 °C, which is much lower than that of In (156.6 °C) and Sn (231.9 °C). In addition, Ga has excellent wetting properties on glass and ceramics. In this research, the purity of Ga and Cu NPs was 99.99% and 99.9%, respectively. All chemical reagents used in this experiment were of analytical grade and without any subsequent treatment.

For the purpose of sintering bonding experiment, samples of SiC$_p$/6063 Al MMCs were cut into pieces of 10 × 15 × 2 mm^3. The bonding surface of the specimen was mechanically polished on 1000 grit grinding paper and then cleaned by ultrasonic cleaners with ethanol for 5 min. The schematic representation of sintering bonded SiC$_p$/6063 Al

MMCs joint is shown in Figure 2. In order to break the oxidation film on SiC$_p$/6063 Al MMCs and improve the bonding properties, a liquid 10 µm thick Ga (equal to 5.9 mg/cm^2) was lightly smeared on the bonding surface by using a warmed (about 40–60 °C) soft cloth impregnated with liquid Ga. The thickness of Ga was controlled by high precision electronic scale. Figure 3 is a photograph of 15 vol.% SiC$_p$/6063 Al MMCs after coating Ga on the bonding surface. As can be seen, Ga is uniformly coated over the surface of the composites. Thereafter, the Ga-coated specimen was quickly transferred to a dry and low temperature environment to solidify the Ga liquid, minimizing intergranular permeability and the solution between Ga and SiC$_p$/6063 Al MMCs. Cu NPs were uniformly dropped on solidified Ga layer. The mass ratio of Ga and Cu was controlled to be 2:1. The binary phase diagram of Cu-Ga demonstrated that these two elements can interact with each other to produce solid solution and intermetallic compounds such as Cu$_3$Ga, Cu$_2$Ga, Cu$_5$Ga$_3$, Cu$_3$Ga$_2$ and CuGa$_2$, depending on the temperature and concentration. The two specimens after Cu NPs dropping were assembled together and then pressed by a preheated steel plate. The sintering bonding process was performed in a resistance furnace for 2 h under a pressure of 3 MPa. During sintering bonding process, rapid heating of the sample is the basic requirement of this experiment since any delay of heating may cause the intergranular diffusion of Ga into SiC$_p$/6063 Al MMCs though grain and phase boundary. The rapid heating process will result in a solid solution of Ga into Al and Cu to avoid intergranular permeability.

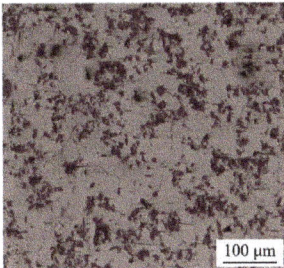

Figure 1. Metallographic structure of 15 vol.% SiC$_p$/6063 Al MMCs manufactured by vacuum stirring casting method.

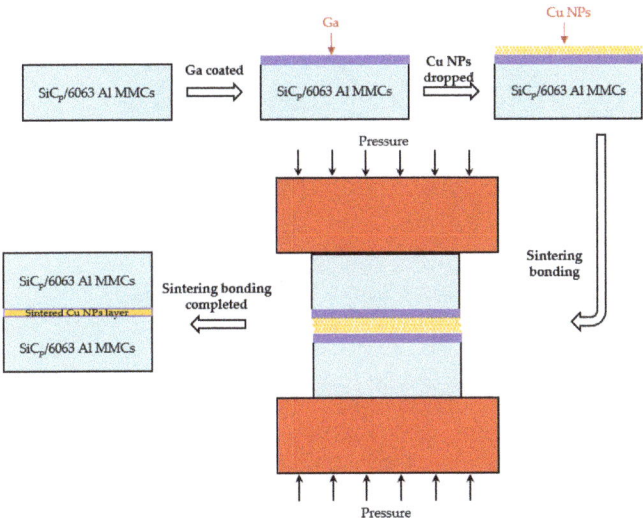

Figure 2. Schematic representation of sintering bonded SiC$_p$/6063 Al MMCs joint.

Figure 3. Photograph of 15 vol.% SiC$_p$/6063 Al MMCs after Ga coating on bonding surface.

Before sintering bonding, the morphological features of Cu NPs were observed by scanning electron microscope (SEM, Carl Zeiss NTS GmbH, Merlin Compact, Jena, Germany). Thermodynamic characteristics of Cu NPs were measured by differential scanning calorimetry (DSC, Q100, TA Instruments, New Castle, DE, USA) at a heating rate of 10 °C/min. After sintering bonding, the shear strength of the lapped joint was evaluated by an electronic universal testing machine (CMT5205, MTS Systems (China) Co. Ltd., Shenzhen, China) with the constant shear rate of 0.2 mm/min. At least 5 sintering bonded specimens related to each group were used for the test, and the average test value was calculated. In addition to that, Vickers hardness (HBRV-187.5, Light-Mach Tech. Co., Ltd., Shanghai, China) was tested near the joint. The morphological character of the sintering bonded interface was investigated by scanning electron microscope (SEM) and optical microscope (OLYMPUS GX51, Olympus Corporation, Tokyo, Japan). X-ray diffraction (XRD, Dmax-RB, Rigaku Corporation, Tokyo, Japan) was utilized to analyze the phase composition in joint fracture. Helium leak mass spectrometer ZQJ-530 (KYKY Technology Development Ltd., Beijing, China) was employed to investigate the gas tightness of the joint.

3. Results

3.1. Characteristics of Cu Nanoparticles

Figure 4 shows the SEM image of Cu NPs. As can be seen in Figure 4, the morphology of Cu NPs is presented in the form of quasi-spherical shapes. Moreover, Cu NPs are homogeneously dispersed without hard agglomerates. The average size of Cu NPs is around 50 nm.

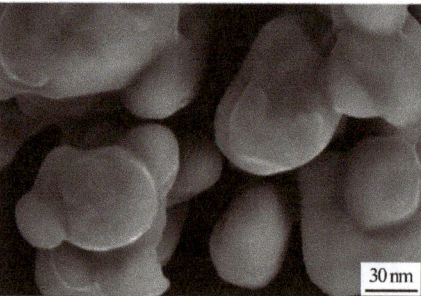

Figure 4. SEM image of Cu nanoparticles.

As is known, the atomic arrangement on surface of nanoparticles varies greatly from the complete lattice inside the crystal, and the atoms at the interface have higher free energy. As a result, nanomaterials are metastable, which is quite sensitive to ambient temperature. To decrease the free energy of the system, atoms on the surface of nanoparticles will experience structural rearrangement and relaxation as the environmental temperature increases [25–27]. Figure 5 shows the DSC analysis of Cu NPs utilized in the experiment. For Cu NPs, a sharp exothermic peak between 120 °C and 160 °C is found in Figure 5.

The second exothermic peak is quite broad compared with the first one, which is between 200 °C to 320 °C, implying the occurrence of a relatively uniform heat flow. The reason for this broad uniformity is that the Cu NPs have gone through a quite mild ripening process, which is a slow exothermic process. Different from the smooth DSC curves of bulk Cu before melting, there is exothermic peaks on DSC curve of Cu NPs prior to melting point of 1083.4 °C. These low-temperature exothermic peaks correspond to the enthalpy release of nanocrystals. The slower the heating rate is, the more heat is released.

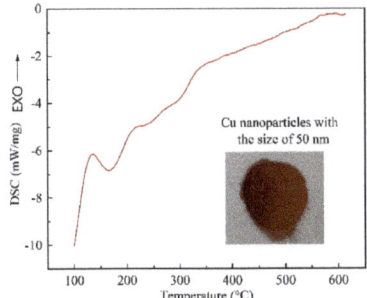

Figure 5. DSC curve of Cu nanoparticles.

In DSC test of metal particles, three types of phenomena can be responsible for the appearance of an exothermic peak, namely, the mutual diffusion of highly unstable atoms on the surface of metal particles during the sintering process, crystallization of amorphous metal particles and recrystallization of strained metal particles by heating. Therefore, in the case of sintering bonding in this research, the appearance of sintering necks by using Cu NPs is the main reason for the exothermic peak. As a result, the DSC analysis suggests that bonding temperature higher than 320 °C can assist the sintering joint performance. According to the thermodynamic melting process, Couchmann and Jesser proposed a relationship between the melting point and the solid particle size dimension, which is expressed as follows [25]:

$$T_m = T_m^e \left[1 - 2\,\sigma_{sl} \cos\theta\,(L\rho_s r)^{-1}\right], \qquad (2)$$

where T_m and T_m^e are the melting point of nano material and bulk material, σ_{sl} the interfacial energy between solid and liquid metal, L the latent heat of fusion, ρ_s the density of solid metal, r the size of particle, θ the contact angle between embedded particle and matrix. Obviously, $\cos\theta$ equal to 1 for the free particles. Equation (2) indicates that the melting point of metal will change as the particle size changes. The change of melting point (T_m-T_m^e) is linearly related to the reciprocal of particle size. Therefore, the smaller the nanoparticle size is, the lower the sintering bonding temperature is. The sintering performance of Cu NPs is one of the key factors for achieving high strength bonding.

3.2. Microstructure Analysis of the Joint and XRD Analysis of the Fracture

The optical microscope image of SiC$_p$/6063 Al MMCs joints after bonding is presented in Figure 6. Since liquid Ga was reported to be effective for bonding of aluminum [28,29], the bonding of SiC$_p$/6063 Al MMCs was carried out by using liquid Ga at different temperatures. As shown in Figure 6a–c, the bonding of SiC$_p$/6063 Al MMCs was realized by using only liquid Ga. Most areas are perfectly bonded with the assistance of liquid Ga. As is known from the Al-Ga binary phase diagram, Ga has the maximum solid solubility in aluminum at the temperature of 26.6 °C, which is around 20.0% (wt.%). The value of solid solubility decreases regardless of the increase or decrease in temperature. As a result, the width of the bonding area was quite small. However, some Ga cannot diffuse into aluminum, as shown in the black area along the bonded line due to the block of a small number of SiC particles and discontinuous oxide layer at the interface. In the area of bonding, these residual Ga are a type of defect, that is unfavorable to the strength of the joints. Figure 6d–f show the joint microstructure sintering bonded with Cu NPs and liquid Ga at temperatures that varied from 440 °C to 460 °C. As can be seen, the bonding of SiC_p/6063 Al MMCs was successfully achieved by using Cu NPs and liquid Ga. Compared with the joint bonded by liquid Ga, the width of the bonding area is much wider due to the existence of the Cu layer. Figure 6d,e show that the morphology of Cu NPs disappears totally, leaving typical sintering characteristics such as micro-voids in the bonding area. Almost the whole nanoparticles fuse together to form continuous sintering Cu layer. However, the Cu NPs sintering bonded at 460 °C shows a different microstructure in the joint, showing isolated Cu particles of relatively large size, as shown in Figure 6f. In the bonding region, the size of Cu particles is much smaller than that of 450 °C, which leads to a fine, brittle structure. At higher sintering bonding temperature, the Cu NPs experience a higher degree of oxidation than that at lower temperature during the bonding, resulting in a sintering barrier for Cu NPs.

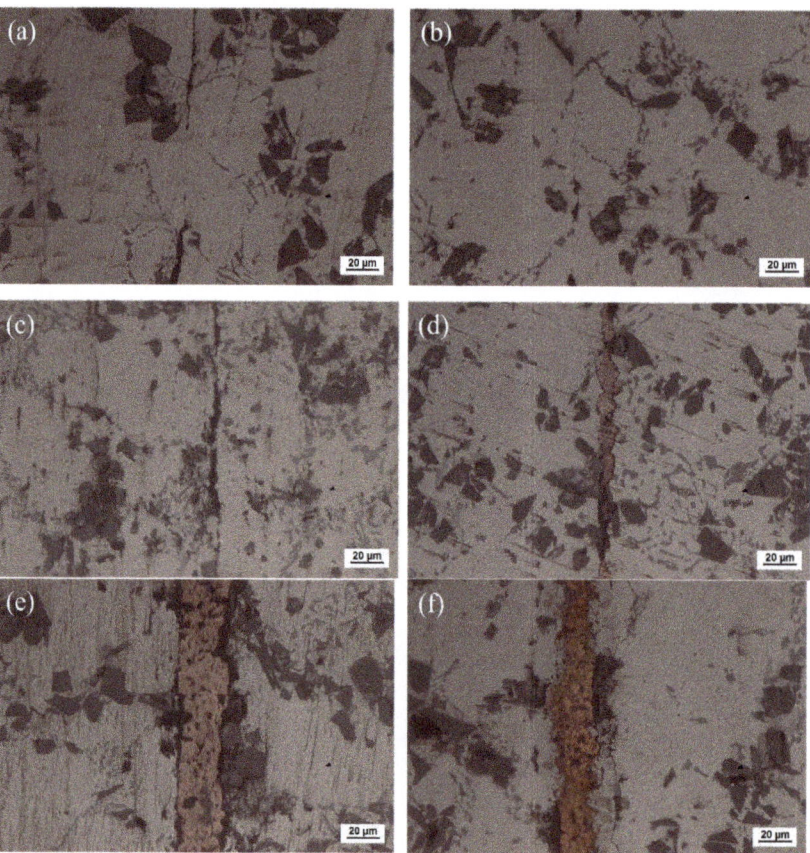

Figure 6. Optical microscope of the joints bonded at: (**a**) 440 °C using liquid Ga; (**b**) 450 °C using liquid Ga; (**c**) 460 °C using liquid Ga; (**d**) 440 °C using Cu NPs and liquid Ga; (**e**) 450 °C using Cu NPs and liquid Ga; (**f**) 460 °C using Cu NPs and liquid Ga.

Figure 7 shows SEM micrograph of the joint sintering bonded at 450 °C using Cu NPs and liquid Ga, and the corresponding energy dispersive X-ray maps showing distribution of elements. As shown in Figure 7a, nearly all of the nanoparticles merged into one during the bonding process, which can be called a long-range sintering process. Meanwhile, some micro-voids were left in the Cu layer after sintering bonding. The size and arrangement of sintering voids affect the mechanical strength of the joint. Moreover, Figure 7a also shows that a new phase appears along the interface between the Cu layer and substrate of SiC_p/6063 Al MMCs. The appearance of this action layer mainly results from the high activity of Cu NPs causing the reaction of Cu and Al at the interface. The Al-Cu binary phase diagram suggests that the most possible phase at that condition is $CuAl_2$ and that will be later confirmed by XRD. Figure 7b–f shows the individual elemental mapping of Al, Cu, Ga, Si and Mg in joint. As can be seen, some Al element diffused into the joint area and a little Cu element diffused into composites as well. The distribution of Mg and Si, which came from SiC_p/6063 Al MMCs, was quite uniform in joint due to the large diffusion coefficient. Moreover, the distribution of Ga in joint was very uniform without aggregation since the diffusion velocity of Ga in Al and Cu NPs was quite high. When Cu NPs reacted with Al, they also reacted with liquid Ga to form solid solution and intermetallic compound. The line scanning profile of elements Al, Cu and Ga in Figure 7g–i also indicated that the mutual diffusion between filler metal and SiCp/6063 Al MMCs proceeded well at bonding condition, and that would be beneficial to the formation of strengthening joint.

Figure 8 shows the SEM micrograph of the joint bonded at 450 °C using only liquid Ga, and the corresponding energy dispersive X-ray maps showing distribution of elements. Different from Figure 7d, element Ga shows an inhomogeneous distribution, as shown in Figure 8c. A large amount of Ga aggregated in the center of joint. This is because the bonding conditions, including temperature and soaking time, were not sufficient for the diffusion of liquid Ga into SiC_p/6063 Al MMCs. Therefore, it is necessary to add some high activity material in joint to consume the residual Ga. In this research, Cu NPs were selected as the high activity material since they easily reacted with both Ga and Al in a relatively low temperature. In addition, Cu NPs have better oxidation resistance compared to other commercial metal nano particles. The distribution of Mg and Si was fairly uniform in the joint, which was similar to Figure 7. The line scanning profiles of Al and Ga also indicated that residual Ga was left in joint, as shown in Figure 8f,g. In the joint area, the residual Ga, as stated earlier, is unfavorable to the strength of the joint. Moreover, the residual Ga in joint may cause the hydrolysis of Al in humid environment, leading to the potential damage of the joint.

The XRD pattern of the joint fracture bonded at different conditions is presented in Figure 9. As shown in Figure 9a, when SiC_p/6063 Al MMCs joint was bonded at 440 °C with only liquid Ga, the results showed that the main phases are Ga and Ga_2O_3 except the basic Al and SiC included in matrix, which further verified the above analysis results. With the addition of liquid Ga and Cu NPs in joint, the main phases in joint fracture consisted of Cu, $CuAl_2$, Cu_9Ga_4 and Ga_2O_3 except the basic Al and SiC, as shown in Figure 9b. Compared with the results in Figure 9a, simple substances of Ga cannot be found in Figure 9b. The reason for that was the reaction between Cu NPs and liquid Ga in sintering bonding condition, which can consume the redundant Ga and formed the intermetallic compound Cu_9Ga_4 in joint. Based on the high activity of Cu NPs, the intermetallic compound $CuAl_2$ was found in Figure 9b, which indicated that the interdiffusion between filler metal and base SiC_p/6063 Al MMCs proceeded well. That intermetallic compound $CuAl_2$ was also observed in Figure 7a, presenting a characteristic of dark grey phase along the interface. In both bonding conditions, Ga_2O_3 was detected while the aluminum oxide was not detected in joint fractures. That indicated that the Ga in joint can protect the aluminum in SiC_p/6063 Al MMCs from the oxidation in atmosphere environment. In addition, Cu oxide was not detected in Figure 9b, which means that Ga can also protect Cu NPs from oxidation as well. Therefore, the bonding of SiC_p/6063 Al MMCs can be realized using liquid Ga or Cu NPs and liquid Ga in air environment.

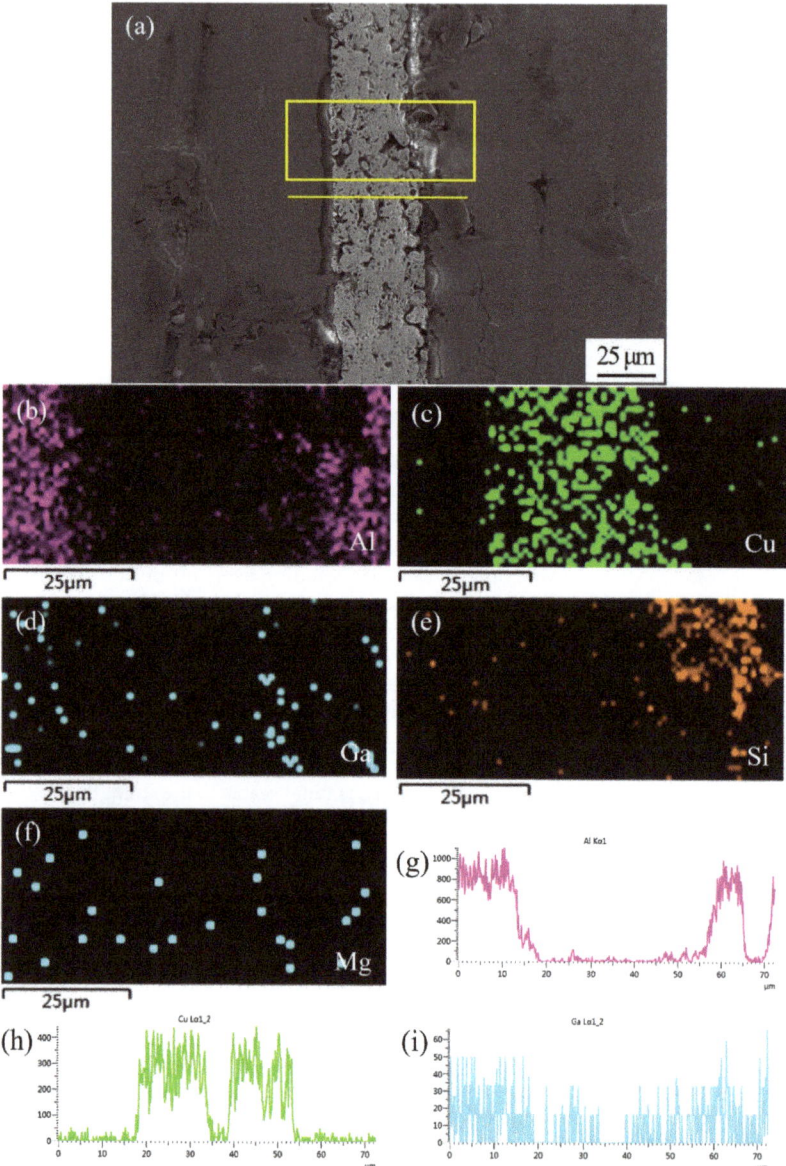

Figure 7. SEM micrograph at the interface of SiC$_p$/6063 Al MMCs joint sintering bonded at 450 °C using Cu NPs and liquid Ga, and the corresponding energy dispersive X-ray maps showing distribution of elements: (**a**) SEM image; (**b**–**f**) individual elemental mapping of Al, Cu, Ga, Si and Mg, respectively; (**g**–**i**) line scanning profile of elements Al, Cu and Ga, respectively.

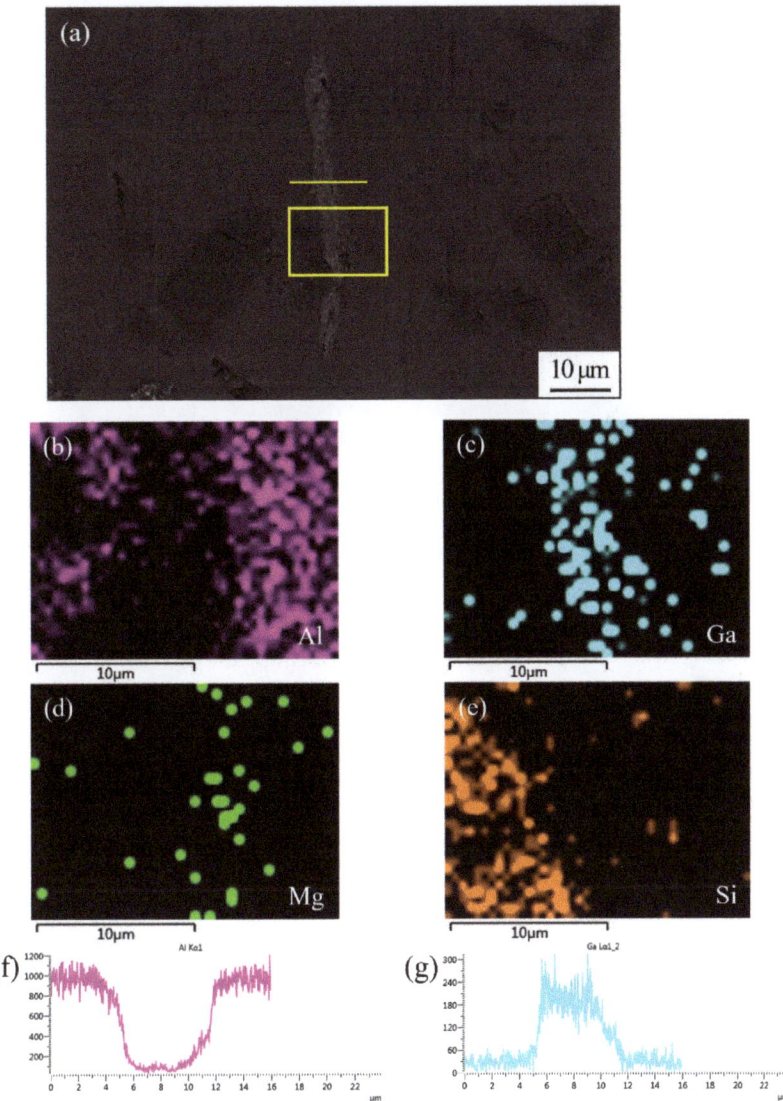

Figure 8. SEM micrograph at the interface of SiC$_p$/6063 Al MMCs joint bonded at 450 °C using liquid Ga, and the corresponding energy dispersive X-ray maps showing distribution of elements: (**a**) SEM image; (**b**–**e**) individual elemental mapping of Al, Ga, Mg and Si, respectively; (**f**–**g**) line scanning profile of elements Al and Ga.

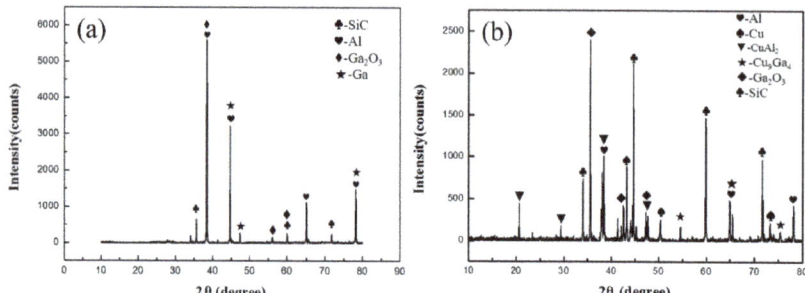

Figure 9. XRD results of the joint fracture bonded using: (**a**) liquid Ga at 440 °C; (**b**) Cu NPs and liquid Ga at 450 °C.

3.3. Mechanical Properties Analysis of the Joint

Figure 10 shows the mechanical properties of bonded SiC_p/6063 Al MMCs joints by using liquid Ga, Cu NPs and liquid Ga at different temperatures under a certain pressure of 3 MPa for 2 h. As shown in Figure 10a, shear strength level of the joint bonded using Ga and Cu NPs as filler metal initially increased and then decreased. Due to the sintering inadequacy of Cu NPs at low temperature, a weak shear strength level of 30.2 MPa was presented at 400 °C. As the bonding temperature increases, the coalescence of Cu NPs leads to a steady interconnection between nanoparticles. Meanwhile, the interdiffusion between Cu NPs and SiC_p/6063 Al MMCs substrate increased significantly with the increasing of temperature. As a consequence, shear strength of the joint increased gradually. When the bonding temperature was 450 °C, the maximum shear strength value of 56.2 MPa was achieved. Although a denser sintered structure can be obtained at a higher bonding temperature, shear strength level of the joint decreased quickly once bonding temperature exceeded 450 °C. At higher bonding temperature, a large amount of brittle intermetallic compounds such as $CuAl_2$ generated along the interface between filler metal and substrate and that was harmful to mechanical property of the joint. Compared to the joint bonded with only liquid Ga, the joint bonded with Cu NPs and liquid Ga as filler metal had higher shear strength when the temperature exceeded 450 °C. Below 450 °C, the joints had lower shear strength compared with that made of liquid Ga on account of sintering inadequacy of Cu NPs.

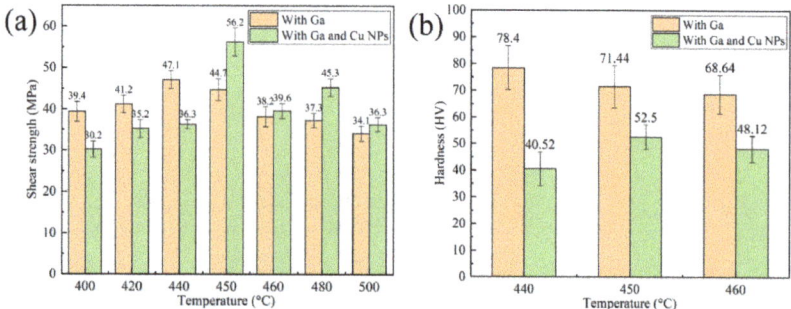

Figure 10. Mechanical properties of the joint bonded at different temperatures: (**a**) Shear strength of joints; (**b**) Hardness of the joints.

Vickers hardness test of the joint was performed on the middle of specimens which were bonded with the temperature range of 440 °C to 460 °C. In Vickers hardness test, the loaded weight and time was set to be 300 g and 5 s, respectively. Each hardness was the average value of five measured specimen. As shown in Figure 10b, the hardness of specimen bonded with only Ga decreased gradually in the temperature range of 440 °C to 460 °C. The maximum hardness was 78.4 HV when the joint was bonded at 440 °C. With the increase of bonding temperature, Ga diffused a longer distance in joint. As a consequence, the content of Ga, which can play the role of solution strengthening, was less and less in the joint center, resulting in the decrease of hardness. The minimum hardness of the joint was 68.64 HV when the joint was bonded at 460 °C. On the other hand, the changing trend of joint hardness was consistent with that of shear strength. The maximum hardness was measured to be 52.5 HV when the joint was bonded at 450 °C. Compared with the joint bonded with only Ga, the joint bonded with Ga and Cu NPs as filler metal had lower hardness in the joint center. The main reason for that was the existence of micro-voids in the Cu layer generated during the sintering bonding process.

3.4. Gas Tightness Test of the Joints

One of important application area of SiC_p/6063 Al MMCs is electronic packaging, in which the assembly component requires strict gas tightness to protect the internal chips from potential damage caused by moist air during service. Therefore, the gas tightness test after bonding experiment was necessary to measure the leak rate of the joint, which is the only leak path of the component. In this research, helium mass spectrum detection technique was applied to measure the leak rate of the joint due to its low effective minimum detectable leak rate (5×10^{-12} Pa·m^3/s) and high sensitivity. The schematic diagram of gas tightness test used in this experiment is displayed in Figure 11. Before bonding experiment, a round hole was machined on one piece of SiC_p/6063 Al MMCs. After bonding experiment, the specimen was placed on a specimen table, in which the blind hole was located just above the detector. Meanwhile, the contact area between specimen and specimen table was sealed with sealant. After that, pure helium was sprayed on the specimen in all directions. The only path for helium entering the detector space was the bonding area.

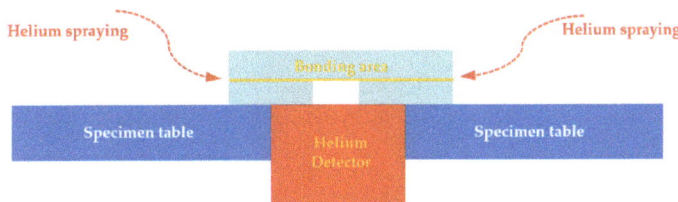

Figure 11. Schematic representation of gas tightness test for sintering bonded joint.

Leak rate test of the joint was immediately performed after the bonding process. In order to test stability of the joint, the same method was carried out after one week. As a qualified component used in electronic packaging field, it is desirable that gas leakage rate keeps at 10^{-10} Pa·m^3/s or below.

Table 1 shows gas tightness test result of the joint sintering bonded by Ga and Cu NPs at temperature range of 440 °C to 500 °C. After bonding, the specimens bonded at temperature range of 440 °C to 480 °C were qualified. Higher bonding temperature of 500 °C increased the gas leakage rate, reaching 10^{-9} Pa·m^3/s. At higher temperature, more brittle intermetallic compounds will be generated at the interface between filler metal and SiC_p/6063 Al MMCs. Due to different coefficient of thermal expansion between Cu layer and SiC_p/6063 Al MMCs, the released stress will lead to damage of brittle intermetallic compounds at interface during the cooling process, causing the gas leakage. The gas tightness test after one week showed that specimens bonded at temperature range of 440 °C to 460 °C were qualified, remaining unchanged compared to the test after bonding immediately. When the specimens were bonded at 480 °C and 500 °C, the leak rates after one week were 10^{-9} Pa·m^3/s and 10^{-8} Pa·m^3/s, respectively. Compared with the leak rate after bonding, the leak rate of the specimens increased by an order of magnitude. The damage of the intermetallic compound layer caused by stress relief during service was the primary reason for increasing of leak rate. Though some micro-voids were observed in microstructure analysis of the joint, gas tightness of the joint sintering bonded with temperature range of 440 °C to 460 °C were qualified whether test was performed immediately after bonding or after one week. It suggests that the sintered micro-voids were disconnected in the Cu layer.

Table 1. Gas tightness of the joint sintering bonded by Ga and Cu NPs at different temperatures.

Temperature (°C)	440	450	460	480	500
Leak rate after bonding (Pa·m^3/s)	10^{-10}	10^{-10}	10^{-10}	10^{-10}	10^{-9}
Leak rate after one week (Pa·m^3/s)	10^{-10}	10^{-10}	10^{-10}	10^{-9}	10^{-8}

4. Conclusions

In this research, 15 vol.% SiC$_p$/6063 Al MMCs were successfully sintering bonded using Cu nanoparticles and liquid Ga as filler metal without any flux in the air environment. The characteristics of Cu nanoparticles were analyzed at first. Based on thermophysical characteristic of Cu nanoparticles, sintering bonding technology was then applied on joining of 15 vol.% SiC$_p$/6063 Al MMCs. The microstructure, shear strength, hardness and gas tightness of the joints were studied to understand the quality of the joint. The following conclusions can be drawn:

(1). For Cu NPs with the average diameter of 50 nm, two exothermic peaks can be found from DSC test. The first exothermic peak was located in the temperature range of 120 °C to 160 °C. The second broad exothermic peak was located in the temperature range of 200 °C to 320 °C. The reason for the later broad uniformity is that the Cu NPs have gone through a mild ripening process, which is a slow exothermic process. The occurrence of sintering necks in Cu NPs is the main reason for the exothermic peak. Sintering performance of Cu NPs is one of the key factors for achieving qualified joint with low gas leakage rate and high strength.

(2). When the joint was bonded at 450 °C, a large amount of pure Ga aggregated in the joint center by using only liquid Ga as filler metal. The addition of Cu NPs can consume the redundant Ga to form the intermetallic compounds in joint at the same bonding temperature. When the joint was bonded at 450 °C by using liquid Ga and Cu NPs, the main phases in joint fracture consisted of Cu, CuAl$_2$, Cu$_9$Ga$_4$ and Ga$_2$O$_3$ besides the basic Al and SiC. The formation of Ga$_2$O$_3$ in joint can protect aluminum in SiC$_p$/6063 Al MMCs and Cu NPs from the oxidation.

(3). The maximum shear strength level of 56.2 MPa can be achieved by applying bonding temperature of 450 °C. At higher temperature, although a denser sintered structure can be obtained in joint, the shear strength level decreases quickly due to the generation of large quantities of brittle intermetallic compounds such as CuAl$_2$ along the interface. Compared with joint bonded with only Ga, the joint bonded with Ga and Cu NPs as filler metal had lower hardness due to the existence of micro-voids in sintering Cu layer.

(4). As the specimens sintering bonded between 480 °C and 500 °C, the released thermal stress will lead to damage of brittle intermetallic compounds along the interface during cooling or service, causing gas leakage. The specimens sintering bonded in temperature range of 440 °C to 460 °C had qualified gas tightness during the service, which can remain in the level of 10^{-10} Pa·m^3/s. The sintering micro-voids in Cu layer had little influence on gas tightness of the joint since they were disconnected.

Author Contributions: Investigation, Z.G., C.Y. and J.F.; resources, D.C. and J.B.; writing—original draft preparation, Z.G. and C.Y.; writing—review and editing, J.B. and J.N.; supervision, P.H. and J.N. All authors have read and agreed to the published version of the manuscript.

Funding: This research was funded by the National Natural Science Foundation of China (No. 51245008), the Science and Technology Project of Henan Province, China (No. 202102210036), China Postdoctoral Science Foundation (No. 2021M692891).

Institutional Review Board Statement: Not applicable.

Informed Consent Statement: Not applicable.

Data Availability Statement: Not applicable.

Conflicts of Interest: The authors declare no conflict of interest.

References

1. Mousavian, R.T.; Khosroshahi, R.A.; Yazdani, S.; Brabazon, D.; Boostani, A.F. Fabrication of aluminum matrix composites reinforced with nano- to micrometer-sized SiC particles. *Mater. Des.* **2016**, *89*, 58–70. [CrossRef]
2. Hu, Q.; Zhao, H.; Li, F. Microstructures and properties of SiC particles reinforced aluminum-matrix composites fabricated by vacuum-assisted high pressure die casting. *Mater. Sci. Eng. A* **2017**, *680*, 270–277. [CrossRef]
3. Qu, X.-H.; Zhang, L.; Wu, M.; Ren, S.-B. Review of metal matrix composites with high thermal conductivity for thermal management applications. *Prog. Nat. Sci. Mater. Int.* **2011**, *21*, 189–197. [CrossRef]
4. Zhu, X.-M.; Yu, J.-K.; Wang, X.-Y. Microstructure and properties of Al/Si/SiC composites for electronic packaging. *Trans. Nonferrous Met. Soc. China* **2012**, *22*, 1686–1692. [CrossRef]
5. Sharma, D.K.; Mahant, D.; Upadhyay, G. Manufacturing of metal matrix composites: A state of review. *Mater. Today Proc.* **2020**, *26*, 506–519. [CrossRef]
6. Zhu, M.-J.; Li, S.; Zhao, X.; Xiong, D.-G. Laser-weldable Sip-SiCp/Al hybrid composites with bilayer structure for electronic packaging. *Trans. Nonferrous Met. Soc. China (Engl. Ed.)* **2014**, *24*, 1032–1038. [CrossRef]
7. Zhang, M.; Wang, C.; Mi, G.; Jiang, P.; Zhang, X. New insight into laser butt welded nonporous SiCp/2A14Al joint: Interfaces, precipitation phase and mechanical properties. *Mater. Charact.* **2021**, *176*, 111082. [CrossRef]
8. Chen, M.-A.; Wu, C.-S.; Zou, Z.-D. Electron beam welding of SiCp/LD2 composite. *Trans. Nonferrous Met. Soc. China* **2006**, *16*, 818–823. [CrossRef]
9. Lei, Y.-C.; Wang, Z.-W.; Chen, X.-Z. Effect of arc-ultrasound on microstructures and mechanical properties of plasma arc welded joints of SiCp/Al MMCs. *Trans. Nonferrous Met. Soc. China* **2011**, *21*, 272–277. [CrossRef]
10. Raja, V.K.B.; Gupta, M. Joining of Metal Matrix Composites. In *Reference Module in Materials Science and Materials Engineering*; Elsevier: Amsterdam, The Netherlands, 2021. [CrossRef]
11. Rodríguez-Reyes, M.; Pech-Canul, M.I.; Rendón-Angeles, J.C.; López-Cuevas, J. Limiting the development of Al_4C_3 to prevent degradation of Al/SiCp composites processed by pressureless infiltration. *Compos. Sci. Technol.* **2006**, *66*, 1056–1062. [CrossRef]
12. Lei, Y.-C.; Yuan, W.-J.; Chen, X.-Z.; Zhu, F.; Cheng, X.-N. In-situ weld-alloying plasma arc welding of SiCp/Al MMC. *Trans. Nonferrous Met. Soc. China* **2007**, *17*, 313–317. [CrossRef]
13. Huang, J.-H.; Dong, Y.-L.; Wan, Y.; Zhao, X.-K.; Zhang, H. Investigation on reactive diffusion bonding of SiCp/6063 MMC by using mixed powders as interlayers. *J. Mater. Process. Technol.* **2007**, *190*, 312–316. [CrossRef]
14. Maity, J.; Pal, T.K.; Maiti, R. Transient liquid phase diffusion bonding of 6061-15 wt% SiCp in argon environment. *J. Mater. Process. Technol.* **2009**, *209*, 3568–3580. [CrossRef]
15. Xiangzhao, Z.; Santuan, Z.; Guiwu, L.; Ziwei, X.; Haicheng, S.; Guanjun, Q. Review on Brazing of High Volume Faction SiCp/Al Composites for Electronic Packaging Applications. *Rare Met. Mater. Eng.* **2017**, *46*, 2812–2819. [CrossRef]
16. Amirizad, A.; Kokabi, A.H.; Gharacheh, M.A.; Sarrafi, R.; Amirkhiz, B.S.; Azizieh, M. Evaluation of microstructure and mechanical properties in friction stir welded A356+15%SiCp cast composite. *Mater. Lett.* **2006**, *60*, 565–568. [CrossRef]
17. Ferchaud, E.; Christien, F.; Barnier, V.; Paillard, P. Characterisation of Ga-coated and Ga-brazed aluminium. *Mater. Charact.* **2012**, *67*, 17–26. [CrossRef]
18. Ludwig, W.; Pereiro-Lopez, E.; Bellet, D. In situ investigation of liquid Ga penetration in Al bicrystal grain boundaries: Grain boundary wetting or liquid metal embrittlement? *Acta Mater.* **2005**, *53*, 151–162. [CrossRef]
19. Yan, J. A review of sintering-bonding technology using Ag nanoparticles for electronic packaging. *Nanomaterials* **2021**, *11*, 927. [CrossRef] [PubMed]
20. Li, J.; Yu, X.; Shi, T.; Cheng, C.; Fan, J.; Cheng, S.; Liao, G.; Tang, Z. Low-temperature and low-pressure Cu-Cu bonding by highly sinterable Cu nanoparticle paste. *Nanoscale Res. Lett.* **2017**, *12*, 1–6. [CrossRef]
21. Li, J.; Liang, Q.; Shi, T.; Fan, J.; Gong, B.; Feng, C.; Fan, J.; Liao, G.; Tang, Z. Design of Cu nanoaggregates composed of ultra-small Cu nanoparticles for Cu-Cu thermocompression bonding. *J. Alloys Compd.* **2019**, *772*, 793–800. [CrossRef]
22. Li, J.; Yu, X.; Shi, T.; Cheng, C.; Fan, J.; Cheng, S.; Li, T.; Liao, G.; Tang, Z. Depressing of Cu-Cu bonding temperature by composting Cu nanoparticle paste with Ag nanoparticles. *J. Alloys Compd.* **2017**, *709*, 700–707. [CrossRef]
23. Liu, J.; Chen, H.; Ji, H.; Li, M. Highly conductive Cu-Cu joint formation by low-temperature sintering of formic acid-treated Cu nanoparticles. *ACS Appl. Mater. Interfaces* **2016**, *8*, 33289–33298. [CrossRef]
24. Mu, G.; Qu, W.; Zhu, H.; Zhuang, H.; Zhang, Y. Low temperature Cu/Ga solid-liquid inter-diffusion bonding used for interfacial heat transfer in high-power devices. *Metals* **2020**, *10*, 1223. [CrossRef]

25. Foster, D.M.; Pavloudis, T.; Kioseoglou, J.; Palmer, R.E. Atomic-resolution imaging of surface and core melting in individual size-selected Au nanoclusters on carbon. *Nat. Commun.* **2019**, *10*. [CrossRef] [PubMed]
26. Goodman, E.D.; Carlson, E.Z.; Dietze, E.M.; Tahsini, N.; Johnson, A.; Aitbekova, A.; Nguyen Taylor, T.; Plessow, P.N.; Cargnello, M. Size-controlled nanocrystals reveal spatial dependence and severity of nanoparticle coalescence and Ostwald ripening in sintering phenomena. *Nanoscale* **2021**, *13*, 930–938. [CrossRef] [PubMed]
27. Ju, S.-P.; Lee, I.J.; Chen, H.-Y. Melting mechanism of Pt-Pd-Rh-Co high entropy alloy nanoparticle: An insight from molecular dynamics simulation. *J. Alloys Compd.* **2021**, *858*. [CrossRef]
28. Shirzadi, A.A.; Wallach, E.R. Novel method for diffusion bonding superalloys and aluminium alloys (USA Patent 6,669,534 B2, European Patent Pending). In Proceedings of the International Conference on New Frontiers of Process Science and Engineering in Advanced Materials, PSEA '04, Kyoto, Japan, 24–26 November 2004; pp. 431–436.
29. Shirzadi, A.A.; Saindrenan, G.; Wallach, E.R. Flux-free diffusion brazing of aluminium-based materials using gallium (patent application: UK 0128623.6). In Proceedings of the Aluminium Alloys 2002 Their Physical and Mechnaical Properties: Proceedings of the 8th International Conference ICAA8, Cambridge, UK, 2–5 July 2002; pp. 1579–1584.

Article

Effect of Cold Rolling on the Evolution of Shear Bands and Nanoindentation Hardness in $Zr_{41.2}Ti_{13.8}Cu_{12.5}Ni_{10}Be_{22.5}$ Bulk Metallic Glass

Abhilash Gunti [1], Parijat Pallab Jana [1], Min-Ha Lee [2] and Jayanta Das [1,*]

[1] Department of Metallurgical and Materials Engineering, Indian Institute of Technology Kharagpur, West Bengal 721302, India; abhilash@iitkgp.ac.in (A.G.); parijat.pallab@iitkgp.ac.in (P.P.J.)
[2] KITECH North America, Korea Institute of Industrial Technology, San Jose, CA 95134, USA; mhlee1@kitech.re.kr
* Correspondence: j.das@metal.iitkgp.ac.in; Tel.: +91-3222-283284; Fax: +91-3222-282280

Abstract: The effect of cold rolling on the evolution of hardness (H) and Young's modulus (E) on the rolling-width (RW), normal-rolling (NR), and normal-width (NW) planes in $Zr_{41.2}Ti_{13.8}Cu_{12.5}Ni_{10}Be_{22.5}$ (Vitreloy 1) bulk metallic glass (BMG) was investigated systematically using nanoindentation at peak loads in the range of 50 mN–500 mN. The hardness at specimen surface varied with cold rolling percentage (%) and the variation is similar on RW and NR planes at all the different peak loads, whereas the same is insignificant for the core region of the specimen on the NW plane. Three-dimensional (3D) optical surface profilometry studies on the NR plane suggest that the shear band spacing decreases and shear band offset height increases with the increase of cold rolling extent. Meanwhile, the number of the pop-in events during loading for all the planes reduces with the increase of cold rolling extent pointing to more homogeneous deformation upon rolling. Calorimetric studies were performed to correlate the net free volume content and hardness in the differently cold rolled specimens.

Keywords: bulk metallic glass; Vitreloy 1; nanoindentation; cold rolling; densification; inhomogeneity

Citation: Gunti, A.; Jana, P.P.; Lee, M.-H.; Das, J. Effect of Cold Rolling on the Evolution of Shear Bands and Nanoindentation Hardness in $Zr_{41.2}Ti_{13.8}Cu_{12.5}Ni_{10}Be_{22.5}$ Bulk Metallic Glass. *Nanomaterials* **2021**, *11*, 1670. https://doi.org/10.3390/nano11071670

Academic Editor: Yang-Tse Cheng

Received: 15 April 2021
Accepted: 22 June 2021
Published: 25 June 2021

Publisher's Note: MDPI stays neutral with regard to jurisdictional claims in published maps and institutional affiliations.

Copyright: © 2021 by the authors. Licensee MDPI, Basel, Switzerland. This article is an open access article distributed under the terms and conditions of the Creative Commons Attribution (CC BY) license (https://creativecommons.org/licenses/by/4.0/).

1. Introduction

Bulk metallic glasses (BMGs) exhibiting high structural strength can be synthesized by quenching from the liquid state at a critical cooling rate of >10^2 K/s [1]. The glassy structure produced at different cooling rates exhibits different free volume contents, leading to the evolution of different configurational states in the potential energy landscape (PEL), and exhibits different fictive temperatures T_f, which affects the mechanical behavior [2,3]. Owing to the absence of long-range order, BMGs are isotropic, but can become anisotropic in terms of elastic, mechanical, and magnetic properties owing to local cooling rate difference fluctuations during vitrification [4] or secondary processing like elastic [5], anelastic [6], homogeneous (creep) [7], and inhomogeneous plastic deformation through compression [8] or high-pressure torsion [9]. Bond-orientation anisotropy (BOA) in BMGs was reported to cause anelastic deformation [8,10]. Furthermore, local bond exchange or the bond reformation during plastic deformation was reported to be the root cause of the observed structural anisotropy [3,7,9].

The plastic deformation of bulk metallic glasses (BMGs) involves two competing processes, including free volume creation and relaxation [11–13]. The free volume creation process includes disordering and dilatation, which led to softening, whereas the relaxation process is linked with diffusional ordering and densification [14,15]. The disordering and softening processes dominate in most of the cases and fail the BMG catastrophically by forming shear bands [11,12,16]. At low temperatures, the run-away of shear bands is dominated by the sluggish diffusion rate, thus the free volume accumulation is significant compared with that of free volume annihilation rate. Therefore, experimental and molecular dynamics simulations have shown that uncontrolled softening is the dominant mechanism during shear banding, as reported by several authors [17–26].

On the other hand, the hardening in BMGs has been reported to occur as a result of the homogeneous nucleation and continuous multiplication of the shear bands [27]. Similarly, the change in the morphology and distribution of the chemically heterogeneous domains in the glassy structure may also contribute to the hardening [28]. Along this line, Lee et al. have reported an increase in the compressive yield strength in Vitreloy 1 upon cold rolling without any change in the local chemical composition nor the deformation induced nanocrystallization [29]. Interestingly, the strain-hardening has also been reported during cyclic compressive loading [30]. Schuh et al. have reported that the hardening in a glassy phase is linked with the formation of new shear bands, which increases the flow stress owing to the deformation induced nanocrystallization in the shear bands [13]. Recently, Wang et al. have also observed the densification in a monolithic BMG upon tensile deformation. Such densification is linked to the diffusional rearrangements leading to rapid annihilation of the free volume, which dominates over the shear flow and the free volume accumulation [31]. Therefore, in-depth studies are required to understand the deformation-induced relaxation and dilatation in the BMGs.

In the present work, monolithic $Zr_{41.2}Ti_{13.8}Cu_{12.5}Ni_{10}Be_{22.5}$ (Vitreloy 1) BMG was cold rolled (CR) to different reductions and the variations of the nanoindentation hardness and Young's modulus on the rolling-width (RW), normal-rolling (NR), and normal-width (NW) planes at different loads were systematically studied. The evolution of the shear bands on the NR plane was studied by optical profilometry. The effect of rolling on the hardness was correlated with the free volume content.

2. Experimental Procedure

Rectangular bars of $Zr_{41.2}Ti_{13.8}Cu_{12.5}Ni_{10}Be_{22.5}$ (Vitreloy 1) BMG with the size of $2 \times 2 \times 4$ mm^3 were polished carefully to achieve a mirror-like surface finish and were cold-rolled (CR) up to 4.5% (CR4.5), 10% (CR10), 20% (CR20), and 31% (CR31), with 0.05 mm reduction during each pass, as shown in Figure 1. The structure of the specimens was characterized using a differential scanning calorimeter (DSC, Perkin Elmer, Waltham, MA, USA, DSC 8000) and X-ray diffraction (XRD), confirming the glassy nature of the as-cast (CR0) and the differently CR specimens. The X-ray diffraction studies were performed using Philips PANalytical X-ray diffraction (XRD) unit (PW3373, The Netherlands) with Cu-Kα radiation. Figure 2 displays the X-ray diffraction (XRD) patterns of CR0 and CR31, confirming the structure of the specimens to be amorphous. DSC experiments were carried out at a heating rate of 20 K/s. As the shear bands are visible on the free surfaces only, the NR plane of the CR specimens was investigated, which remained untouched during rolling to construct the 3D contour of the surface. The 3D contour of the surfaces was constructed using the interference pattern, as constructed by the 1376 pixel × 1032 pixel high resolution colour charge-coupled device camera attached to Bruker Contour GT 3D-optical surface profilometer (OSP). The instrument has a wide dynamic range with lateral resolution of 0.3 μm and RMS of < 0.03 nm. The shear band spacing and the offset height were estimated using Vision64 analysis software (Billerica, MA, USA).

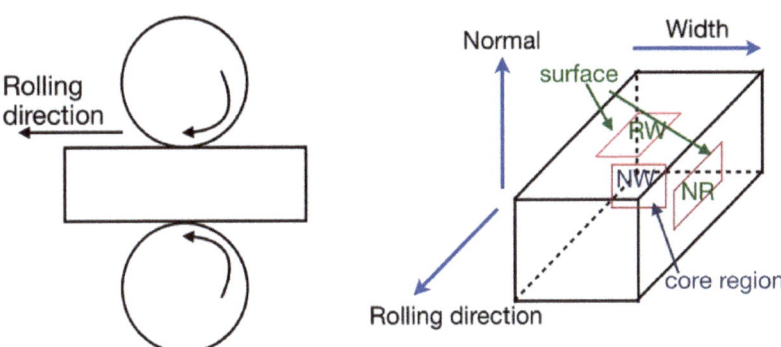

Figure 1. Schematic illustration of cold rolling and regions of normal-rolling (NR), rolling-width (RW), and normal-width (NW) planes for nanoindentation measurements.

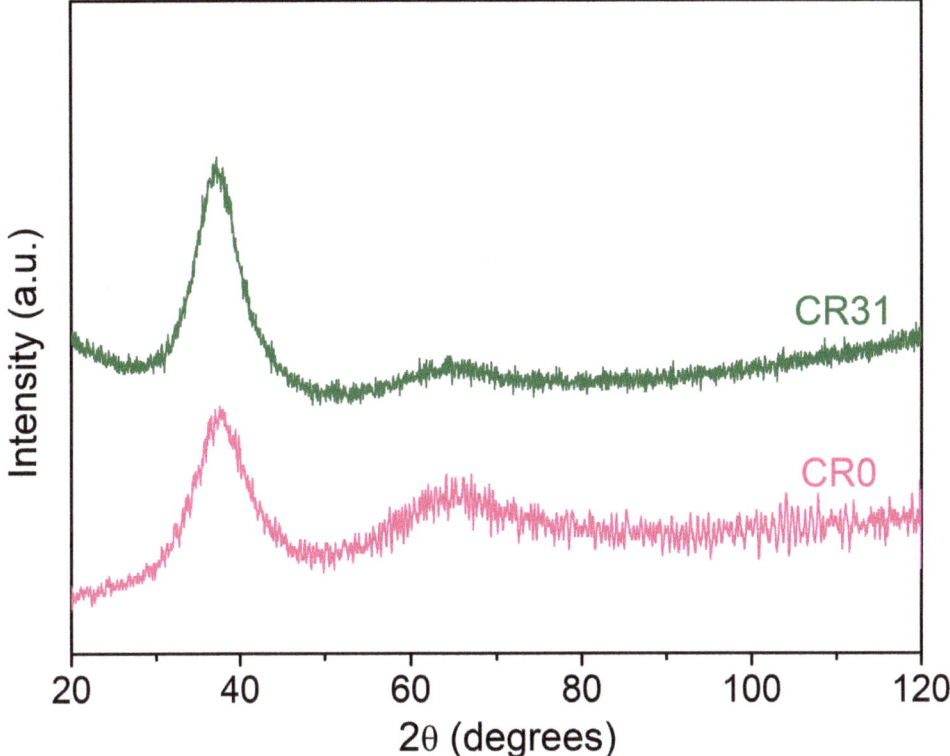

Figure 2. X-ray diffraction (XRD) patterns of CR0 and CR31 showing amorphous humps confirming the glassy nature of the samples.

Similarly, an optical microscope attached to the nanoindenter was used to choose the flat surface in between the two nearby shear bands for the indentation studies on the NR plane. A minimum distance was maintained between two indentations following ASTM standard, which is 20 times of the indentation depth, to avoid an overlapping of the two nearby plastically deformed zones. As no shear bands were observed on the RW and NW planes, the location for nanoindentation tests was chosen randomly, whereas the as-cast and

CR samples were cut into two halves along the NW plane for the nanoindentation studies for NW planes, as depicted in Figure 1. The nanoindentation tests were performed using Hysitron TI950 TriboIndenter, Minneapolis, MN with a standard three-sided Berkovich indenter at room temperature in control loading rate (CLR) mode with varying maximum applied load (P_{max}) at 50 mN, 100 mN, 200 mN, and 500 mN. Instrumented nanoindentation carried out by pressing a Berkovich indenter into the sample surface with of 30 s of loading followed by load holding of 40 s at peak load and unloading of 10 s, as illustrated in Figure 3. The nanoindenter load resolution is ±1 nN and displacement resolution is ±0.02 nm. The thermal drift was set to be ±0.05 nm/s. The contact area between the diamond indenter and the specimen was calibrated using a fused-quartz standard [32]. A field emission scanning electron microscope (FESEM, Zeiss Merlin Gemini 2, Zeiss, Jena, Germany) was used to study the nanoindentation impressions in the as-cast and differently CR samples. The hardness (H) and elastic modulus (E) values were calculated from P-h curves using the Oliver and Pharr model [32].

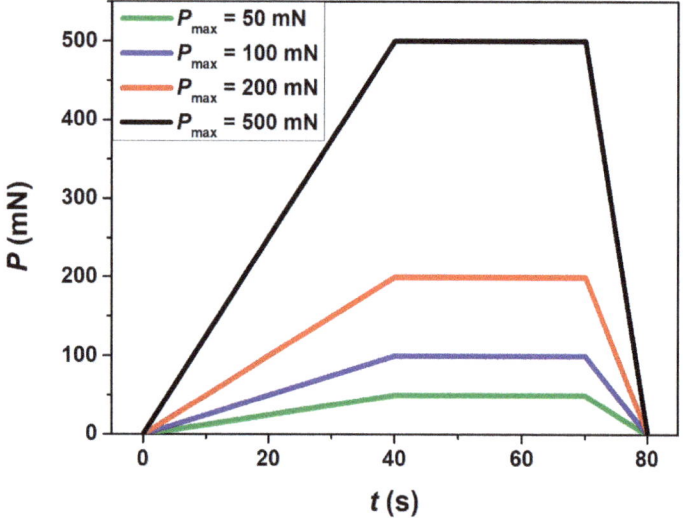

Figure 3. Schematic illustration of load (p)–time (t) plot used in the nanoindentation experiments.

3. Results

3.1. DSC Studies and Free Volume Estimation

Figure 4a displays the DSC traces for CR0, CR4.5, CR10, CR20, and CR31. The onset of glass transition (T_g^{onset}) was estimated to be 609.8 ± 8.0 K (CR0), 611 ± 3.7 K (CR4.5), 616.6 ± 7.5 K (CR10), 619.1 ± 3.8 K (CR20), and 619.9 ± 5.5 K (CR31). The estimated T_g^{onset} values of CR specimens are higher than that of CR0. The onset of the crystallization (T_x) was measured to be 752.5 ± 0.1K (CR0), 759.9 ± 0.1 K (CR4.5), 760.1 ± 0.2 K (CR10), 759.2 ± 0.1 K (CR20), and 756.7 ± 0.1 K (CR31), respectively, whereas the enthalpy change (ΔH) during crystallization was estimated to be −78.78 ± 0.46 J/g (CR0), −78.53 ± 0.09 J/g (CR4.5), −78.6 ± 0.33 J/g (CR10), −77.93 ± 0.63 J/g (CR20), and −81.08 ± 0.57 J/g (CR31), pointing to similar ΔH values of the as-cast and CR specimens. Hence, the glassy structure remains similar upon cold rolling without any hint of nanocrystallization. Such a conclusion was also confirmed by transmission electron microscopic investigation, and crystallization did not occur upon cold rolling [33]. Several researchers have shown an increase of ΔH values with an increase in the extent of cold rolling [11–14,34]. In contrary, Flores et. al. have shown that ΔH values varied with prior deformation [35].

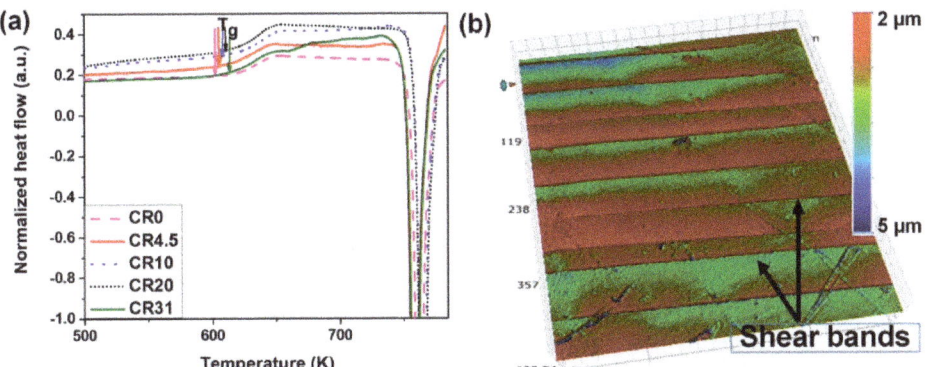

Figure 4. (a) Differential scanning calorimeter (DSC) plots of CR0, CR4.5, CR10, CR20, and CR31 showing the glass transition event followed by crystallization. (b) 3D optical profile showing the evolution of shear bands with step-like features on the surface of CR10.

According to Beukel and Sietsma [16], the reduced free volume (x) is expressed as $x = v_f/(\gamma v^*)$, where the average free volume per atom is v_f, v^* represents the critical free volume for atomic diffusion, and the overlap factor is indicated by γ. For simplicity, the reduced free volume (x) is named as free volume from here onwards in this work. The x value varies with the temperature when a metallic glass is subjected to continuous heating during DSC experiments. The glassy structure relaxes and attains the equilibrium upon heating at a constant rate below T_g [16]. The stored enthalpy releases during the reduction in the free volume and, therefore, the structural relaxation process is featured by an exothermic event. On the other hand, the rise in x with temperature after relaxation leads to an increase in the enthalpy and was characterized by an endothermic reaction, which is related to the glass transition [34]. The onset of the endothermic reaction, pointing to the rise in x, is called the onset of glass transition temperature or T_g^{onset}. At T_g^{onset}, x reaches the equilibrium free volume x_{eq} (T_g^{onset}), which can be expressed as follows:

$$x_{eq}(T) = \frac{(T - T_o)}{B} \tag{1}$$

where T is the temperature at which free volume is measured; T_o is the Vogel–Fulcher–Tammann (VFT) temperature (i.e., 390 K); and B is the product of T_o and the fragility parameter, reported to be 9282 [16,36].

A proportional relation exists between the ΔH and the free volume change (Δx) during both the structural relaxation and the glass transition event [6,34]:

$$\Delta H = A \cdot \Delta x \tag{2}$$

where A is the proportionality constant. The structural relaxation process in BMGs occurs at temperatures in between room temperature and T_g^{onset}. The change in the enthalpy throughout the structural relaxation process is related to the variation of the net free volume Δx, which is a measure of the difference between $x(T_{RT})$ at 297 K and the free volume at T_g^{onset} in Vitreloy 1. Thus, the value of $x(T_{RT})$ was estimated to be 0.0234, 0.0222, 0.0245, 0.0241, and 0.0229 using Equations (1) and (2) for CR0, CR4.5, CR10, CR20, and CR31, respectively. The $x(T_{RT})$ value of CR0 is similar to that reported by Masuhr et al. [37]. The value of $x(T_{RT})$ of CR10 and CR20 increased, whereas the same value of CR4.5 and CR31 decreased compared with that of CR0. Both the increase and decrease of free volume content upon plastic deformation of BMGs were reported earlier [11–14,21,38,39].

3.2. Characterization of Shear Bands on the NR Plane

Three-dimensional (3D)-OSP studies were performed to quantitatively estimate the shear bands on the free surface of the NR plane, as the shear bands are visible on the free surface, which remained untouched during rolling. For example, a 3D-OSP image of shear bands formed on CR4.5 is shown in Figure 4b. The shear band spacing (λ) and the offset height (δ) were measured in each different CR sample by analyzing the 3D-OSP images, and their relative frequency distribution data shown in Figure 5a,b are very well fitted using the lognormal function, which has a regression coefficient of >0.9 in all cases. The shear band density increased with the increase of cold rolling percentage. It was observed that the λ value decreased and δ increased with the increase in the extent of cold rolling. The average value of shear-band spacing (λ_{avg}) was estimated to be 37.6 ± 17.1 µm, 23.3 ± 19.7 µm, 26.8 ± 15.6 µm, and 19.3 ± 12.1 µm for CR4.5, CR10, CR20, and CR31, respectively, whereas the average value of shear band height (δ_{avg}) was estimated to be 1.7 ± 1.2 µm, 2.6 ± 1.8 µm, 3.4 ± 0.9 µm, and 2.8 ± 2.1 µm for CR4.5, CR10, CR20, and CR31, respectively.

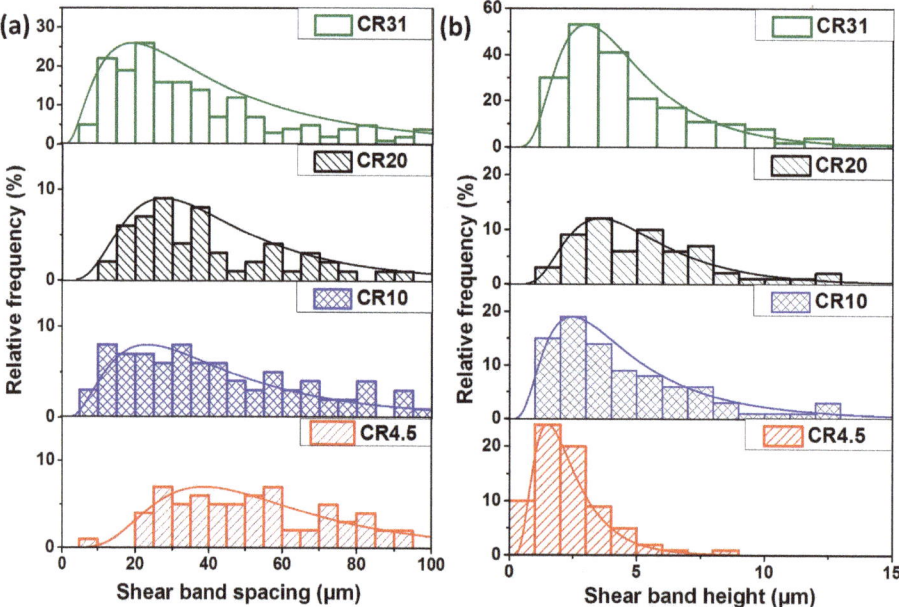

Figure 5. Histogram plot showing the distribution of (**a**) shear band spacing and (**b**) shear band offset height in CR specimens.

3.3. Nanoindentation Studies at Various P_{max} (50 mN–500 mN)

Figure 6a–e and Figures S1–S3 show the load (*P*) versus displacement (*h*) plots of CR0, CR4.5, CR10, CR20, and CR31 at different loading rates of 1.66, 3.33, 6.67, and 16.67 mN/s loaded up to P_{max} = 50 mN, 100 mN, 200 mN, and 500 mN, respectively, for NR, RW, and NW planes. The indentation hardness (*H*) was estimated using the following equation:

$$H = (P_{max}/A_c) \tag{3}$$

where A_c is the corrected contact area, which is equal to $C_1 h_c^2 + C_2 h_c + C_3 h_c^{1/2} + C_4 h_c^{1/4} + C_5 h_c^{1/8} + C_6 h_c^{1/16}$, and the C_n terms are constants, as described earlier [32,40]. The h_c values were corrected as $h_c = h_{max} - 0.75(P_{max}/S)$, where *S* is the unloading stiffness and

h_{max} is the maximum penetration depth. The elastic modulus (E) of the specimen was measured using the following equation:

$$\frac{1}{E_r} = \frac{1-\nu^2}{E} + \frac{1-\nu_i^2}{E_i} \qquad (4)$$

where the reduced modulus $E_r = 0.5\sqrt{\pi/A}\,(dp/dh)$, E_i is the modulus of the indenter, and ν_i is the Poisson's ratio of the indenter. The Poisson's ratio of Vitreloy 1 sample is $\nu = 0.37$ [41].

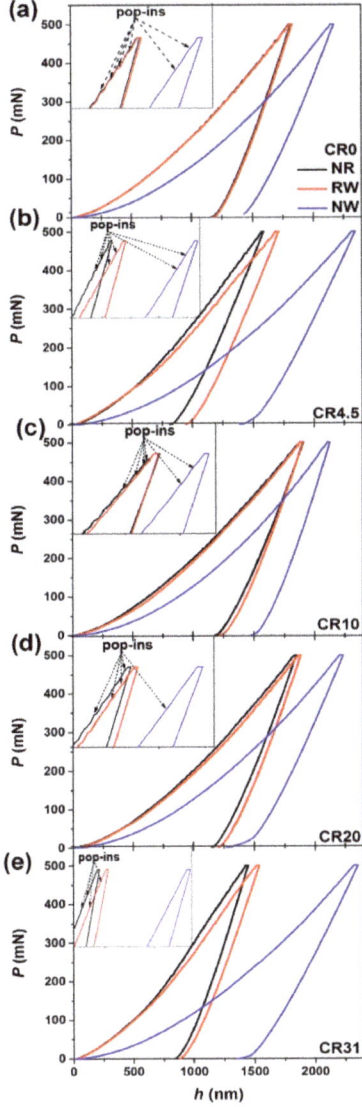

Figure 6. P-h plots at $P_{max} = 500$ mN for NR, RW, and NW planes: (**a**) CR0, (**b**) CR4.5, (**c**) CR10, (**d**) CR20, and (**e**) CR31; pop-in events are more pronounced in CR0, which gradually decreases with the increase of cold rolling percentage.

It was noticed that the h_{max} value for a given load is higher on the NW plane than that of the NR and RW planes. In addition, the h_{max} values are higher in CR10 and CR20, and lower in CR4.5 and CR31 for a given load than that of CR0 for the NR and RW planes, as shown in Figure 6a–e. Figure 7 displays the variation H and E with the cold rolling percentage for P_{max} of 50 mN, 100 mN, 200 mN, and 500 mN for the NR, RW, and NW planes. The estimated H values of CR0, CR4.5, CR10, CR20, and CR31 were 9.6 ± 0.1 GPa, 14.8 ± 0.3 GPa, 8.9 ± 0.2 GPa, 9.5 ± 0.3 GPa, and 15.9 ± 0.1 GPa, respectively, for the NR plane, whereas the H values of the RW plane were estimated to be 9.6 ± 2.1 GPa, 11.8 ± 0.5 GPa, 8.1 ± 0.2 GPa, 8.9 ± 0.6 GPa, and 13.9 ± 0.7 GPa for CR0, CR4.5, CR10, CR20, and CR31, respectively. Furthermore, the H values were estimated to be lower for the NW plane as 5.9 ± 0.2 GPa, 5.7 ± 0.1 GPa, 5.5 ± 0.1 GPa, 6.3 ± 0.3 GPa, and 5.5 ± 0.1 GPa in CR0, CR4.5, CR10, CR20, and CR31, respectively. The values of H and E in CR0, CR4.5, CR10, CR20, and CR31 at various P_{max} values in between 50 and 500 mN for the NR, RW, and NW planes are presented in Tables 1–3, respectively. Even though the hardness fluctuation on a particular plane at a given cold rolling percentage is small, as evident from the above hardness error values, the variation of H and E with cold rolling percentage on the NR and RW planes is significant and shows a similar trend at all P_{max} values, as shown in Figure 7a–d. The effect of cold rolling percentage on the H and E values is lesser on the NW plane than that of the NR and RW planes. Therefore, the glassy structure was modified upon cold rolling, leading to the variation of H and E values with the increase of cold rolling percentage in Vitreloy 1.

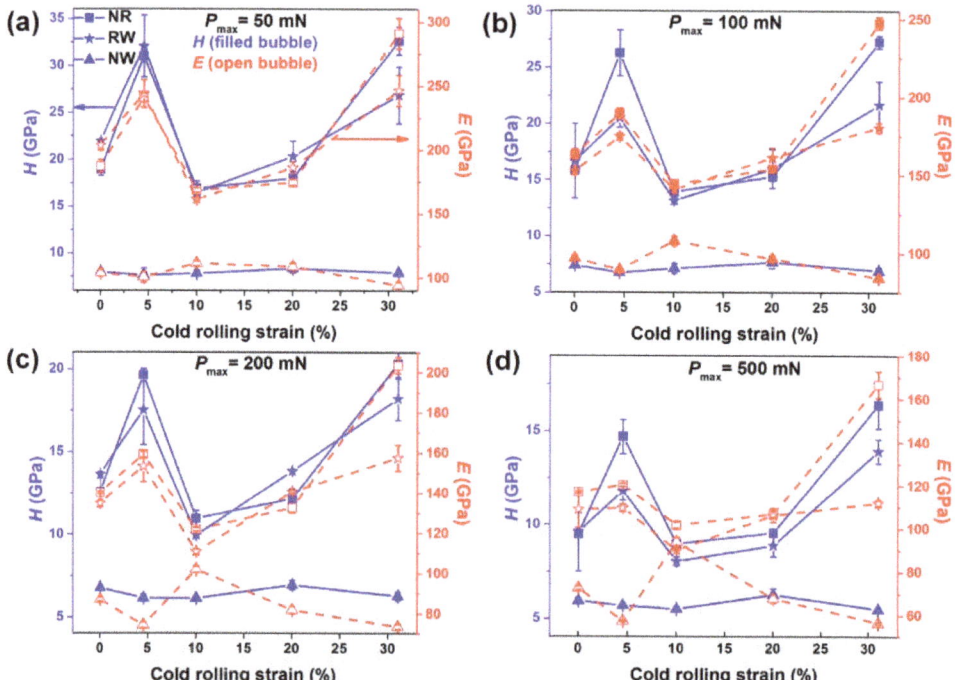

Figure 7. Hardness (H) and Young's modulus (E) variation with cold rolling percentage in the NR, RW, and NW planes at P_{max} of (**a**) 50 mN, (**b**) 100 mN, (**c**) 200 mN, and (**d**) 500 mN.

Table 1. H and E values of the NR plane in CR0, CR4.5, CR10, CR20, and CR31 at different P_{max} values in between 50 and 500 mN.

	H (GPa) NR				E (GPa) NR			
P_{Max} (mN)	50 mN	100 mN	200 mN	500 mN	50 mN	100 mN	200 mN	500 mN
CR0	18.9 ± 0.6	15.9 ± 0.1	12.6 ± 0.9	9.5 ± 0.1	187 ± 5	163 ± 3	140 ± 1	117 ± 1
CR4.5	31.0 ± 0.6	26.3 ± 2.0	19.6 ± 0.4	14.7 ± 0.9	241 ± 4	189 ± 4	159 ± 2	121 ± 1
CR10	16.9 ± 0.8	13.9 ± 0.5	10.9 ± 0.5	9.0 ± 0.1	169 ± 2	144 ± 2	122 ± 1	102 ± 2
CR20	18.0 ± 0.3	15.3 ± 0.4	12.2 ± 0.6	9.5 ± 0.1	175 ± 3	154 ± 1	133 ± 2	107 ± 2
CR31	32.6 ± 1.6	27.3 ± 0.6	20.3 ± 0.9	16.4 ± 1.2	291 ± 12	248 ± 4	204 ± 4	167 ± 6

Table 2. H and E values of the RW plane in the as-cast and differently cold rolled samples at different maximum loads of 50 mN, 100 mN, 200 mN, and 500 mN.

	H (GPa) RW				E (GPa) RW			
P_{Max} (mN)	50 mN	100 mN	200 mN	500 mN	50 mN	100 mN	200 mN	500 mN
CR0	21.9 ± 0.2	16.7 ± 3.3	13.8 ± 0.1	9.6 ± 2	204 ± 4	153 ± 3	135 ± 1.5	109 ± 8
CR4.5	32.1 ± 3.3	20.5 ± 0.8	17.5 ± 2.1	11.8 ± 0.5	244 ± 11	175 ± 2	153 ± 8	110 ± 2
CR10	16.5 ± 0.2	13.2 ± 0.3	9.9 ± 0.1	8.0 ± 0.2	161 ± 2	141 ± 3	111 ± 1	90 ± 0.3
CR20	20.3 ± 1.6	16.0 ± 1.8	13.6 ± 0.1	8.9 ± 0.6	186 ± 5	161 ± 6	141 ± 1	106 ± 3
CR31	26.8 ± 3.0	21.7 ± 2.1	18.2 ± 1.3	13.9 ± 0.7	246 ± 12	181 ± 3	158 ± 6	112 ± 1

Table 3. H and E values of the NW plane in CR0, CR4.5, CR10, CR20, and CR31 at various P_{max} values in between 50 and 500 mN.

	H (GPa) NW				E (GPa) NW			
P_{Max} (mN)	50 mN	100 mN	200 mN	500 mN	50 mN	100 mN	200 mN	500 mN
CR0	7.9 ± 0.3	7.4 ± 0.3	6.8 ± 0.1	5.9 ± 0.2	104 ± 2	97 ± 2	87 ± 1	73 ± 1
CR4.5	7.6 ± 0.7	6.8 ± 0.1	6.2 ± 0.2	5.7 ± 0.1	100 ± 4	90 ± 0.2	74 ± 2	58 ± 0.4
CR10	7.9 ± 0.3	7.1 ± 0.4	6.2 ± 0.1	5.5 ± 0.1	111 ± 2	108 ± 3	102 ± 2	94 ± 2
CR20	8.4 ± 0.5	7.7 ± 0.5	6.9 ± 0.3	6.3 ± 0.3	109 ± 3	96 ± 2	82 ± 2	68 ± 2
CR31	7.9 ± 0.3	6.9 ± 0.3	6.3 ± 0.3	5.5 ± 0.1	94 ± 3	84 ± 2	74 ± 1	57 ± 0.5

Figure 6 and Figures S1–S3 display the analogous nature of the P-h curves on the NR and RW planes, with more pop-in events in the case of CR0 owing to more trapped in free volume upon cooling than that of the interior of the NW plane, which exhibited a more relaxed glassy structure. Meanwhile, a more parabolic nature of P-h curves with fewer serrations was observed on the NW plane in CR0 with less pop-in depth, revealing a more homogeneous deformation in the NW plane than that of the NR and RW planes. The lower number of pop-in events during nanoindentation on the NW plane than on the NR and RW planes indicates a low number of shear band activation and their propagation. Such reduced pop-in events due to the cooling rate difference between the sample surfaces and interior portion of Zr-based BMG rods have been observed earlier [42]. Likewise, all the cold rolled samples have shown similar behavior on the different planes at a given cold rolling strain.

3.4. Studies on the Indentation Impression under SEM

Figure 8a shows the indentation impressions, which are located 70–100 μm away from each other on the surface of CR0. The indentation impression under P_{max} = 500 mN in between two shear bands in the case of CR10 and CR31 is shown in Figure 8b,c, respectively. It is worth mentioning that δ_{avg} varied between 1.7 μm (CR4.5) and 2.8 μm (CR31) for different CR specimens, exhibiting a much larger length scale than that the indentation impressions for the NR plane. As the location for indentation was chosen between the flat region between two shear bands, the error on the measurement of h_c and subsequently on H and E must be neglected. Furthermore, Figure 8d shows the indentation impression on CR0 at P_{max} = 500 mN, which shows the formation of very finely spaced shear bands formed around as well as beneath the impression, suggesting inhomogeneous deformation. Meanwhile, a reduction in the number of shear bands around and beneath the indentation impressions was noticed, as shown in Figure 8e,f. Such a disappearance becomes more pronounced with the increase of cold rolling extent, implying the evolution of homogeneous deformation in cold rolled samples.

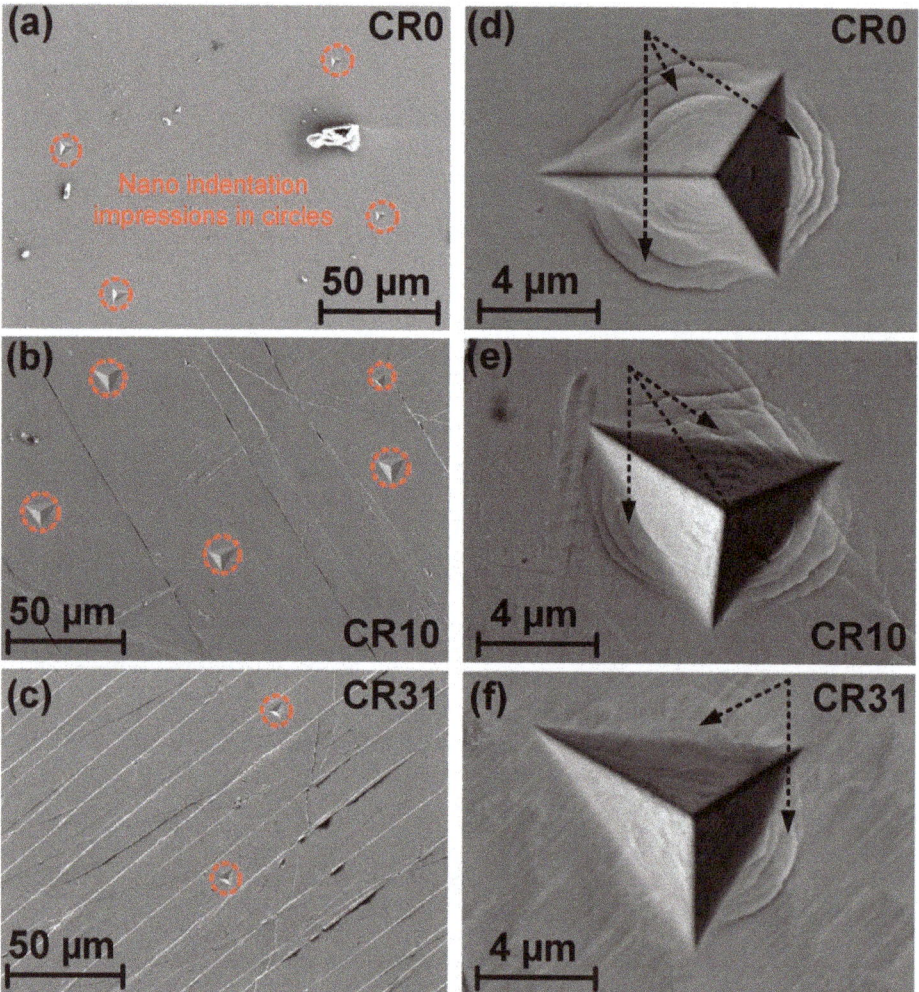

Figure 8. SEM secondary electron (SE) images showing nanoindentation marks in (**a**) CR0 along with evolved shear bands on the NR plane upon cold rolling in (**b**) CR10 and (**c**) CR31. The magnified image of indentation impression at P_{max} = 500 mN showing shear bands in (**d**) CR0, (**e**) CR10, and (**f**) CR31.

4. Discussion

4.1. Effect of Cold Rolling on Elastic-Plastic Response

The P-h plots of a specific sample for a given plane (either NR, RW, or NW) coincided with each other for all the P_{max} values in the range of 50–500 mN, which supports the repeatability and reproducibility of the experiments. The nature of elastic-plastic deformation beneath the indenter can be evaluated using the h_f/h_{max} ratio, which lies between 0 and 1 [43]. $h_f/h_{max} = 0$ implies a complete elastic deformation, whereas $h_f/h_{max} = 1$ points to a fully plastic deformation. Figure 9 shows the h_f/h_{max} ratio at different P_{max} of all the CR specimens for the NR, RW, and NW planes. The CR0 and different CR specimens showed $h_f/h_{max} < 0.7$, indicating the deformation to be elastic-perfectly plastic. The effect of cold rolling percentage on h_f/h_{max} exhibits a similar trend to that of H variation with cold rolling, as depicted in Figure 7a–d, whereas h_f/h_{max} values lay between 0.6 and 0.7, depicting a more plastic nature in the NW plane than that for the NR and RW planes for all samples. Furthermore, more plastic behavior was observed in CR4.5 and CR31 than that of CR0, CR10, and CR20 for the RW and NR planes. Pharr et al. have shown, using experimental and finite element simulations, that pileup is not a significant factor when $h_f/h_{max} < 0.7$ and, hence, the Oliver–Pharr data analysis procedure can give reasonable results [32]. In the case of $h_f/h_{max} > 0.7$, the pileup can result in wrong estimation of H owing to an erroneous contact area deduced from indentation load-displacement data [44,45]. In the present study, h_f/h_{max} values < 0.7, indicating calculated H values, are realistic in nature for all samples for the NR, RW, and NW planes.

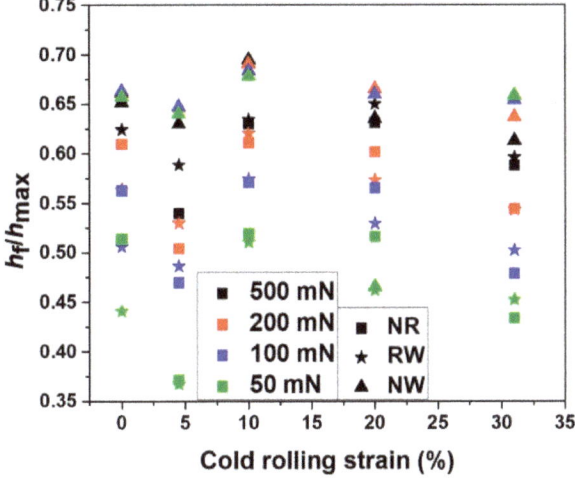

Figure 9. The h_f/h_{max} ratio with cold rolling percentage for the NR, RW, and NW planes at P_{max} of 50 mN, 100 mN, 200 mN, and 500 mN.

The affected volume during nanoindentation was estimated by $(3*h_f)^3$ for all the specimens at all P_{max} for NR, RW, and NW, as shown in Table 4 [46]. The affected volume during nanoindentation is more than 0.4^3 µm^3, 0.7^3 µm^3, 1.3^3 µm^3, and 2.6^3 µm^3 for P_{max} of 50 mN, 100 mN, 200 mN, and 500 mN, respectively, for the NR and RW planes for all the samples. The hardened CR4.5 and CR31 show less affected volume during indentation than the CR0 and softened CR10 and CR20. The affected volume during nanoindentation in the central core of CR specimen on NW plane is estimated to be $>1.2^3$ µm^3 at all P_{max} values. Such an affected volume during nanoindentation validates a homogeneous deformation and corroborates the measured H and E values. A schematic potential energy landscape (PEL) model illustrates the available energy for glasses state: the lowest energy minima configuration for crystal representing corresponding atomic configurations or inherent states, as shown in Figure 10. No compositional variation was noticed on the three perpendicular planes of the specimens; therefore, the variations of H and E are linked to the evolution of different structural states only. Such structural fluctuations may evolve during synthesis owing to cooling rate differences between surface planes (NR and RW) and the interior core of the NW plane. NR and RW planes have shown similar H and

E values, representing a similar glassy structure to that of the interior NW plane, which exhibited lower H and E values. Such an observation of different structural state formation was also reported in BMGs owing to cooling rate differences between the surface and core region of the sample [42]. Even though no compositional variation was observed in the specimens [33], the variation of H and E are linked to the different structural states that evolved upon cold rolling. Therefore, the evolution of structural fluctuations must have occurred with the increase of cold rolling strain on the NR and RW planes, which resulted in hardening in CR4.5 and CR31, whereas softening was observed in CR10 and CR31, as shown in Figure 4. Therefore, H and E values on the NW plane have shown only slight variation with increasing cold rolling strain indicating less fluctuations of the structural states.

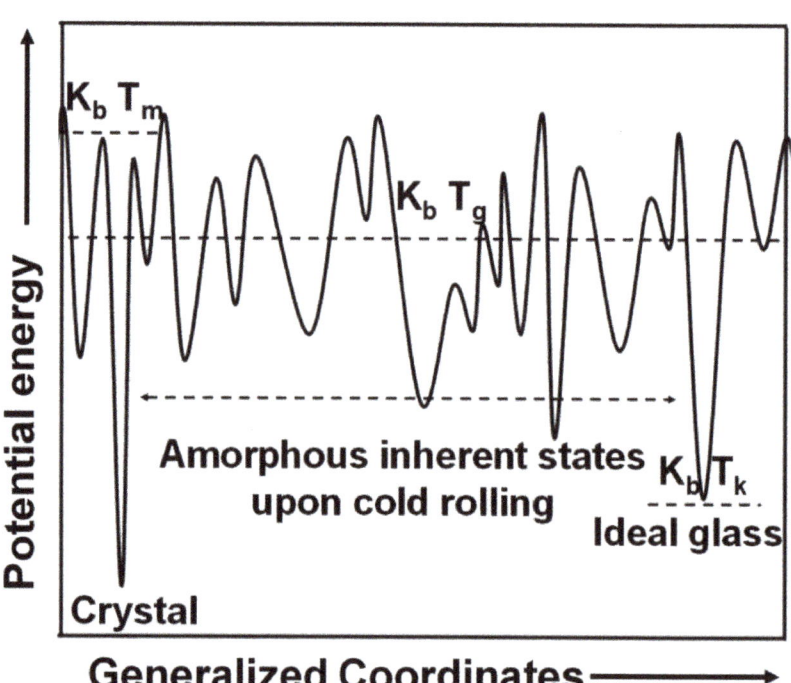

Figure 10. Schematic potential energy landscape (PEL) model illustrating different metastable energy states in the glassy phase.

Table 4. The affected deformed volume in CR specimens during nanoindentation on the NR, RW, and NW planes at different P_{max}.

	Affected Volume ($\mu m)^3$											
	NR				RW				NW			
P_{max} (mN)	50	100	200	500	50	100	200	500	50	100	200	500
CR0	0.6^3	1.0^3	1.8^3	3.6^3	0.5^3	0.9^3	1.7^3	3.3^3	1.2^3	1.8^3	2.6^3	4.3^3
CR4.5	0.4^3	0.8^3	1.3^3	2.6^3	0.4^3	0.8^3	1.4^3	3.1^3	1.2^3	1.7^3	2.6^3	4.4^3
CR10	0.6^3	1.1^3	2.0^3	3.6^3	0.6^3	1.2^3	2.1^3	3.8^3	1.2^3	1.8^3	2.7^3	4.5^3
CR20	0.6^3	1.1^3	1.8^3	3.5^3	0.5^3	1.0^3	1.7^3	3.7^3	1.2^3	1.8^3	2.7^3	4.2^3
CR31	0.4^3	0.7^3	1.3^3	2.6^3	0.4^3	0.8^3	1.4^3	2.8^3	1.3^3	1.8^3	2.6^3	4.3^3

4.2. Serrated to Smooth Flow on NR, RW, and NW Planes upon Cold Rolling

Strain bursts or pop-in events during loading portion of the P-h curve were observed owing to the nucleation of shear bands and/or their propagation, whereas elbow nature at the end part of the unloading P-h curve was also observed. The elbow-like step in the P-h curve is due to phase transition or softer regions or grain boundaries or pores in crystalline specimens. Such "pop-in" events were observed to be more in CR0 and a smoother flow curve was achieved in CR specimens, as depicted in Figure 6 at P_{max} = 500 mN for the NR, RW, and NW planes. A similar observation was made for all other P_{max} values, as depicted in Figures S1–S3. In addition, P-h curves exhibited more serration behavior for the NR and RW planes in CR0 and all CR samples than that for the NW plane at all P_{max} values.

The plastic event in a glassy phase is associated with the evolution of two different spatial regions, i.e., the shear transformation zones (STZs) and the residual glassy matrix [13]. The macroscopic plastic events proceed with the nucleation and proliferation of new shear bands. Such a phenomenon can be explained from the viewpoint of the percolation of the SRO and the atomic rearrangements in the glassy structure. If the local $\dot{\varepsilon}$ reaches a critical value, then the atomic rearrangements cannot further accommodate the applied strain. Hence, the differences of strain rates among the surrounding matrix and STZs would still remain, which results an inhomogeneous (non-Newtonian) flow and serrations during nanoindentation as reflected in the P–h curves, as shown for CR0 in Figure 6. The nature of the P-h curves for NR and RW planes have changed from ideal stair step-like behavior for CR0 and CR4.5 (lower cold rolling reductions) to a more parabolic nature with increasing cold rolling extent for all loading rates in the NR, RW, and NW planes, pointing to plastic deformation being more homogeneous upon cold rolling for all planes. Similar behavior of reduced pop-in events with a cooling rate difference between surfaces to the interior portion of the transverse direction was observed in Zr-based BMG rods [42]. Such features were supported by the disappearance of shear bands around the indentation impressions in CR31, as shown in Figure 8f. The reduction of the shear bands around the indentation impressions in CR samples also confirms the homogeneous deformation behavior upon cold rolling. Similarly, high-pressure torsion (HPT) processed BMGs also showed more homogeneous deformation [47].

4.3. Effect of Cold Rolling on Structure, H and E

The deformation of BMGs could be understood by the free volume theory as proposed by Spaepen, and Argon has suggested the STZ model involving clusters of atoms undergoing cooperative shear displacements [11,12,14]. The shear stress and indentation hardness are correlated as follows: $H \approx 3\sqrt{3}\tau$. According to the free volume model, the evolved shear strain rate ($\dot{\gamma}$) is influenced by the temperature (T), applied shear stress (τ), free volume accumulation, and viscosity (η), which is defined by the ratio of shear stress and shear strain rate as follows [11,14]:

$$\dot{\gamma} = 2c_f \alpha_o k_{f,O} \frac{\varepsilon_o v_o}{\Omega} \sinh\left(\frac{\tau \varepsilon_o v_o}{2k_B}\right) \exp\left(\frac{-\Delta G}{k_B T}\right) \tag{5}$$

$$\eta = \frac{\tau}{\dot{\gamma}} = \frac{\tau}{2f \sinh(\tau\Omega/2k_B T)} \cdot \exp\left(\frac{\chi v^*}{v_f} + \frac{\Delta G^m}{k_B T}\right) \tag{6}$$

$$H = \frac{6\sqrt{3}k_B T}{\Omega} \sinh^{-1}\left[\frac{\dot{\gamma}}{2f \cdot \gamma_0} \cdot \exp\left(\frac{\chi v^*}{v_f} + \frac{\Delta G^m}{k_B T}\right)\right] \tag{7}$$

where Ω indicates the atomic volume, α_o is a coefficient featuring the quantity of the material undergoing shear transformation, $\varepsilon_o v_o$ denotes a flow event activation volume, τ is the shear stress, k_B is the Boltzmann constant, ΔG is the activation energy for defect migration, and $k_{f,o}$ is associated with the Debye frequency.

The atomic radii of the elements Zr, Ti, Cu, Ni, and Be are 155 pm, 140 pm, 135 pm, 135 pm, and 112 pm, respectively, and the individual atomic volume of these elements was estimated to be 0.0156, 0.115, 0.0103, 0.0103, and 0.00588 nm³, respectively. Therefore, the average atomic volume (i.e., Voronoi volume) in Vitreloy 1 is 0.01165 nm³ [48]. The activation volume of the flow event ($\varepsilon_o v_o$) was estimated using nanoindentation data. The strain rate sensitivity (m) was estimated using the log(σ)–log($\dot{\varepsilon}$) plot, where $m = \frac{d\ln(H)}{d\ln(\dot{\varepsilon})}$, activation volume $V = \frac{3\sqrt{3}(k_B)T}{mH}$, H is the indentation hardness at temperature T, and k_B is the Boltzmann constant. Assuming the tip radius of the Berkovich indenter is 25 nm, which remains constant for all experiments, τ_{max} was estimated as $\tau_{max} = \frac{0.47PR}{h}$. By considering ΔG, α_o, k_B, T, and $k_{f,o}$ to be constant, the $\frac{\dot{\gamma}}{\gamma_0}$ ratio was estimated to be 1, 0.93, 0.30, 0.71, and 0.61, for CR0, CR4.5, CR10, CR20, and CR31, respectively. The variation of $\dot{\gamma}$ indicates a change in the flow defect concentration, which will either increase or decrease the free volume content in order to accommodate the applied strain into the glassy phase. The application of shear stress during cold rolling alters PEL by reducing or eliminating the barrier to a neighboring/different metastable state [33]. Hence, the evolved $\dot{\gamma}$ in the CR specimens largely differs from that of as-cast BMG, further confirming the modification of the SRO cluster upon cold rolling.

As insignificant variation of H and E values with cold rolling percentage was noticed at the central core of the specimens on the NW plane, Spaepen's free volume model was adopted to understand the observed behavior for the NR and RW planes. According to Spaepen's model, the rate of free volume accumulation and annihilation can be estimated using the following equation [11].

$$\left(\frac{dx}{dt}\right)_+ = \frac{f}{\gamma}\exp\left(-\frac{\Delta G^m}{kT}\right)\exp\left(-\frac{1}{x}\right)\left[\cosh\left(\frac{\tau\Omega}{2kT}\right) - 1\right]\frac{2kT}{xSv^*} \quad (8)$$

$$\left(\frac{dx}{dt}\right)_- = -fx^2\left[\exp\left(-\frac{1}{x}\right) - \exp\left(-\frac{1}{x_{eq}}\right)\right]\exp\left(-\frac{\Delta G^m - \sigma_m V}{kT}\right) \quad (9)$$

where $S = \frac{2\mu(1+\vartheta)}{3(1-\vartheta)}$, μ is shear modulus, ϑ is Poisson's ratio, ΔG is the activation energy, f is an attempt frequency, k is Boltzmann's constant, Ω is atomic volume, v^* is critical volume, γ is the geometrical factor, and the mean stress $\sigma_m = 0$ during rolling operation. The values of the above parameters are $T = 300$ K, $\gamma = 0.15$, $f = 5.415 \times 10^{12}$ s^{-1}, $\Omega = 2.424 \times 10^{-29}$, $\mu = 35.3$ GPa, $v^* = 1.67 \times 10^{-29}$ m³, $\Delta G^m = 10^{-19}$ J, and $\vartheta = 0.37$ [48]. The values of activation volume (V) and free volume (x and x_{eq}) were obtained from the nanoindentation and DSC studies, respectively. Equation (8) indicates that the free volume accumulation depends on the applied shear stress, whereas Equation (9) indicates that free volume annihilation is linked with the diffusional atomic movement, which is further accelerated by the applied stress. Our calculation suggests that the free volume accumulation rate is 2.29×10^{-19} (CR0), 6.74×10^{-20} (CR4.5), 4.15×10^{-6} (CR10), 2.20×10^{-14} (CR20), and 6.42×10^{-20} (CR31). On the other hand, the free volume annihilation rates in CR0, CR4.5, CR10, CR20, and CR31 were estimated to be 5.09×10^{-20}, 1.80×10^{-15}, 5.56×10^{-20}, 1.888×10^{-18}, and 2.98×10^{-19}, respectively. Plastic deformation in BMGs occurs inhomogeneously by forming shear bands at temperature <T_g and the coalescence of the free volume occurs gradually; therefore, the BMGs exhibit a dynamic response during nanoindentation.

Usually, the microhardness of BMGs decreases with prior deformation, as observed in Pd-base, Zr-base, and Cu-base glassy alloys [12,21,33,49]. Several authors have also reported an increase in hardness in BMGs upon deformation [13,28–32,34]. Furthermore, the increase of the yield strength of Vitreloy 1 upon cold rolling or laser shock peening was reported by Lee et al. and Cao et al., respectively [30,34,50]. The rise and fall of average free volume content with increasing cold rolling percentage were linked to the competition between the average free volume annihilation rate and creation rate, as studied using positron annihilation spectroscopy and DSC studies [33,35]. Furthermore, Pan and co-workers have proved that work hardening occurred as a result of densification in tensile pre-strained Zr-based BMGs using micro and nanoindentation, when free volume annihilation rate dominated the free creation rate [31,51].

Equations (5) and (6) indicate that the viscosity and indentation hardness decrease with the increase of free volume content. Turnbull and Cohen suggested that the excess free volume in MGs decreases the atomic bonding energy by increasing the average interatomic distances [52]. As the value of E is linked to the atomic bond strength, the increase of the E in CR4.5 and CR31 points to an increase of the bond strength compared with that of CR0. Meanwhile, the decrease of E in CR10 and CR20 points to a decrease in the bond strength compared with that of CR0 [53,54]. Thus, short range ordered (SRO) clusters in the glassy phase must have been modified into a newer atomic configuration metastable state upon cold rolling in the PEL model, as proposed by us [33]. In the case of CR10 and CR20, the decreas of the hardness is linked to a higher free volume content, as reported earlier [12,21,49]. Hence, the synergetic effect of free volume accumulation and annihilation during cold rolling is responsible for producing a relaxed glassy structure in CR4.5 and CR31 and a dilated glassy structure in CR10 and CR20, which eventually resulted in enhanced hardness or reduced hardness in those specimens, respectively. Figure 11a illustrates the schematic of the atomic cluster in an as-cast glassy phase, which is modified as a result of the alteration of the free volume content and subsequent dilation or densification at the later stage of cold rolling, as shown in Figure 11b,c, respectively. The relaxed glassy phase in CR4.5 and CR31 exhibited densification-induced hardening during nanoindentation. However, such deformation-induced densification was also reported earlier in the case of silicate and polymeric glasses [55–57]. Cold rolling induced structural relaxation, accompanied by reduced free volume content, as reported by us earlier [34]. Thus, the free volume accumulation and annihilation play crucial roles during cold rolling in Vitreloy 1.

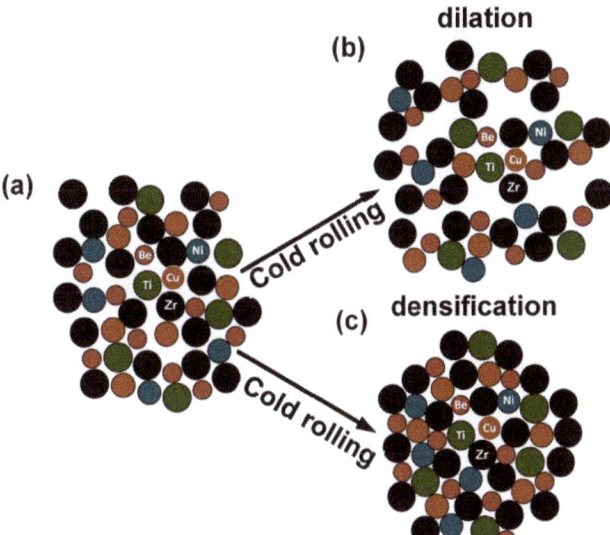

Figure 11. Schematic illustration of (a) the atomic arrangement in the glassy structure, (b) local dilation causing enhanced free volume content, as well as (c) densification during cold rolling.

5. Conclusions

The following conclusions were drawn from the present study:

- The hardness at the specimen surface varied with cold rolling percentage (%) and the variation is similar on the RW and NR planes at all different peak loads in the range of 50 mN–500 mN, whereas the same is insignificant for the core region of the specimen

on the NW plane. H and E values increase in CR4.5 and CR31, whereas CR10 and CR20 become softer than CR0 upon cold rolling on the RW and NR planes.
- 3D optical surface profilometry studies on the NR plane suggest that the shear band spacing decreases from 57.7 μm (CR4.5) to 24.6 μm (CR31), and the shear band offset height increases from 2.4 μm (CR4.5) to 4.4 μm (CR31) with the increase in the extent of cold rolling. Meanwhile, the number of pop-in events during nanoindentation for all the planes reduces with the increase in the extent of cold rolling, and the disappearance of shear bands around indentation impression points to more homogeneous deformation in cold rolled BMGs.
- The nanoindentation, DSC studies, and model calculations suggest that the annihilation rate dominated over the free volume creation rate, which produce a relaxed and dense glassy structure in CR4.5 and CR31, exhibiting enhanced H and E.

Supplementary Materials: The following are available online at https://www.mdpi.com/article/10.3390/nano11071670/s1, Figure S1: P-h plots at P_{max} = 200 mN along along NR, RW and NW; Figure S2: P-h plots at P_{max} = 100 mN along along NR, RW and NW; Figure S3: P-h plots at P_{max} = 50 mN along NR, RW and NW.

Author Contributions: Conceptualization, A.G. and J.D.; validation, A.G. and P.P.J.; formal analysis, A.G.; investigation, A.G. and P.P.J.; data curation, A.G. and P.P.J.; writing—original draft preparation, A.G.; writing—review and editing, M.-H.L. and J.D.; supervision, J.D.; project administration, J.D.; funding acquisition, M.-H.L. and J.D. All authors have read and agreed to the published version of the manuscript.

Funding: This research was funded by SRIC IIT Kharagpur, grant number SGIRG and the APC was funded by Korea Institute of Industrial Technology (KITECH), grant number PJE21030.

Data Availability Statement: The data presented in this study are available in this article.

Acknowledgments: The authors thank Central Research Facility at IIT Kharagpur for providing experimental facilities.

Conflicts of Interest: The authors declare no conflict of interest. The funders had no role in the design of the study; in the collection, analyses, or interpretation of data; in the writing of the manuscript, or in the decision to publish the results.

References

1. Suryanarayana, C.; Inoue, A. *Bulk Metallic Glasses*; CRC Press: New York, NY, USA, 2011.
2. Kumar, G.; Neibecker, P.; Liu, Y.H.; Schroers, J. Critical fictive temperature for plasticity in metallic glasses. *Nat. Commun.* **2013**, *4*, 1536. [CrossRef] [PubMed]
3. Egami, T. Formation and deformation of metallic glasses: Atomistic theory. *Intermetallics* **2006**, *14*, 882–887. [CrossRef]
4. Sinning, H.R.; Leonardsson, L.; Cahn, R.W. Irreversible anisotropic length changes in fe40ni40b20 and a search for reversible length changes in several metallic glasses. *Int. J. Rapid Solidif.* **1985**, *1*, 175–197.
5. Poulsen, H.F.; Wert, J.A.; Neuefeind, J.; Aki, V.H.; Daymond, M. Measuring strain distributions in amorphous materials. *Nat. Mater.* **2004**, *4*, 33–36. [CrossRef]
6. Ott, R.T.; Kramer, M.J.; Besser, M.F.; Sordelet, D.J. High-energy X-ray measurements of structural anisotropy and excess free volume in a homogenously deformed Zr-based metallic glass. *Acta Mater.* **2006**, *54*, 2463–2471. [CrossRef]
7. Dmowski, W.; Tong, Y.; Iwashita, T.; Yokoyama, Y.; Egami, T. Universal mechanism of thermomechanical deformation in metallic glasses. *Phys. Rev. B* **2015**, *91*, 60101. [CrossRef]
8. Shahabi, H.S.; Scudino, S.; Kaban, I.; Stoica, M.; Rütt, U.; Kühn, U.; Eckert, J. Structural aspects of elasto-plastic deformation of a Zr-based bulk metallic glass under uniaxial compression. *Acta Mater.* **2015**, *95*, 30–36. [CrossRef]
9. Revesz, A.; Schafler, E.; Kovacs, Z. Structural anisotropy in a $Zr_{57}Ti_5Cu_{20}Al_{10}Ni_8$ bulk metallic glass deformed by high pressure torsion at room temperature. *Appl. Phys. Lett.* **2008**, *92*, 011910. [CrossRef]
10. Dmowski, W.; Egami, T. Structural anisotropy in metallic glasses induced by mechanical deformation. *Adv. Eng. Mater.* **2008**, *10*, 1003–1007. [CrossRef]
11. Spaepen, F. A microscopic mechanism for steady state inhomogeneous flow in metallic glasses. *Acta Metall.* **1977**, *25*, 407–415. [CrossRef]
12. Heggen, M.; Spaepen, F.; Feuerbacher, M. Creation and annihilation of free volume during homogeneous flow of a metallic glass. *J. Appl. Phys.* **2005**, *97*, 1–8. [CrossRef]
13. Schuh, C.A.; Hufnagel, T.C.; Ramamurty, U. Mechanical behavior of amorphous alloys. *Acta Mater.* **2007**, *55*, 4067–4109. [CrossRef]

14. Argon, A.S. Plastic deformation in metallic glasses. *Acta Metall.* **1979**, *27*, 47–58. [CrossRef]
15. Steif, P.S.; Spaepen, F.; Hutchinson, J.W. Strain localization in amorphous metals. *Acta Metall.* **1982**, *30*, 447–455. [CrossRef]
16. van den Beukel, A.; Sietsma, J. The glass transition as a free volume related kinetic phenomenon. *Acta Metall. Mater.* **1990**, *38*, 383–389. [CrossRef]
17. Schuh, C.A.; Lund, A.C. Atomistic basis for the plastic yield criterion of metallic glass. *Nat. Mater.* **2003**, *2*, 449–452. [CrossRef] [PubMed]
18. Furukawa, A.; Tanaka, H. Inhomogeneous flow and fracture of glassy materials. *Nat. Mater.* **2009**, *8*, 601–609. [CrossRef]
19. Spaepen, F. Must shear bands be hot? *Nat. Mater.* **2006**, *8*, 601–609. [CrossRef]
20. de Hey, P.; Sietsma, J.; van den Beukel, A. Structural Disordering In Amorphous $Pd_{40}Ni_{40}P_{20}$ Induced By High Temperature Deformation. *Acta Mater.* **1998**, *46*, 5873–5882. [CrossRef]
21. Bei, H.; Xie, S.; George, E.P. Softening caused by profuse shear banding in a bulk metallic glass. *Phys. Rev. Lett.* **2006**, *96*, 105503. [CrossRef]
22. Pan, J.; Chen, Q.; Liu, L.; Li, Y. Softening and dilatation in a single shear band. *Acta Mater.* **2011**, *59*, 5146–5158. [CrossRef]
23. Jian, W.R.; Wang, L.; Yao, X.H.; Luo, S.N. Tensile and nanoindentation deformation of amorphous/crystalline nanolaminates: Effects of layer thickness and interface type. *Comput. Mater. Sci.* **2018**, *154*, 225–233. [CrossRef]
24. Jian, W.R.; Wang, L.; Li, B.; Yao, X.H.; Luo, S.N. Improved ductility of $Cu_{64}Zr_{36}$ metallic glass/Cu nanocomposites via phase and grain boundaries. *Nanotechnology* **2016**, *27*, 175701. [CrossRef] [PubMed]
25. Deng, C.; Schuh, C.A. Atomistic mechanisms of cyclic hardening in metallic glass. *Appl. Phys. Lett.* **2012**, *100*, 251909. [CrossRef]
26. Zhao, D.; Zhu, B.; Wang, S.; Niu, Y.; Xu, L.; Zhao, H. Effects of pre-strain on the nanoindentation behaviors of metallic glass studied by molecular dynamics simulations. *Comput. Mater. Sci.* **2021**, *186*, 110073. [CrossRef]
27. Das, J.; Tang, M.B.; Kim, K.B.; Theissmann, R.; Baier, F.; Wang, W.H.; Eckert, J. "Work-Hardenable" Ductile Bulk Metallic Glass. *Phys. Rev. Lett.* **2005**, *94*, 205501. [CrossRef]
28. Kim, K.B.; Das, J.; Venkataraman, S.; Yi, S.; Eckert, J. Work hardening ability of ductile $Ti_{45}Cu_{40}Ni_{7.5}Zr_5Sn_{2.5}$ and $Cu_{47.5}Zr_{47.5}Al_5$ bulk metallic glasses. *Appl. Phys. Lett.* **2006**, *89*, 071908. [CrossRef]
29. Lee, M.H.; Lee, K.S.; Das, J.; Thomas, J.; Kühn, U.; Eckert, J. Improved plasticity of bulk metallic glasses upon cold rolling. *Scr. Mater.* **2010**, *62*, 678–681. [CrossRef]
30. Packard, C.E.; Witmer, L.M.; Schuh, C.A. Hardening of a metallic glass during cyclic loading in the elastic range. *Appl. Phys. Lett.* **2008**, *92*, 171911. [CrossRef]
31. Wang, Z.T.; Pan, J.; Li, Y.; Schuh, C.A. Densification and strain hardening of a metallic glass under tension at room temperature. *Phys. Rev. Lett.* **2013**, *111*, 135504. [CrossRef]
32. Oliver, W.C.; Pharr, G.M. An improved technique for determining hardness and elastic modulus using load and displacement sensing indentation experiments. *J. Mater. Res.* **1992**, *7*, 1564–1583. [CrossRef]
33. Jana, P.P.; Gunti, A.; Das, J. Improvement of intrinsic plasticity and strength of $Zr_{55}Cu_{30}Ni_5Al_{10}$ metallic glass by tuning the glass transition temperature. *Mater. Sci. Eng. A* **2019**, *762*, 138102. [CrossRef]
34. Slipenyuk, A.; Eckert, J. Correlation between enthalpy change and free volume reduction during structural relaxation of $Zr_{55}Cu_{30}Al_{10}Ni_5$ metallic glass. *Scr. Mater.* **2004**, *50*, 39–44. [CrossRef]
35. Flores, K.M.; Sherer, E.; Bharathula, A.; Chen, H.; Jean, Y.C. Sub-nanometer open volume regions in a bulk metallic glass investigated by positron annihilation. *Acta Mater.* **2007**, *55*, 3403–3411. [CrossRef]
36. Launey, M.E.; Kruzic, J.J.; Li, C.; Busch, R. Quantification of free volume differences in a $Zr_{44}Ti_{11}Ni_{10}Cu_{10}Be_{25}$ bulk amorphous alloy. *Appl. Phys. Lett.* **2007**, *91*, 051913. [CrossRef]
37. Masuhr, A.; Busch, R.; Johnson, W.L. Thermodynamics and Kinetics of the $Zr_{41.2}Ti_{13.8}Cu_{10}Ni_{12.5}Be_{22.5}$ Bulk Metallic Glass Forming Liquid: Glass Formation from a Strong Liquid. *J. Non Cryst. Solids.* **1999**, *250–252*, 566–571. [CrossRef]
38. Cao, Q.P.; Li, J.F.; Zhou, Y.H.; Jiang, J.Z. Microstructure and microhardness evolutions of $Cu_{47.5}Zr_{47.5}Al_5$ bulk metallic glass processed by rolling. *Scr. Mater.* **2008**, *59*, 673–676. [CrossRef]
39. Liu, J.W.; Cao, Q.P.; Chen, L.Y.; Wang, X.D.; Jiang, J.Z. Shear band evolution and hardness change in cold-rolled bulk metallic glasses. *Acta Mater.* **2010**, *58*, 4827–4840. [CrossRef]
40. Maity, T.; Roy, B.; Das, J. Mechanism of lamellae deformation and phase rearrangement in ultrafine β-Ti/FeTi eutectic composites. *Acta Mater.* **2015**, *97*, 170–179. [CrossRef]
41. Wei, Y.; Lei, X.; Huo, L.S.; Wang, W.H.; Greer, A.L. Towards more uniform deformation in metallic glasses: The role of Poisson's ratio. *Mater. Sci. Eng. A* **2013**, *560*, 510–517. [CrossRef]
42. Plummer, J.D.; Goodall, R.; Figueroa, I.A.; Todd, I. A study of mechanical homogeneity in as-cast bulk metallic glass by nanoindentation. *J. Non Cryst. Solids.* **2011**, *357*, 814–819. [CrossRef]
43. Bolshakov, A.; Pharr, G.M. Influences of pileup on the measurement of mechanical properties by load and depth sensing indentation techniques. *J. Mater. Res.* **1998**, *13*, 1049–1058. [CrossRef]
44. Tsui, T.Y.; Oliver, W.C.; Pharr, G.M. Influences of stress on the measurement of mechanical properties using nanoindentation: Part I. Experimental studies in an aluminum alloy. *J. Mater. Res.* **1996**, *11*, 752–759. [CrossRef]
45. Bolshakov, A.; Oliver, W.C.; Pharr, G.M. Influences of stress on the measurement of mechanical properties using nanoindentation: Part II. Finite element simulations. *J. Mater. Res.* **1996**, *11*, 760–768. [CrossRef]
46. Nemecek, J. (Ed.) *Nanoindentation in Materials Science*; BoD—Books on Demand: Rijeka, Croatia, 2012.

47. Ebner, C.; Pauly, S.; Eckert, J.; Rentenberger, C. Effect of mechanically induced structural rejuvenation on the deformation behaviour of CuZr based bulk metallic glass. *Mater. Sci. Eng. A* **2020**, *773*, 138848. [CrossRef]
48. Stachurski, Z.H. *Fundamentals of Amorphous Solids: Structure and Properties*; John Wiley-VCH: Verlag, Germany, 2014.
49. Li, N.; Chen, Q.; Liu, L. Size dependent plasticity of a Zr-based bulk metallic glass during room temperature compression. *J. Alloys Compd.* **2010**, *493*, 142–147. [CrossRef]
50. Cao, Y.; Xie, X.; Antonagalia, J.; Winiarski, B.; Wang, G.; Shin, Y.C.; Withers, P.J.; Dahmen, K.A.; Liaw, P.K. Laser shock peening on Zr-based bulk metallic glass and its effect on plasticity: Experiment and modelling. *Sci. Rep.* **2015**, *5*, 10789. [CrossRef]
51. Pan, J.; Wang, Y.X.; Guo, Q.; Zhang, D.; Greer, A.L.; Li, Y. Extreme rejuvenation and softening in a bulk metallic glass. *Nat. Commun.* **2018**, *9*, 560–569. [CrossRef]
52. Turnbull, D.; Cohen, M.H. On the free-volume model of the liquid-glass transition. *J. Chem. Phys.* **1970**, *52*, 3038–3041. [CrossRef]
53. Yang, F.; Geng, K.; Liaw, P.K.; Fan, G.; Choo, H. Deformation in a $Zr_{57}Ti_5Cu_{20}Ni_8Al_{10}$ bulk metallic glass during nanoindentation. *Acta Mater.* **2007**, *55*, 321–327. [CrossRef]
54. Liu, L.; Chan, K.C. Plastic deformation of Zr-based bulk metallic glasses under nanoindentation. *Mater. Lett.* **2005**, *59*, 3090–3094. [CrossRef]
55. Bridgman, P.W.; Simon, I. Effects of Very High Pressures on Glass. *J. Appl. Phys.* **1953**, *24*, 405–413. [CrossRef]
56. Rouxel, T.; Ji, H.; Hammouda, T.; Moréac, A. Poisson's ratio and the densification of glass under high pressure. *Phys. Rev. Lett.* **2008**, *100*, 225501. [CrossRef] [PubMed]
57. Drozdov, A.D. Stress-induced densification of glassy polymers in the subyield region. *J. Appl. Polym. Sci.* **1999**, *74*, 1705–1718. [CrossRef]

Article

Experimental, Theoretical and Simulation Studies on the Thermal Behavior of PLA-Based Nanocomposites Reinforced with Different Carbonaceous Fillers

Giovanni Spinelli [1,*], Rosella Guarini [1], Rumiana Kotsilkova [1], Evgeni Ivanov [1,2] and Vittorio Romano [3]

[1] Open Laboratory on Experimental Micro and Nano Mechanics (OLEM), Institute of Mechanics, Bulgarian Academy of Sciences, Acad. G. Bonchev Str. Block 4, 1113 Sofia, Bulgaria; rgrosagi@gmail.com (R.G.); kotsilkova@yahoo.com (R.K.); ivanov_evgeni@yahoo.com (E.I.)
[2] Research and Development of Nanomaterials and Nanotechnologies (NanoTech Lab Ltd.), Acad. G. Bonchev Str. Block 4, 1113 Sofia, Bulgaria
[3] Department of Industrial Engineering, University of Salerno, Via Giovanni Paolo II, 132, 84084 Fisciano, Italy; vromano@unisa.it
* Correspondence: spinelligiovanni76@gmail.com; Tel.: +359-2-979-6476

Citation: Spinelli, G.; Guarini, R.; Kotsilkova, R.; Ivanov, E.; Romano, V. Experimental, Theoretical and Simulation Studies on the Thermal Behavior of PLA-Based Nanocomposites Reinforced with Different Carbonaceous Fillers. *Nanomaterials* **2021**, *11*, 1511. https://doi.org/10.3390/nano11061511

Academic Editor: Riccardo Rurali

Received: 18 May 2021
Accepted: 3 June 2021
Published: 7 June 2021

Publisher's Note: MDPI stays neutral with regard to jurisdictional claims in published maps and institutional affiliations.

Copyright: © 2021 by the authors. Licensee MDPI, Basel, Switzerland. This article is an open access article distributed under the terms and conditions of the Creative Commons Attribution (CC BY) license (https://creativecommons.org/licenses/by/4.0/).

Abstract: Many research efforts have been directed towards enhancing the thermal properties of polymers, since they are classically regarded as thermal insulators. To this end, the present study focuses on the thermal investigation of poly(lactic acid) (PLA) filled with two types of carbon nanotubes (trade names: TNIMH4 and N7000), two type of graphene nanoplatelets (trade names: TNIGNP and TNGNP), or their appropriate combination. A significant increase in the thermal conductivity by 254% with respect to that of unfilled polymer was achieved in the best case by using 9 wt% TNIGNP, resulting from its favorable arrangement and the lower thermal boundary resistance between the two phases, matrix and filler. To theoretically assist the design of such advanced nanocomposites, Design of Experiments (DoE) and Response Surface Method (RSM) were employed, respectively, to obtain information on the conditioning effect of each filler loading on the thermal conductivity and to find an analytical relationship between them. The numerical results were compared with the experimental data in order to confirm the reliability of the prediction. Finally, a simulation study was carried out with Comsol Multiphysics® for a comparative study between two heat sinks based on pure PLA, and to determine the best thermally performing nanocomposite with a view towards potential use in heat transfer applications.

Keywords: biodegradable polymers; graphene; carbon nanotubes; nanocomposites; thermal transport properties; design of experiments; multiphysics simulations

1. Introduction

Nowadays, the high demand for custom-made products requires the development of new materials. Polymers have been identified as promising candidates for this aim, to the point where their current impact on our lives is almost unquantifiable. In fact, polymer-based products have been favorably adopted everywhere: synthetic fibers are also increasingly being used for clothing production, plastic bags are adopted for multiple, epoxy glues (and not only them) are widely present in the field of adhesives, fiberglass or carbon-based reinforced composites are being used as structural parts, and so on, with the list being potentially endless [1]. Noteworthy is the recent study on the possibility of using fibers obtained from polyethylene terephthalate (PET) waste bottles to improve the ductility of concrete [2], as well as the efforts in selecting appropriate nanoclays to combine with polyamide (PA) fibers in order to improve the flame-retardant and tensile properties of knitted fabrics [3]. Due to the increasing interest in environmentally friendly materials over the years, poly(lactic acid) (PLA) has aroused a great deal of attention, and therefore it is being intensely investigated, in both industry and academia [4,5]. The creation of non-woven membrane supports made with bamboo-fiber-reinforced poly(lactic acid) composites for an energy-efficient and sustainable technology is of interest [6], as well as the design of poly(lactic) acid bio-composites that include three types of silk fibers and wool protein microparticles for targeted biomedical applications [7]. More recently, with the advent of new fabrication processes based on additive manufacturing (AM) technologies, PLA, as a result of its vegetable-based nature, which confers excellent biocompatibility, sustainability and biodegradability, has become increasingly widespread, to the extent that it represents the primary natural raw material used in extrusion-based 3D printing techniques (popularly known as fused deposition modeling, FDM) [8]. This technology relies on a continuous filament consisting of a thermoplastic polymer that is fed by the extrusion head to the nozzle, where it melts, at a selected temperature, to then subsequently re-solidify on a building plate, thus creating, layer-by-layer in a pre-determined path, the designed 3D object [9,10]. Compared to classical fabrication processes with a subtractive nature, additive manufacturing technology presents remarkable benefits in terms of cost effectiveness, reduced processing waste, light weight, and versatility in the manufacturing of complex structures, which are increasingly being used in several fields [11]. To cite just some of them, AM has been adopted as a fast prototyping technology in the early stages of the design and development of components in aerospace and defense fields [12], for the manufacture of biomaterials suitable in medical applications and in dental care [13,14], and to produce polymer-based packages with electromagnetic (EM) shielding properties or heat exchangers, which are essential in the electronic industry [15]. Nevertheless, despite all of the progress made so far in terms of the development of AM techniques, the selection of suitable materials remains a critical bottleneck for their full implementation [16,17]. Thermal dissipation and electrical insulation are challenging issues that have still not been overcome when it comes to polymer-based composites. Therefore, different research efforts have been devoted to their improvement in terms of both thermal and electrical conductivity [18–22]. The negligible values of thermal conductivity typical for polymers are due to their structures; the random structure of molecule chains strongly limits the thermal transport due to the phonons, and as result, low values of thermal conductivity are observed at the macroscale level. Although the thermal conductivity can be affected by different intrinsic features of polymers, to engineer the polymeric matrix through the dispersion of highly conductive fillers inside it, it is recognized as a valid method for enhancing the overall thermal performance of the resulting materials, regardless of the thermoplastic or thermosetting nature of the host polymer [23]. Table 1 summarizes the thermal conductivity values of some polymers commonly used in the AM field, as well as for some classic fillers, which are grouped according to their physical type, i.e., carbon-based, metallic, and ceramic fillers [24]. Information on other polymers is reported in the handbook written by Yang [25].

Table 1. Thermal conductivity (at room temperature) of some well-known polymers for AM and of some dispersive fillers classically adopted for improving their thermal behavior [24].

Polymer		
Polymer Name	Acronym	Thermal Conductivity [W/m·K]
High density polyethylene	HDPE	0.44
Polyphenylsulfone	PPSU	0.35
Ethylene Vinyl Acetate	EVA	0.34
Acrylonitrile Butadiene Styrene	ABS	0.33
Poly(Butylene Terephthalate)	PBT	0.29
Nylon-6	PA6	0.25
Polyether Ether Ketone	PEEK	0.25
Poly(dimethylsiloxane)	PDMS	0.25

Table 1. Cont.

Polymer		
Polymer Name	Acronym	Thermal Conductivity [W/m·K]
* Poly(lactic) Acid	PLA	0.20
Polymethylmethacrylate	PMMA	0.21
Polyvinyl Chloride	PVC	0.19
Polyvinylidene Difluoride	PVDF	0.19
Polystyrene	PS	0.14
Polypropylene	PP	0.12
Filler		
Group	Type	Thermal Conductivity [W/m·K]
Carbon-based Fillers	Carbon nanotubes	2000÷6000 (longitudinal)
Carbon-based Fillers	Graphite	100÷400 (in plane)
Carbon-based Fillers	Carbon Black (CB)	6÷174
Metallic Fillers	Copper (Cu)	483
Metallic Fillers	Silver (Ag)	450
Metallic Fillers	Gold (Au)	345
Metallic Fillers	Aluminum (Al)	204
Ceramic Fillers	Boron nitride (BN)	250÷300
Ceramic Fillers	Aluminum oxide (Al_2O_3)	30

* Value from the experimental characterization carried out in the present study.

On the basis of the analysis of these data, it is evident that the thermal conductivity of the fillers, which is always dependent on the material structure, is significantly higher than that of the neat polymers.

Heat transfer can be associated with purely phonon mechanisms, or with the combined effect of phonons and electrons. Therefore, carbon-based and metallic fillers, the reticular structures of which contain freely moving electrons, present a dual heat conduction mechanism that leads to higher thermal conductivity compared to the values exhibited by ceramic fillers which are characterized only by a phonon transmission [26]. Carbon-based fillers (used either as single fillers or in combination) are considered ideal candidates for obtaining advanced composites, not only due to their remarkable thermal and electrical conductivities, but also due to their extraordinary corrosion resistance and thermal expansion coefficients, which are lower than other reinforcement particles. A scientifically recognized theory explaining both the thermal and the electrical conduction in composite structures is the so-called percolation theory, which is based on the creation of suitable electrically/thermally conductive paths: as soon as the filler concentration reaches a suitable value (percolation threshold), continuous paths (network) will be established, and the heat/current can flow through them due to the lower thermal/electrical resistance pathways, which are otherwise not present in the host polymers. As a result, a significant improvement in the thermal and electrical conductivity is observed [27]. 3D-printing silicone acrylate-based formulations (Polydimethylsiloxane, PDMS) with enhanced thermal conductivity due to the introduction of boron nitride (BN) as conductive filler were proposed and investigated by Pezzana et al. [28]. A 3D-printed composite with alumina and oriented carbon nanofibers (CFs) with enhanced thermal conductivity compare to that of cast composites was discussed by Ji et al. [29]. PLA-based nanocomposites filled with four different metal particles were prepared by 3D-printing and then characterized in order to investigate the influence of the additive microstructures, along with the printing parameter settings, on the thermal conductivity of the resulting structures. [30]. Even though the properties of the polymers were significantly improved due to the addition of various conductive fillers, a comprehensive knowledge of the overall performance has still not yet been fully achieved, since the observed results are usually far from those theoretically expected. In fact, many literature studies indicate that several factors, including the intrinsic features of filler particles, such as their geometrical shape and aspect ratio (AR), their concentration, their spatial dispersion in the polymer matrix, and the adhesion and interaction at the interface between the constituent phases, strongly affect the thermal and electrical behavior (among other things) of the composites [20,31]. In a previous study [32], the authors investigated formulations based exclusively on a single reinforcement type, MWCNTs or GNPs, in order to analyze the influence of their different geometrical features on the final performance of the resulting materials. Here, the investigation is enriched with the additional analysis of hybrid systems (multiphase nanocomposites), including both types of fillers (MWCNTs/GNPs), in some selected weight ratios, and up to a maximum of 9% total infill weight. Another novelty is represented by the Design of Experiments (DoE), which was performed to identify the most influential design parameters, and the Response Surface Method (RSM), which was applied to find the analytic relationship between them and the parameter of interest, i.e., thermal conductivity. Moreover, simulation studies carried out with the commercial software package Multiphysics (COMSOL Multiphysics®) are presented with the aim of predicting the thermal behavior of this novel nanocomposite, designed as a heat sink for potential heat transfer applications.

2. Materials and Methods

2.1. Materials

The base polymer used in this study for the compounding formulation was Ingeo™ Biopolymer PLA-3D850 (Nature Works, Minnetonka, MN, USA), which is particularly indicated for the manufacturing of 3D printer monofilament, since it is characterized by fast crystallization rate, good adhesion to the build plate, and high printing speed, as well as less warping or curling, low odor emission, and much more. Among its main physical properties, it is worth mentioning its glass transition temperature (Tg) of 55–60 °C and peak melt temperature of 165–180 °C (both measured in agreement with the D3418 standard of the American Society for Testing and Materials, ASTM), as well as its melt mass-flow rate (MFR) of 7–9 g/10 min (according to the D1238 ASTM standard). The aim was to develop a non-conventional material for additive manufacturing and to obtain, thanks to the potentialities of this technology, heat sinks with an ad hoc design for heat transfer applications.

Four types of carbon nanofiller were chosen to manufacture the nanocomposites investigated in the present study. In brief: (i) industrial graphene TNIGNP (from Times Nano, Chengdu, China); (ii) industrial MWCNTs—TNIMH4 (from Times Nano, Chengdu, China); (iii) graphene—TNGNP (from Times Nano, Chengdu, China); and (iv) MWCNTs —N7000 (NC7000™ series, Nanocyl® SA, Sambreville, Belgium).

The names used—(i) industrial graphene nanoplatelets (TNIGNP), and (ii) industrial multiwall carbon nanotubes (TNIMH4)—were provided by the producer (Times Nano, Chengdu, China). The term industrial indicates that their production was carried out in large quantities and at a low price in contrast to higher-quality and more expensive nanofillers, which are produced in small quantities under laboratory or semi-industrial conditions (e.g., TNGNP and N7000). In this study, both industrial fillers and higher-quality fillers were used to compare the characteristics, properties, advantages and disadvantages of the polymeric nanocomposite materials obtained from them.

In particular, the two types of graphene nanoplatelets (GNPs) and the two grades of multiwall carbon nanotubes (MWCNTs) were selected here based on their low price versus good technical specifications. The size, shape, aspect ratio, specific surface area and functionalization of nanoparticles were varied in order to estimate the essential nanofiller characteristics governing the thermal, electrical and other physical properties of the nanocomposites [32]. MWCNTs differ mainly with respect to aspect ratio (1000 for TNIMH4 and 150 for N7000), OH content (2.48% for TNIMH4 and absent for N7000), and surface area (110 m^2/g and 250 m^2/g for TNHIMH4 and N7000, respectively). Instead, GNPs differ principally with respect to aspect ratio (240 for TNIGNP and about 500 for TNGNP), volume resistivity (<0.15 Ω·cm for TNIGNP and about 10^{-4} for N7000), and purity (>90%

and >99.5% for TNIGNP and TNGNP, respectively). Other technical characteristics are summarized in the schematic representation presented in Figure 1.

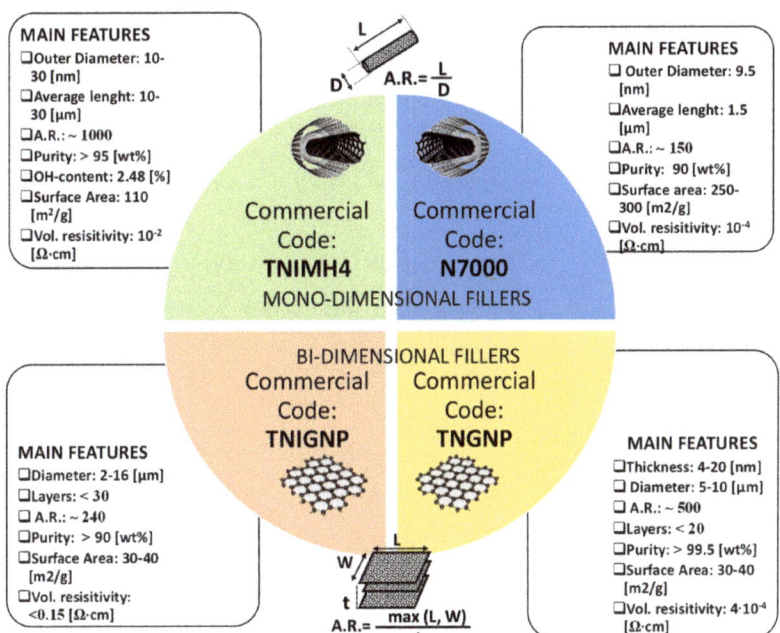

Figure 1. Trade names and technical specifications of each of the fillers used in this study. Source: technical data sheets available on the website of the manufacturing companies.

The definition of the aspect ratio (i.e., A.R.) for both 1-dimensional filler (MWCNTs) and 2-dimensional ones (GNPs) is also schematically reported in Figure 1. The aspect ratio of an object is defined as the ratio of its longest dimension to its shortest one. For rodlike fillers such as carbon nanotubes or nanofibers, the A.R. coincides with the ratio of the length (L) to the diameter D (A.R. = L/D), whereas for planar particles, such as graphene and its derivatives, it is given by the ratio of the largest lateral dimension (length, L or width, W) to the thickness t (A.R. = max (L, W)/t).

2.2. Preparation of Nanocomposites and Test Samples

Two sets of nanocomposites were prepared by varying the type and content of the fillers (1.5–9 wt%). The first set was prepared by melt mixing industrial graphene (TNIGNP) and industrial multi-walled carbon nanotubes (TNIMH4) with PLA 3D850 in pellet form. The second set of nanocomposites was obtained using the same host matrix (PLA 3D850) milled to a powder but filled with the other types of graphene nanoplatelets (TNGNP) and MWCNTs (N7000). The mono (PLA/MWCNT and PLA/GNP) and bi-filler nanocomposites (PLA/MWCNT/GNP) were processed by melt extrusion at 170–180 °C using a twin-screw extruder (COLLIN Teach-Line ZK25T, Maitenbeth, Germany), at a screw speed of 40 rpm. The PLA and nanofillers were dried for 4 h at 80 °C in a vacuum oven, and masterbatches of 9 wt% filler were initially extruded, and then further diluted with PLA by a second extrusion run to produce mono-filler composites with varying filler contents. The bi-filler composites with 3, 6 and 9 wt% total filler content and various GNP: MWCNT ratios were fabricated by mixing the GNP/PLA and the MWCNT/PLA masterbatches with the neat PLA in appropriate amounts.

The concentrations were identified during the pre-planning phase in order to perform a Design of Experiment (see Section 2.3.4. Design of Experiments (DoE) for Thermal Characterization), initially based on uniformly distributed parameters at four levels in the interval [0÷9] wt%, and eventually with further intermediate values being considered in a second stage, if necessary. The maximum amount of 9 wt% was dictated by a compromise in terms of overall the performances exhibited by the resulting nanocomposites and their easier processability, on the basis of our previous study [32].

Table 2 summarizes the compositions studied in the present study. Circular-shaped specimens with a diameter of 16 mm and a thickness of 3 mm were hot pressed from extruded pellets, polished and then used for electrical and thermal conductivity tests.

Table 2. List of mono-filler and bi-filler composites on the base of PLA, industrial graphene (TNIGNP) and industrial MWCNTs (TNIMH4) as well as graphene TNGNP and MWCNT of type N7000.

Composition	GNP Content [wt%]	MWCNT Content [wt%]	PLA Content [wt%]	GNP/MWCNT [Ratio]
PLA	-	-	100	-
Mono-phase nanocomposites based on GNPs (TNIGNP or TNGNP)				
PLA/1.5% GNP	1.5	-	98.5	-
PLA/3% GNP	3	-	97	-
PLA/6% GNP	6	-	94	-
PLA/9% GNP	9	-	91	-
Mono-phase nanocomposites based on MWCNTs (TNIMH4 or N7000)				
PLA/1.5% MWCNT	-	1.5	98.5	-
PLA/3% MWCNT	-	3	97	-
PLA/6% MWCNT	-	6	94	-
PLA/9% MWCNT	-	9	91	-
Multiphase (hybrid) nanocomposites based on GNPs/MWCNTs (TNIGNP/TNIMH4 or TNGNP/N7000)				
PLA/1.5% GNP/1.5% MWCNT	1.5	1.5	97	1:1
PLA/3% GNP/3% MWCNT	3	3	94	1:1
PLA/4.5% GNP/1.5% MWCNT	4.5	1.5	94	3:1
PLA/6% GNP/3% MWCNT	6	3	91	2:1
PLA/1.5% GNP/4.5% MWCNT	1.5	4.5	94	1:3
PLA/3% GNP/6% MWCNT	3	6	91	1:2

2.3. Experimental and Numerical Methods

2.3.1. Morphological Analysis

The dispersion state of the nanofillers inside the PLA-based nanocomposites, as well as their morphological features, was investigated by means of transmission electron microscopy (TEM) and scanning electronic microscopy (SEM). More specifically, the transmission electron microscopy (TEM) analysis was carried out using an FEI TECNAI G12 Spirit-Twin (LaB6 source, FEI Company, Hillsboro, OR, USA) working with an acceleration voltage of 120 kV with a magnification variable between 22 and 300 kX in combination with an FEI Eagle-4k charged coupled device camera (CCD). Prior to the investigation, sections of the specimens suitable for analysis were cut, at room temperature, using using a Leica EM UC6/FC6 ultramicrotome, and they were then positioned on 400 mesh TEM copper grids. With respect to the SEM, a JSM-6700F apparatus (JSM-6700F, Jeol, Akishima, Japan) was used on suitable fragments of the whole specimens obtained through fragile fractures in liquid nitrogen. Prior to the investigation, the samples were chemically etched and then gold-sputtered following a method already described by Spinelli et al. [33,34], and therefore omitted here.

2.3.2. DC Electrical Conductivity Analysis

The bulk conductivity (at room temperature of 20 °C) of each formulated composition was tested using a pico-ammeter (Keithley 2400, Keithley Instruments Inc., Beaverton, OR, USA), which acted simultaneously as both a source and a meter, in accordance with the electrical schematization reported in Figure 2a). Three samples were measured for each nanocomposite, and the average values of the measurements are reported as results in the electrical section of the present study. During the test, the electrical resistance (R_{mis}) of the material was measured by applying Ohm's first law between the DC voltage applied to the samples (Vm) and the measured current (Im) flowing in it ($R_{mis} = Vm/Im$). The electrical conductivity, σ_{DC} [S/m], of the bulk sample was determined using Ohm's second law:

$$\sigma_{DC} = \frac{1}{R_{mis}} \cdot \frac{H}{\pi \cdot (D_4/2)^2} \quad (1)$$

where H is the sample thickness and D_4 is the sample diameter of the metalized measuring electrodes. In fact, in order to ensure Ohmic contacts, both sample surfaces, see Figure 2a,b, were covered with a silver paint having a volume resistivity of 0.001 Ω·cm (Alpha Silver Coated Copper Compound Screening, RS 186–3600, Corby, UK). Moreover, in order to guarantee the exclusive measurement of bulk currents, a guard ring was applied on the top side of the sample in order to drain any surface current towards the mass of the system, especially for composites with high electrical conductivity.

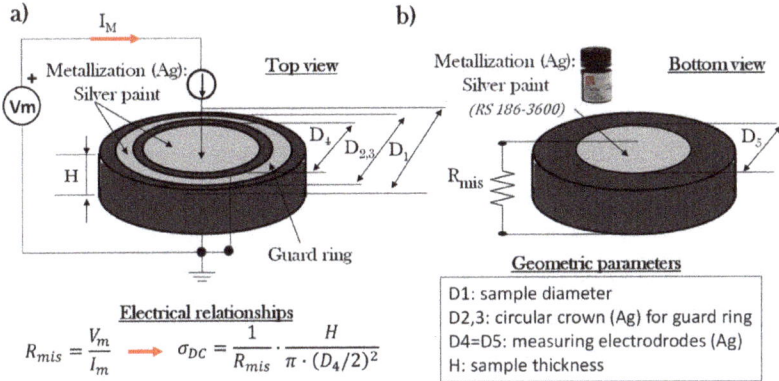

Figure 2. Setup for the measurement of the DC bulk conductivity of circular-shaped specimens with top view in (**a**) and bottom view in (**b**), respectively. Geometric details and the electrical relationships adopted for its determination are also reported.

2.3.3. Thermal Characterization

A Hot Disk 2500 thermal constant analyzer (Hot-Disk AB TPS 2500, Göteborg, Sweden) was used to measure the thermal conductivity of the GNP- and MWCNT-based polymer composites using the transient plane source method (i.e., TPS) in accordance with the specifications of ISO 22007-2-2015 (International Organization for Standardization) [35]. Before measurements, the samples were polished using sandpaper to obtain very flat surfaces. The measurements were performed by placing the TPS element (3 mm diameter), which acts simultaneously as a heater and a temperature sensor, between two similar slabs of material (Figure 3). The sensor supplied a heat pulse of 0.01 W for 40 s to the sample at room temperature, and the associated change in temperature was recorded. The thermal conductive parameters, including the thermal conductivity and the thermal diffusivity of the samples, were measured according to the theory already described in detail in Spinelli et al. [30], and only briefly summarized here.

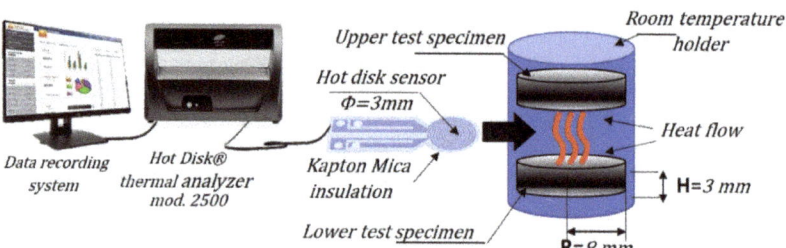

Figure 3. Experimental setup for the thermal characterization performed by using a Hot Disk® thermal constant analyzer.

The experimental measurements were correlated with the time-dependent resistance of the TPS element, and were evaluated in accordance with the following expression:

$$R(t) = R_0[1 + \beta \Delta T(\tau)]. \qquad (2)$$

In Equation (2), above, R_0 represent the initial value for the resistance of the TPS before the transient recording (about 4 Ω at room temperature), and β is its temperature coefficient, whereas $\Delta T(\tau)$ is the temperature increase recorded over time, calculable by means of the formula:

$$\Delta T(\tau) = \frac{P_0}{(\pi^{3/2} r \lambda)} D(\tau) \qquad (3)$$

where the dimensionless time $\tau = (t \cdot \alpha / r^2)^{1/2}$ is dependent on the measurement time t, the thermal diffusivity α, the radius of the sensor r, the input heating power P_0, and the thermal conductivity λ. Finally, $D(\tau)$ is a Bessel-based dimensionless shape function that also accounts for the number of concentric circles forming the hot disk sensor [34].

Once the temperature evolution has been determined by means of Equation (2) and by fitting it to Equation (3), it is possible to derive the thermal features of the material under test.

2.3.4. Design of Experiments (DoE) for Thermal Characterization

Design of Experiments (DOE) belongs to the applied statistics branch introduced for evaluating so-called cause-effects, i.e., the relationship between the factors conditioning a process/product and its output or performance function (*PF*) of interest [36,37].

As schematically shown in Figure 4, during the initial pre-planning design phase, the most common elements to be considered include the controllable input factors (i.e., X_i, which can be arbitrarily modified), the uncontrollable input factors (i.e., N_i, which cannot be changed due to noise sources and obvious tolerances on *Xi* variables), and the output of the process (the performance function of interest, P.F.) The latter is the result of the joint action of controllable input factors (X = (X_1, X_2, ..., X_p)) and uncontrollable input factors (N = (N_1, N_2, ..., N_q)), i.e., P.F. = f (X, N). In the presence of multiple design parameters, DoE is a powerful statistical tool that makes it possible to identify the most influential one among them. Moreover, DoE helps to choose the best combination of them for optimizing the selected P.F., or at least to contain its deviation in response to the action of uncontrollable factors (robust design, RD) [38].

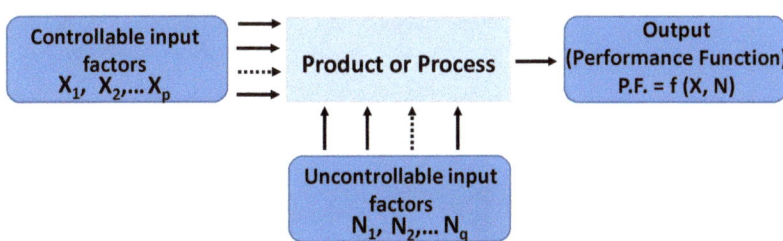

Figure 4. Schematic illustration of a system (product or process) at the design stage.

An approach based on DoE was successfully adopted to produce polymer microfibers with carbon nanotubes through electrospinning experiments [39], and has recently been combined with artificial intelligence (AI) to develop the sustainable electrochemical synthesis of zeolitic imidazolate frameworks (ZIF-8) [40].

In this paper, DoE is performed to analyze the influence of the two main controllable input factors, i.e., the nanofiller loading ($wt\%_{MWCNTs}$, $wt\%_{GNPs}$) on the thermal conductivity (i.e., λ) of the resulting nanocomposites, which represents the targeted performance function in this analysis. For this purpose, a specific Matlab® routine was designed. As a matter of course, a discretization level for the controllable input factors should be properly set. Uniformly distributed parameter values are advisable at first, whereas further intermediate points may be considered in order to refine the model at a later stage [41]. In any case, in order to obtain an effective predictive model, a good space-filling of the experimental region data must be ensured. In light of the above, in the present study, the input variable vector was:

$$\bar{x} = (wt\%_{MWCNTs}, wt\%_{GNPs}) \in \mathbb{R}^2 \tag{4}$$

where a discretization over four levels was selected for the two controllable input factors in the interval [0÷9] wt%, i.e.,:

$$wt\%_{MWCNTs\_1} = 0, wt\%_{MWCNTs\_2} = 3, wt\%_{MWCNTs\_3} = 6, wt\%_{MWCNTs\_4} = 9 \tag{5}$$

and:

$$wt\%_{GNPs\_1} = 0, wt\%_{GNPs\_2} = 3, wt\%_{GNPs\_3} = 6, wt\%_{CNPs\_4} = 9 \tag{6}$$

As result, the variable space is compact:

$$\mathcal{D} = wt\%_{MWCNTs} \times wt\%_{GNPs} \subset \Re^2 \tag{7}$$

whereas the P.F. is estimated for each ordered pair (x_1, x_2) of the controllable input factors, i.e.,:

$$(x_1, x_2) = (wt\%_{MWCNTs}, wt\%_{GNPs}) \in \mathcal{D}. \tag{8}$$

By following the aforementioned methodological steps, DoE on the basis of $2^4 = 16$ points $\in \mathcal{D}^* \subset \mathcal{D}$ generates scattered data for the P.F., which is required for carrying out sensitivity analysis by means of Dex Scatter Plot (DSP), Main Factor Plot (MFP) and Response Surface Methodology (RSM). Their meaning, along with their relative results, will be illustrated in the corresponding Results sections.

2.3.5. Multiphysics Simulations of Thermal Properties

In the present work, we study the thermal behavior of designed nanocomposites in the case of their use as thermal dissipators in heat transfer applications. The considered samples, modeled as cylinders with a diameter of 16 mm and a height of 3 mm, include one of pure polymer (PLA), with the others being the best-performing nanocomposites, in terms of thermal properties, investigated here, i.e., PLA including 9% by weight of TNIGNP (as schematically shown in Figure 5a). For this purpose, a mathematical model was developed, and a numerical analysis based on the Finite Element Method (FEM) was performed using the commercial software COMSOL Multiphysics®. The main model definitions selected for the numerical analysis are summarized in Figure 5b.

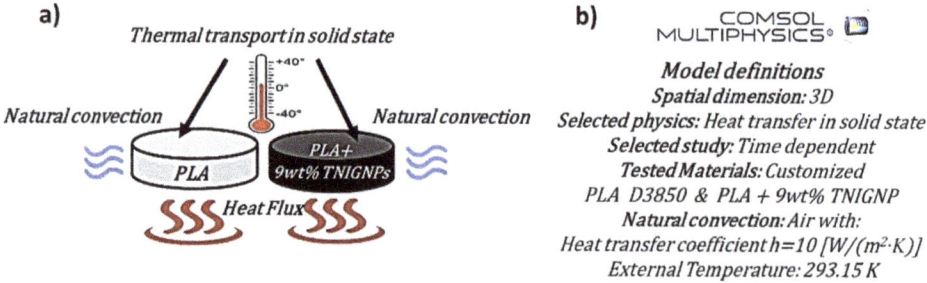

Figure 5. Schematic illustration (a) of the simulation studies carried out on pure PLA and on the best (in terms of thermal properties) composite (PLA + 9 wt% TNIGNPs). The main model definitions adopted in COMSOL are also reported (b).

With respect to potential applications as heat sinks, the heating of the sample was simulated by applying a heat flux (900 W/m²) to the bottom surface and then cooling in still air at a temperature of 293.15 K. The heat is transferred from the hot surface to the inside the sample by conduction (Figure 6a) and to the surrounding air by natural convection, a consequence of the different density between the hot air close to the heated surface and the cold air surrounding the lateral surface and the upper surface of the sample (Figure 6b).

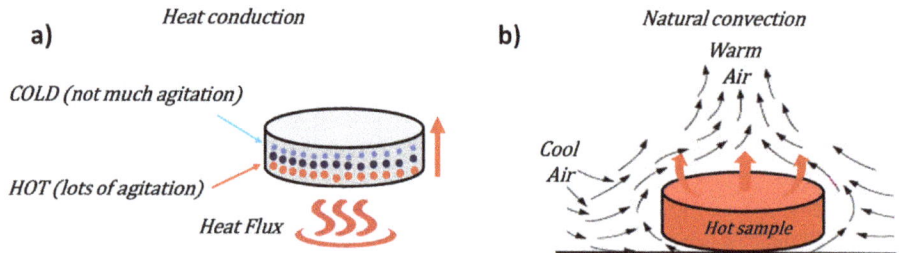

Figure 6. Schematic illustration of the two main modes of thermal energy transfer: (a) heat conduction and (b) natural convection.

The mathematical theory of heat conduction was formulated by Fourier, and the basic equation (in differential form), bearing his name, governs conduction heat transfer in accordance with the following vectorial equation:

$$q = -\lambda \cdot \nabla T \qquad (9)$$

where the involved parameters are as follows (including the SI unit):
- q is the heat flux, i.e., the rate of heat flow per unit area [W/m²];

- λ is the intrinsic thermal conductivity of the material, assumed to be constant in the present work [W/m·K];
- ∇T is the temperature gradient [K/m].

In several simple applications, Fourier's law can be expressed in one-dimensional form in a generic i direction, in which case Equation (9) can be written as:

$$q_i = -\lambda \cdot \frac{dT}{di} \text{ with } i = x, y \text{ or } z \text{ axis} \quad (10)$$

It should be noted that the algebraic sign reported in Equation (9) and in Equation (10) is used when the heat flux occurs in the opposite direction to the temperature gradient, as occurs in the case in question, in which the temperature decreases in the direction in which the radius increases. In the case of interest for the temperature profile inside the medium, the mathematical formulation, considering both the law of energy conservation and Fourier, leads to the universally known differential equation for heat conduction for a solid at constant pressure, written in vectorial form:

$$\nabla \cdot (\lambda \nabla T) + \dot{q} = \rho \cdot c_p \cdot \frac{\partial T}{\partial t} \quad (11)$$

where:
- ρ is the density of the material [kg/m^3];
- c_p is the specific heat of material [J/kg·K];
- $\dot{q}$ is the heat generated per unit volume [W/m^3].

Adapted to the study of our specific nanocomposite solids, the thermal energy Equation (11) can be written in cylindrical coordinates for a differential volume $2\pi r \Delta r \Delta z$, as [42]:

$$\frac{1}{r}\frac{\partial}{\partial r}\left[r\left(\lambda \frac{\partial T}{\partial r}\right)\right] + \frac{\partial}{\partial z}\left(\lambda \frac{\partial T}{\partial z}\right) = \rho c_p \frac{\partial T}{\partial t} \quad (12)$$

As initial conditions, the material is assumed to be at a uniform temperature T_0; as boundary conditions, symmetry on the axis for $r = 0$ is assumed, whereas a loss of heat because of natural convection is considered at the external boundaries ($r = R$) and at the top surface ($z = 3$ mm); finally, a heat flux is imposed at the bottom surface ($z = 0$), as summarized in Table 3.

Table 3. Conditions for the solving the thermal energy equation.

Initial and Boundary Condition	Equation	Validity
$t = 0$	$T = T_0$	$\forall r, \forall z$
$r = 0$	$\frac{\partial T}{\partial r} = 0$	$\forall z, \forall t > 0$
$r = R$	$-K\frac{\partial T}{\partial r} = h(T - T_\infty)$	$\forall z, \forall t > 0$
$z = 0$	$-K\frac{\partial T}{\partial r} = q$	$\forall r, \forall t > 0$
$z = H$	$-K\frac{\partial T}{\partial r} = h(T - T_\infty)$	$\forall r, \forall t > 0$

The heat transfer coefficient by natural convection h [W/m^2·K] is defined by Newton's law of cooling:

$$Q = h \cdot S \cdot \Delta T \quad (13)$$

it represents the proportionality factor between the heat flow Q [W] and the ΔT [K], which causes the convective transport between a hot solid surface S [m^2] and the surrounding air.

3. Results

3.1. Morphological Analysis

Given their crucial roles, it is of particular importance to investigate the basic morphological features of carbon nanotubes and graphene, as well as their dispersion state (also when combined together at low loadings of both fillers) before presenting the electrical and thermal performance of the resulting nanocomposites in which they have been dispersed. Representative TEM images of all four types of carbonaceous fillers (in powder form) used in the present study are shown in Figure 7 with a magnification of 200 nm for MWCNTs and 2 µm for GNPs, respectively.

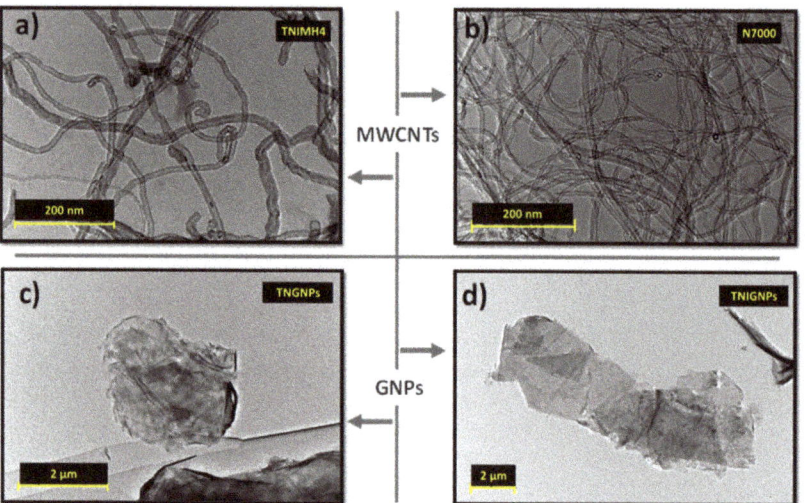

Figure 7. TEM images showing fillers powder of multiwalled carbon nanotubes of the types (**a**) TNIMH4 and (**b**) N7000, and graphene nanoplatelets of the types (**c**) TNGNPs and (**d**) TNIGNPs, respectively.

On the basis of their quick analysis, it is worth noting the nanotube waviness and the larger size of TNIMH4 with respect to N7000 in terms of both average length and diameter, as well as the relevant dimension of planar fillers (GNPs) compared to the monodimenisonal ones (MWCTNs). Figure 8 presents the TEM images of the aforementioned fillers when incorporated at the maximum investigated loading (9 wt%) in the polymer matrix.

It is noticeable at first sight that the issue of agglomeration between the fillers is more evident for MWCNTs of type TNIMH4 and GNPs of type TNGNP, whereas N7000 and TNIGNP particles appear well dispersed, since the aggregates were limited both in terms of number and dimension. The phenomena of bundles or stable aggregates may occur due to the remarkable intermolecular interaction between the fillers. Consequently, in addition to a non-uniform dispersion, the agglomeration effect reduces the aspect ratio of the nanofillers, which in turn affects the overall properties of final nanocomposites by worsening the performance compared to that expected [43]. Finally, Figure 9 illustrates some selected SEM images with respect to the unfilled PLA and nanocomposites with a total charge of 9 wt% including some hybrid systems based on the simultaneous combination of GNPs and MWCNTs in the selected weight ratio.

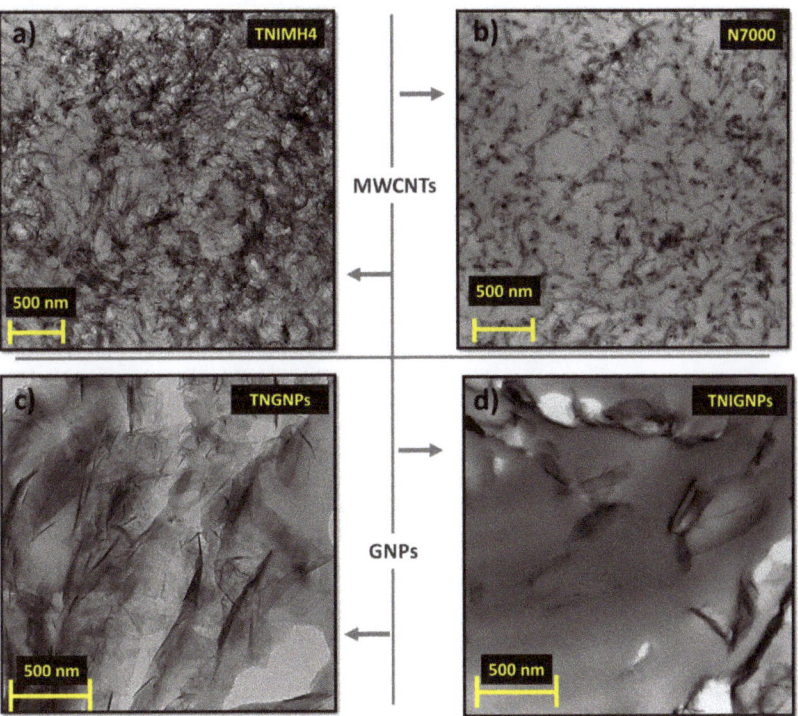

Figure 8. TEM images showing multiwalled carbon nanotubes of the types (**a**) TNIMH4 and (**b**) N7000, and graphene nanoplatelets of the types (**c**) TNIGNPs and (**d**) TNGNPs, respectively, dispersed in the polymer with 9 wt% loading.

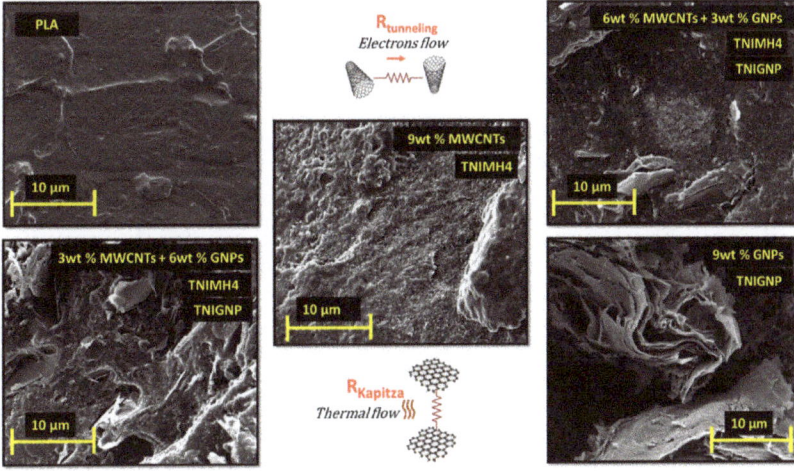

Figure 9. SEM images of pure PLA and different nanocomposites with a total charge of 9 wt% including hybrid systems made with both fillers (TNIMH4 and TNIGNP) in some selected concentrations.

A good adhesion and interaction at the matrix/filler interface is observed, since no significant porosity can be identified; only some unavoidable cavities in the case of graphene-based composites. Instead, what is important to point out is that the uniform and homogenous arrangement of carbon nanotubes seems to be more favorable for the creation of the percolating network within the polymer, and therefore, MWCNTs appear to be better indicated for improving the electrical performance of the resulting nanocomposites given the improved electron tunneling between the conductive particles. In contrast, the clearly visible stacked arrangement of the graphene nanoplatelets is certainly suited for a more highly efficient phononic heat flow compared to the nanotubes, due to a lowering of Kapitza resistance, i.e., the interfacial thermal resistance between the two phases, GNPs/matrix. This is because the bi-dimensional surfaces of GNPs are more easily wetted by the polymer. Hybrid systems represent a recent design attempt to achieve a suitable balance in terms of both the electrical and thermal properties of nanocomposites.

3.2. DC Electrical Conductivity

The addition of conductive nanofillers inside the polymers conditions a whole series of physical properties, including the thermal and electrical ones. Changes in the DC electrical conductivity (σ_{DC}) of the composites based on PLA/MWCNTs (both types, TNIMH4 and N7000) and PLA/GNPs (both types, TNIGNP and TNGNP) as a function of the filler loading are reported in Figure 10a,b, respectively. The progressive introduction of MWCNTs and GNPs nanofillers determines, especially at higher filler loadings, a remarkable improvement in the electrical conductivity of the nanocomposites with respect to the value exhibited by the pure PLA (5.9×10^{-10} S/m). In the best case (i.e., 9 wt% N7000), a conductivity value of about 2 S/m was obtained. For all series of nanocomposites, regardless of the type of nanofillers dispersed, and as expected on the basis of percolation theory, the trend of the electrical conductivity followed a power law dependence described by the following equation:

$$\sigma_{DC} = \sigma_0 \cdot (v - EPT)^t \quad for \; v > EPT \tag{14}$$

where σ_0 is a pre-exponential factor that is dependent on the intrinsic electrical conductivity of the fillers, their resistance of contact and the topology of the percolation cluster [44,45], v is the filler content, and t is a critical exponent that accounts the morphological arrangement of the filler in the percolating structure [27]. From a graphical point of view, the evolution of electrical conductivity of nanofilled composites versus filler concentration can be divided into three main phases. At the start, the conductivity, due to the small amount of additives, assumes values comparable to that of the neat polymer. Later, it begins to progressively increase, because first electrical junctions start to form, and a tunneling effect occurs between close neighbor particles. In the last phase, the increase in the filler content forms continuous paths (at percolation threshold) for the electron flow, thus enhancing the overall electrical conductivity, which in turn evolves towards a plateau value that is imposed by the tunneling resistance (R_{tunnel}) between the filler particles, which can be evaluated according to the following expression

$$R_{tunnel} = \frac{h^2 d}{Ae^2\sqrt{2m\lambda}} exp^{(\frac{4\pi d}{h}\sqrt{2m\lambda})} \tag{15}$$

where h is Plank's constant, A and d are the surface area of the filler and the interparticle distance involved in the tunneling effect, respectively, e and m are the charge and the mass of electrons, and λ is the height of the barrier (generally, few eV) due to the insulating behavior of the host polymer [46–48].

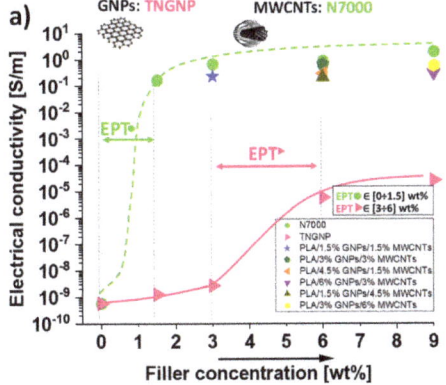 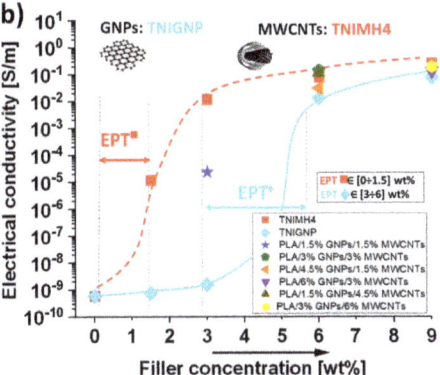

Figure 10. (a) DC electrical conductivity of PLA reinforced with different types of carbon-based fillers: TNGNP or N7000 and their combination (a) and TNIGNP or TNIMH4 and their combination (b).

Moreover, in the case of PLA enriched with mono-dimensional fillers, such as carbon nanotubes (N7000 and TNIMH4), the electrical conductivity already increased significantly at a filler concentration of 1.5 wt%, indicating a much lower percolation threshold (EPT) for this type of nanofiller compared to that observed for bi-dimensional ones, i.e., graphene nanoplatelets (TNIGNP, TNGNP). For these latter ones, a filler loading in the range [3÷6] wt% is required before a sharp insulator–conductor transition in the behavior of PLA is observed. With respect to hybrid systems, it is worth noting that, due to the synergistic effect between MWCNTs and GNPs, the electrical percolation threshold in such formulations was achieved with a 3 wt% total charge (with a filler weight ratio 1:1). This indicates that multiphase systems can be a valid option to improve the ability of GNP particles to easily form percolation paths, since the effectiveness of their dispersion in the polymer is improved. As a consequence, percolation is achieved with a lower filler content than that required when GNPs are used exclusively as filler for the preparation of the composites, thus enhancing their processability. Once again, the N7000 carbon nanotubes achieved the best performance (with respect to the filler TNIMH4), since better conductive networks are also established when combined with GNPs.

All of results of this electrical characterization for each of the investigated nanocomposites are collected in Table 4.

Table 4. Results of the electrical characterization.

FILLER:	TNIGNP (GNP), TNIMH4 (MWCNT)		TNGNP (GNP), N7000 (MWCNT)	
Composites	Electrical Conductivity [S/m]		Electrical Conductivity [S/m]	
PLA	5.9×10^{-10}	$(\pm 1.30 \times 10^{-11})$	5.9×10^{-10}	$(\pm 1.30 \times 10^{-11})$
1.5%GNP	7.8×10^{-10}	$(\pm 1.82 \times 10^{-11})$	1.3×10^{-9}	$(\pm 6.78 \times 10^{-10})$
3%GNP	1.6×10^{-9}	$(\pm 1.5 \times 10^{-10})$	2.8×10^{-9}	$(\pm 4.60 \times 10^{-10})$
6%GNP	0.0133	$(\pm 2.12 \times 10^{-3})$	6.04×10^{-6}	$(\pm 1.13 \times 10^{-7})$
9%GNP	0.07932	$(\pm 5.88 \times 10^{-3})$	2.7×10^{-5}	$(\pm 2.01 \times 10^{-6})$
1.5%MWCNT	1.2×10^{-5}	$(\pm 5.94 \times 10^{-6})$	0.17	$(\pm 2.50 \times 10^{-2})$
3%MWCNT	0.0121	$(\pm 2.74 \times 10^{-3})$	0.69	$(\pm 2.73 \times 10^{-2})$
6%MWCNT	0.0783	$(\pm 1.01 \times 10^{-3})$	0.89	$(\pm 2.63 \times 10^{-2})$
9%MWCNT	0.26404	$(\pm 1.81 \times 10^{-2})$	2.00	$(\pm 1.21 \times 10^{-2})$
1.5%GNP + 1.5%MWCNT	2.41×10^{-5}	$(\pm 2.21 \times 10^{-6})$	0.24207	$(\pm 9.98 \times 10^{-2})$
1.5%GNP + 4.5%MWCNT	0.11393	$(\pm 1.31 \times 10^{-2})$	0.21099	$(\pm 4.32 \times 10^{-2})$
3%GNP + 3%MWCNT	0.13928	$(\pm 1.44 \times 10^{-2})$	0.67	$(\pm 5.34 \times 10^{-2})$
4.5%GNP + 1.5%MWCNT	0.03153	$(\pm 1.60 \times 10^{-2})$	0.30336	$(\pm 1.19 \times 10^{-2})$
3%GNP + 6%MWCNT	0.19565	$(\pm 2.53 \times 10^{-2})$	0.58949	$(\pm 1.55 \times 10^{-2})$
6%GNP + 3%MWCNT	0.13234	$(\pm 2.47 \times 10^{-2})$	0.30879	$(\pm 6.19 \times 10^{-2})$

3.3. Thermal Conductivity of PLA-Based Nanocomposites

The thermal conductivity of a polymer can be affected by a great number of intrinsic features, and can be improved with the addition of conductive fillers into the matrix. Both MWCNTs and GNPs, given their extraordinary intrinsic thermal conductivity, are carbon-based nanofillers that are widely recognized to have a very strong influence on the thermal performance of the composites in which they are dispersed [49,50]. The thermal conductivity is shown as a function of the filler concentration up to 9 wt%, for monophase and hybrid nanocomposites realized with TNGNP and N7000 or TNIGNP and TNIMH4 carbonaceous fillers in Figure 11a,b, respectively. As expected for a pure polymer, unfilled PLA shows a low thermal conductivity value of 0.20 (W/m·K), whereas remarkable increases are observed with the progressive addition of both types of filler, especially with addition of the graphene type. In fact, at the highest investigated filler loading (9 wt%), thermal conductivities of 0.725 (W/m·K) and 0.662 (W/m·K) were measured for PLA including TNIGNPs and TNGNPs, and 0.436 (W/m·K) and 0.341 (W/m·K) in the case of PLA filled with N7000 and TNIMH4, respectively. Therefore, a remarkable increment of the thermal conductivity of 254% with respect to that of the unfilled polymer was achieved in the best case (9 wt% TNIGNPs).

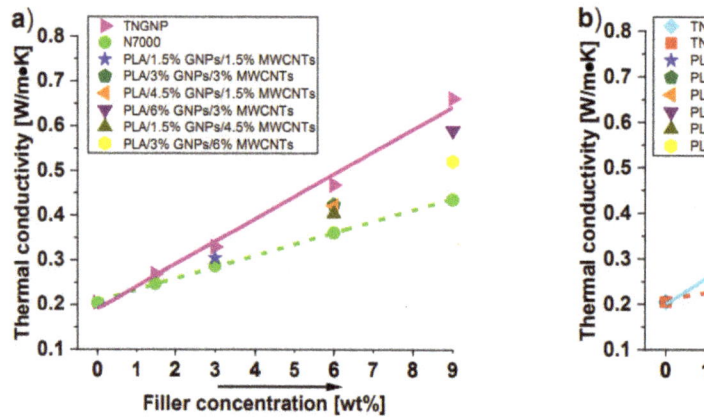
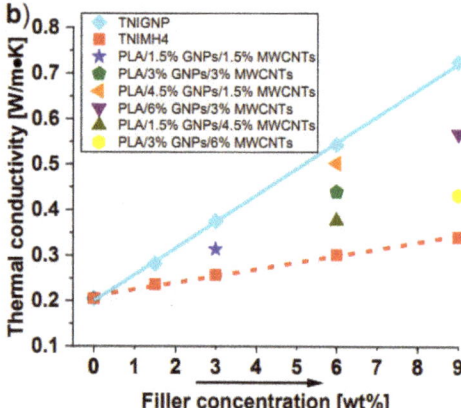

Figure 11. Thermal conductivity of PLA nanocomposites filled with TNGNP or N7000 and their combination in (**a**) and Table 4, and their combination in (**b**).

Intermediate values were observed for the thermal conductivities of the nanocomposites designed by combining both fillers in some selected weight ratios. In any case, such values are lower than those measured for monophase nanocomposites exclusively filled with GNP nanoparticles. For hybrid systems, as is evident from the analysis of Figure 12, the thermal conductivity is enhanced with increasing addition of graphene (regardless the type), thus confirming its key role, compared to that of carbon nanotubes, in improving the thermal transport mechanism in such nanocomposite systems.

All of the results for this thermal characterization, including thermal diffusivity values for each of the investigated nanocomposites, are reported in Table 5.

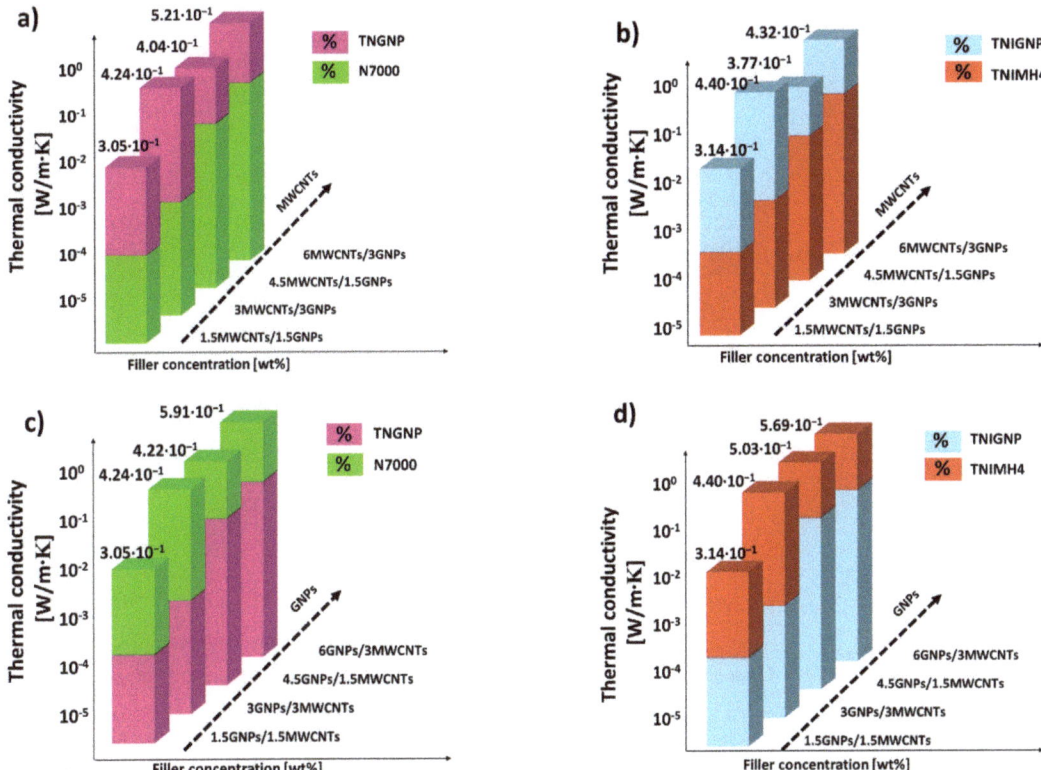

Figure 12. Thermal conductivity of hybrid systems including both types of filler (MWCNTs and GNPs) of type N7000 and TNGNP in (**a,c**) and TNIMH4 and TNIGNP in (**b,d**), respectively.

3.4. Design of Experiment (DoE): Dex Scatter Plot (DSP) and Main Factor Plot (MFP) for Thermal Conductivity

The DoE approach leads to the Dex Scatter Plot (DSP) and Main Factor Plot (MFP), as shown in Figure 13a–d with reference to nanocomposites based on N7000 and TNGNP or TNIMH4 and TNIGNP or fillers, respectively. In brief, and for the sake of clarity, let us remember that a DSP chart shows on vertical axis the scattered experimental data of the P.F. (i.e., the thermal conductivity in the present study), which constitutes the dependent variable, while the independent variables are reported on the horizontal axis (filler loadings, ordered pairs of Equation (8)). From a graphic point of view, the DSP indicates when the P.F. reflects changes in controllable input factors, highlighting the most influential one among them, as well as providing information on its influence (improvement or worsening) [51,52]. To complement DSP information, the main factor plot (MFP) is usually reported in order to evaluate the differences between the mathematical averages for one or more input factors. From a technical point of view, it is possible to evaluate the influence of the controllable factor on the P.F. by examining the slope of the segment that joins the average points of the range of values for the performance function corresponding to the minimum and maximum levels of each factor. For a given variable, a horizontal line (parallel to the x-axis) indicates a null effect on the P.F., whereas the presence of a slope is indicative of a certain influence that can also be quantified and then compared to those exhibited by other input factors [53].

Table 5. Thermal results for each formulation investigated in the present study.

Composites	Thermal Conductivity [W/m·K]		Mean Thermal Diffusivity [mm²/s]	
PLA 3D850 (Polymer), TNIGNP (GNPs), TNIMH4 (MWCNTs)				
PLA	2.05×10^{-1}	($\pm 3.91 \times 10^{-3}$)	$0{,}15 \times 10^{-1}$	($\pm 6.15 \times 10^{-2}$)
1.5%GNP	2.82×10^{-1}	($\pm 7.21 \times 10^{-4}$)	2.10×10^{-1}	($\pm 1.18 \times 10^{-2}$)
3%GNP	3.75×10^{-1}	($\pm 4.11 \times 10^{-3}$)	3.94×10^{-1}	($\pm 1.78 \times 10^{-2}$)
6%GNP	5.44×10^{-1}	($\pm 7.41 \times 10^{-3}$)	4.34×10^{-1}	($\pm 1.96 \times 10^{-3}$)
9%GNP	7.25×10^{-1}	($\pm 4.24 \times 10^{-2}$)	6.29×10^{-1}	($\pm 6.15 \times 10^{-3}$)
1.5%MWCNT	2.37×10^{-1}	($\pm 1.27 \times 10^{-3}$)	1.75×10^{-1}	($\pm 4.01 \times 10^{-3}$)
3%MWCNT	2.58×10^{-1}	($\pm 2.85 \times 10^{-3}$)	1.88×10^{-1}	($\pm 4.86 \times 10^{-3}$)
6%MWCNT	3.02×10^{-1}	($\pm 2.40 \times 10^{-3}$)	2.29×10^{-1}	($\pm 3.52 \times 10^{-3}$)
9%MWCNT	3.41×10^{-1}	($\pm 2.95 \times 10^{-3}$)	2.41×10^{-1}	($\pm 3.06 \times 10^{-3}$)
1.5%GNP + 1.5%MWCNT	3.14×10^{-1}	($\pm 3.94 \times 10^{-3}$)	2.33×10^{-1}	($\pm 8.06 \times 10^{-3}$)
1.5%GNP + 4.5%MWCNT	3.77×10^{-1}	($\pm 1.03 \times 10^{-2}$)	2.99×10^{-1}	($\pm 5.81 \times 10^{-3}$)
3%GNP + 3%MWCNT	4.40×10^{-1}	($\pm 1.20 \times 10^{-2}$)	3.61×10^{-1}	($\pm 2.08 \times 10^{-2}$)
4.5%GNP + 1.5%MWCNT	5.03×10^{-1}	($\pm 4.88 \times 10^{-3}$)	3.91×10^{-1}	($\pm 2.93 \times 10^{-2}$)
3%GNP + 6%MWCNT	4.32×10^{-1}	($\pm 8.34 \times 10^{-3}$)	3.57×10^{-1}	($\pm 7.88 \times 10^{-3}$)
6%GNP+ 3%MWCNT	5.69×10^{-1}	($\pm 1.72 \times 10^{-2}$)	4.86×10^{-1}	($\pm 8.58 \times 10^{-3}$)
PLA 3D850 (Polymer), TNGNP (GNPs), N7000 (MWCNTs)				
PLA	2.05×10^{-1}	($\pm 3.91 \times 10^{-3}$)	0.15×10^{-1}	($\pm 6.15 \times 10^{-2}$)
1.5%GNP	2.69×10^{-1}	($\pm 2.31 \times 10^{-3}$)	2.08×10^{-1}	($\pm 7.18 \times 10^{-3}$)
3%GNP	3.30×10^{-1}	($\pm 5.08 \times 10^{-3}$)	3.35×10^{-1}	($\pm 9.07 \times 10^{-2}$)
6%GNP	4.68×10^{-1}	($\pm 1.50 \times 10^{-2}$)	5.08×10^{-1}	($\pm 2.50 \times 10^{-2}$)
9%GNP	6.62×10^{-1}	($\pm 1.85 \times 10^{-2}$)	6.24×10^{-1}	($\pm 5.72 \times 10^{-2}$)
1.5%MWCNT	2.47×10^{-1}	($\pm 2.44 \times 10^{-2}$)	2.57×10^{-1}	($\pm 1.15 \times 10^{-2}$)
3%MWCNT	2.88×10^{-1}	($\pm 1.81 \times 10^{-2}$)	2.90×10^{-1}	($\pm 6.45 \times 10^{-2}$)
6%MWCNT	3.61×10^{-1}	($\pm 8.62 \times 10^{-3}$)	2.64×10^{-1}	($\pm 6.12 \times 10^{-3}$)
9%MWCNT	4.36×10^{-1}	($\pm 7.83 \times 10^{-3}$)	3.44×10^{-1}	($\pm 2.17 \times 10^{-2}$)
1.5%GNP + 1.5%MWCNT	3.05×10^{-1}	($\pm 6.47 \times 10^{-3}$)	2.24×10^{-1}	($\pm 3.25 \times 10^{-3}$)
1.5%GNP + 4.5%MWCNT	4.04×10^{-1}	($\pm 1.68 \times 10^{-2}$)	2.97×10^{-1}	($\pm 1.28 \times 10^{-2}$)
3%GNP + 3%MWCNT	4.24×10^{-1}	($\pm 8.60 \times 10^{-3}$)	3.28×10^{-1}	($\pm 6.19 \times 10^{-3}$)
4.5%GNP + 1.5%MWCNT	4.22×10^{-1}	($\pm 2.18 \times 10^{-2}$)	5.15×10^{-1}	($\pm 7.26 \times 10^{-2}$)
3%GNP + 6%MWCNT	5.21×10^{-1}	($\pm 1.67 \times 10^{-2}$)	4.21×10^{-1}	($\pm 2.60 \times 10^{-2}$)
6%GNP + 3%MWCNT	5.91×10^{-1}	($\pm 1.88 \times 10^{-2}$)	4.96×10^{-1}	($\pm 4.48 \times 10^{-2}$)

By analyzing these plots, it is possible to highlight the individual and combined influence of each filler on the overall thermal conductivity of the resulting nanocomposites. For formulations based on both N7000 and TNIGNP or TNIMH4 and TNIGNP fillers (Figure 13a,c or Figure 13b,d), an enhancement of the thermal properties is evident due to the progressive increase in graphene loading, as demonstrated by the positive slope of the MFP segments, which show coefficients $\alpha = 0.1709$ and $\alpha = 0.2263$ when TNGNP and TNGINP are used as reinforcement, respectively. Due to the higher value of the α-coefficient, the latter are thus confirmed to be the best indicated for improving thermal transport in composite materials. The gradual introduction of an amount of TNHMH4 carbon nanotubes as a replacement for graphene loading led to a worsening of the thermal conductivity, as demonstrated by the negative slope of the MFP, with a coefficient $\alpha = -0.0621$, whereas a sort of balancing effect ($\alpha = 0.0011$) was observed when using N7000 series.

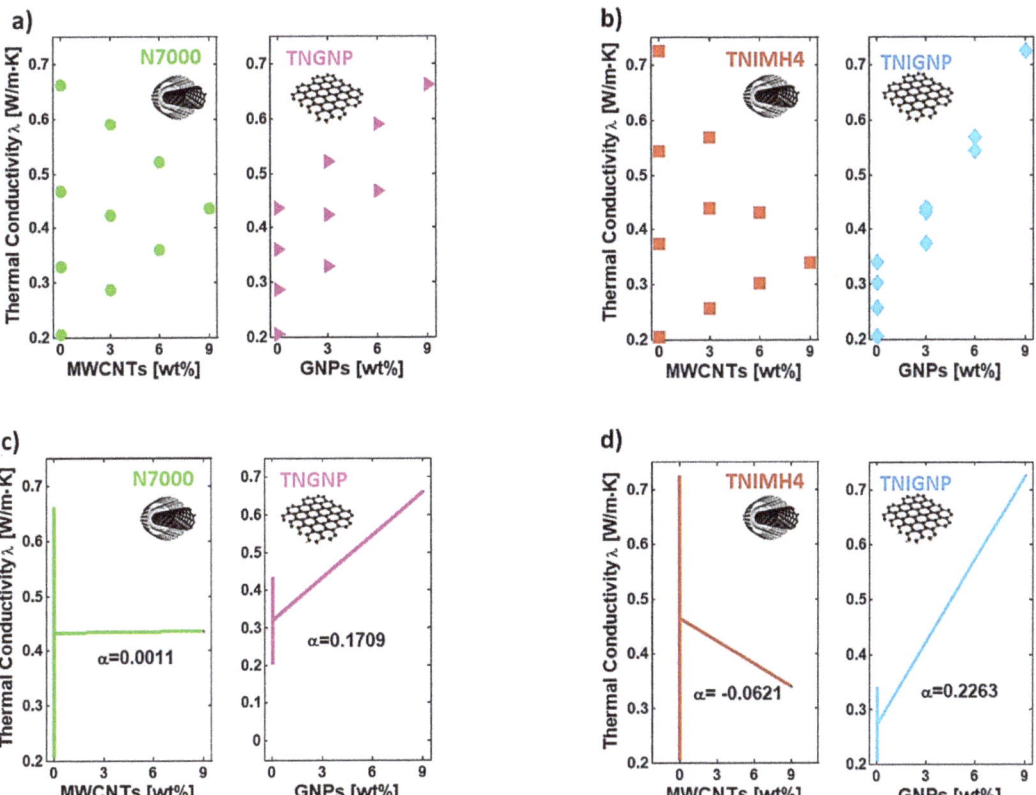

Figure 13. Dex Scatter Plot (DSP) and Main Factor Plot (MFP) for the experimental data of the thermal conductivity related to PLA reinforced with TNIMH4 and TNIGNP in (**a**,**b**) and with N7000 and TNGNP in (**c**,**d**), respectively.

3.5. Response Surface Methodology (RSM)

Although introduced for first the time in the early 1950s by Box and Wilson [54], even today, Response Surface Methodology (RSM), based on Design of Experiments, represents a mathematical method for predicting the relationship between several controlled factors and experimentally observed results. Since the form of the P.F. is not known, the main aim of RSM analysis is to predict the topography of the dependent variable (response surface, R.S.) in order to identify local maxima and minima, as well as the region in which the most effective response occurs in the face of controllable input changes. In general, R.S. can be mathematically expressed as:

$$R.S. = f(X_1, X_2, \ldots X_n) + \varepsilon \tag{16}$$

where f is the relationship between the R.S. and the independent input variables (X_i) and ε is the experimental error having a normal distribution with a null mean and a constant variance, as classically observed in statistical modeling. In this scenario, polynomial models are adopted for the surface prediction, and in particular, a first-order (linear) or second-order (quadratic) model, like that used in the current study, are normally sufficient for estimating the response of the most problems, especially if the problem is based on the variability of only two input variables (i.e., $wt\%_{MWCNTs}$, $wt\%_{GNPs}$, in our case) [55,56]. From a mathematical point of view, the polynomial quadratic model (n = 2) can be represented by the following expression:

$$R.S. = \beta_0 + \sum_{i=2}^{n} \beta_i x_i + \sum_{i=2}^{n} \beta_{ii} x_i^2 + \sum_{i=1}^{n-1} \sum_{j=i+1}^{n} \beta_{ij} x_i x_j \qquad (17)$$

where x_i, x_j are the coded independent input variables, β_0 is the coefficient of intercept whereas β_i, β_{ii} and β_{ij} are the linear, quadratic and interaction regression coefficients, respectively, which are determined by the least squares method. Here, with a particular interest in the thermal conductivity (i.e., λ), the aim is to derive an empirical model that correlates this property with the weight percentage of filler content, i.e., $\lambda = f(wt\%_{MWCNTs}, wt\%_{GNPs}) = f(x_1, x_2)$, for short. In accordance with Equation (17), the following quadratic polynomial approximates the values of the dependent variable λ:

$$\lambda = f(x_1, x_2) = \beta_0 + \beta_1 x_1 + \beta_2 x_2 + \beta_{12} x_1 x_2 + \beta_{11} x_1^2 + \beta_{22} x_2^2 \qquad (18)$$

The regression coefficients of the RSM are reported in Table 6, whereas a 3D plot of the R.S. with respect to the thermal conductivity of the nanocomposites including N7000 and TNGNP or TNIMH4 and TNIGNP are reported in Figure 14a,b, respectively.

Table 6. RSM regression coefficients for the quadratic response of the thermal conductivity λ.

Coefficient	β_0	β_1	β_2	β_{12}	β_{11}	β_{22}
Value for λ: (N7000-TNGNP):	+0.2073	+0.0270	+0.0331	+0.0022	-1.7143×10^{-4}	+0.0019
Value for λ (TNIMH4-TNIGNP):	+0.2021	+0.0231	+0.0591	−0.0023	-9.0714×10^{-4}	-1.4286×10^{-4}

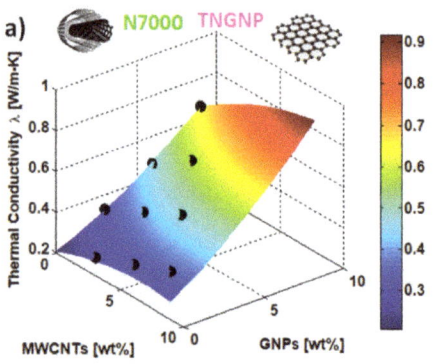
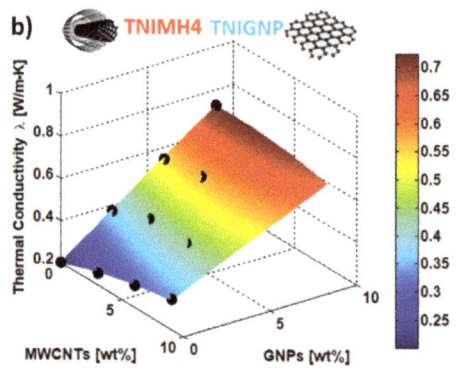

Figure 14. Response surface plot (full quadratic model) for the thermal conductivity depending upon the controllable input variables in the present study, i.e., the weight percentage of the fillers: N7000 and TNGNP or TNIMH4 and TNIGNP in (a,b), respectively. The black markers are the experimental data regarding thermal characterization.

It should be noted from analysis of the graph in Figure 14, that the R.S. approaches the experimental data (black markers) very well, thus evidencing the validity of the regression model for estimating the properties of a material/performance with respect to the conditioning parameters. The design of new advanced hybrid materials on the basis of simulation studies and numerical tools such as RSM may lead to the optimization of such materials without the need to test physical specimens, thus reducing their development time and avoiding the expensive costs necessitated by trial-and-error experiments. Moreover, the best composition and structure of novel materials can be achieved by coupling simulation-based approaches with statistical tools. Consequently, this will also make it possible to recognize the most sensitive controllable input factors in order to assist the design-choices in the manufacturing stage.

3.6. Simulation Results of Thermal Transport in Solid-State Heat Sinks

Heat sinks are passive heat exchangers designed to dissipate the heat generated by electronic devices, transferring it, by natural or forced convection, to a surrounding fluid medium (generally air), or in any case away from the device, in order to regulate the increase in temperature resulting from the operation frequency in the electronics industry. In recent years, advanced polymers have been studied for this purpose, and the advent of additive manufacturing has enabled the fabrication of 3D objects in more desirable complex shape, favoring the design of ad hoc heat sinks.

3.6.1. Thermal Analysis

Figure 15 shows a comparative analysis of the average temperature calculated on the lower and upper surfaces of the two considered heat sinks (pure PLA and PLA containing 9 wt% TNIGP), the thermal behaviors of which were simulated in the present study. It can be observed that, during the thermal transient, the temperature on each surface increases rapidly and, at the same time, the temperature difference between the lower and upper surfaces also increases; subsequently, at around 600 s, the temperature profiles tend asymptotically toward constant steady-state values, so that the temperature differences between the lower and upper surfaces also asymptotically tend toward a constant value; in fact, under steady-state conditions, the heat flow rate supplied at the lower surface is the same as that dissipated by natural convection from the upper surface and from the lateral surface.

Specifically, Figure 15a shows in the foreground the temperature profiles for both inferior and superior surfaces of the disc based on pure PLA and, in the insert of the same figure, the same profiles for that containing 9 wt% TNIGNP. The temperature difference evaluated at 1500 s between the surfaces of the individual heat sinks goes from a value of 10 K for pure PLA to 3 K for 9 wt% TNIGNP according to Equation (9), suggesting that, when subjected to the same heat flow, the higher the thermal conductivity of the solid, the lower the internal temperature gradient.

Analyzing Figure 15b, it can be observed that, under steady-state conditions at about 1500 s, the temperature difference between the lower surfaces of the two samples, through which the heat flow is supplied, remains constant at about 7 K, while the average temperatures of upper surfaces are comparable. This indicates that there is better heat transport in the graphene-based heat sink, as it is able to dissipate heat more efficiently than the disc made from unfilled polymer. The corresponding 3D temperature maps are shown in Figure 15c,d, in which the red arrows represent the direction and intensity of the conductive flux that crosses the samples under analysis and which will be analyzed in detail in the next section (Section 3.6.3. Heat Flux (Conductive/Convective) Analysis).

Continuing the thermal analysis, Figure 16 presents the temperature profiles as a function of thickness in correspondence with the symmetry axis and the respective temperature maps of the simulated disk samples (PLA in Figure 16a,c and 9 wt% TNIGNP in Figure 16b,d, respectively). Starting from the initial time, $t = 0$ s, at which point the sample is at an ambient temperature of 293.15 K, following the application of heat flow, times were chosen at intervals of 100 s during the transient phase up to 400 s, and then progressively increased as steady-state conditions approach 1500 s, a time beyond which no significant temperature differences were found. With reference to the lower ($z = 0$ mm) and upper ($z = 3$ mm) surfaces, the temperature values estimated at 900 s and 1500 s and their difference for the sample of pure PLA and for that with 9 wt% TNIGNP are shown in Table 7, below.

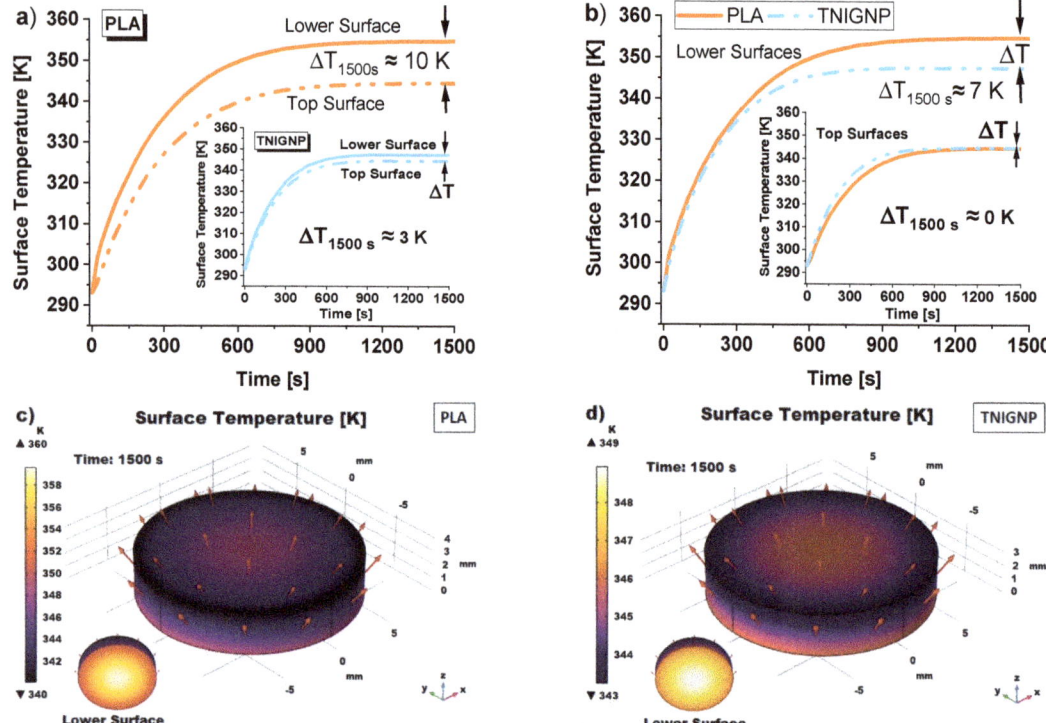

Figure 15. Average surface temperature profiles vs. time for heat sinks realized with (**a**) pure PLA and PLA including 9 wt% TNIGNP in the inset. A comparison of the mutual surfaces of the discs is reported in (**b**). 3D simulated views of the heat distribution on the surfaces of PLA-based disc (**c**) and on the surfaces of TNIGNP-based disc (**d**). In these last representations, red arrows indicate the direction and intensity of conductive heat flux.

Table 7. Lower/upper surface temperatures at 900 s and 1500 s.

Heat Sink	Lower/Upper Surface	T (900 s) [K]	T (1500 s) [K]	ΔT [K]
PLA	z = 0 mm	358.4	359.6	1.2
	z = 3 mm	347.8	349.0	0.1
TNIGNP	z = 0 mm	348.8	348.9	0.1
	z = 3 mm	345.8	345.9	0.1

It is worth noting that the expected drop in temperature through the thickness as the distance from the heat source increases (z = 0) occurs gradually for the heat sink made with TNIGNP, while it is more pronounced in that containing only PLA, as is visually evident from the temperature maps of the surfaces of the simulated heat sinks designed with PLA and TNIGNP, shown in Figure 16c,d, respectively. Once again, the thermal profile for thermal dissipative materials containing graphene-based nanoparticles is more uniform.

3.6.2. Total Internal Energy Analysis

A comparative analysis between the thermal profiles of the two heat sink discs, each of which was stressed with the same thermal flux on the lower surface, is presented in in Figure 17, which compares the internal energy trend during heating, showing that the internal energy follows the temperature profile reported in Figure 15a. The internal energy of each sample increases progressively until it reaches its own final value under steady-state conditions at 1500 s, while the difference between the internal energy of the PLA sample and that of the 9 wt% NTIGNP sample reaches a value of approximately 14,883 J/kg.

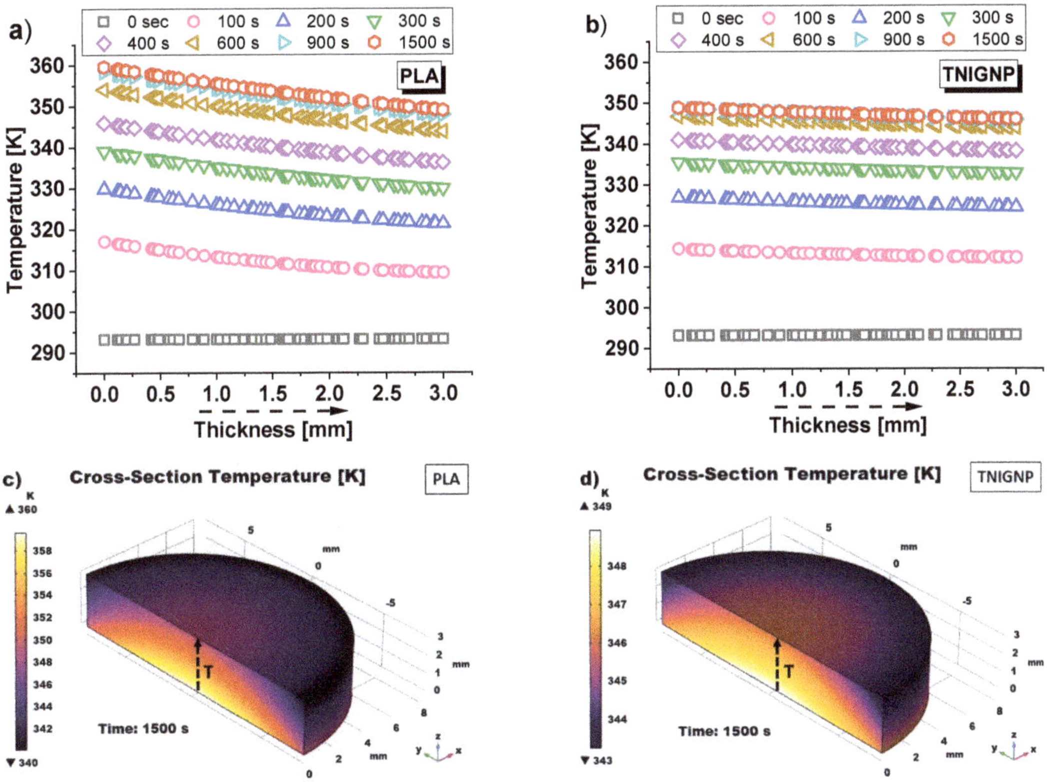

Figure 16. Temperature profiles (evaluated along the symmetry axis) for heat sinks made with PLA (**a**) and TNIGNP (**b**). 3D views of the cross-section temperature profiles are reported in (**c**,**d**) with an indication (black dashed arrows) of the direction in which the results were estimated.

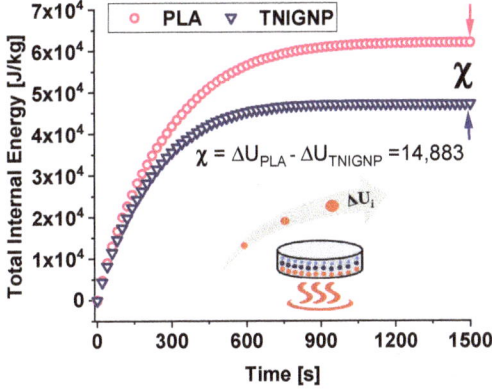

Figure 17. Variation of total internal energy due to the heating of the disc when it acts as a heat sink.

In the case of PLA, the internal energy variation is greater than that estimated for the disc containing 9% by weight of TNIGNP due to the different level of trapped warmth, and the resulting temperature profile, within the solid.

Furthermore, in Figure 18a,b, which show the 3D maps of temperature and total internal energy for PLA and TNIGNP, respectively, evaluated at t = 1500 s, a visual analysis of the variation along the radial direction is presented, as well as with the thickness.

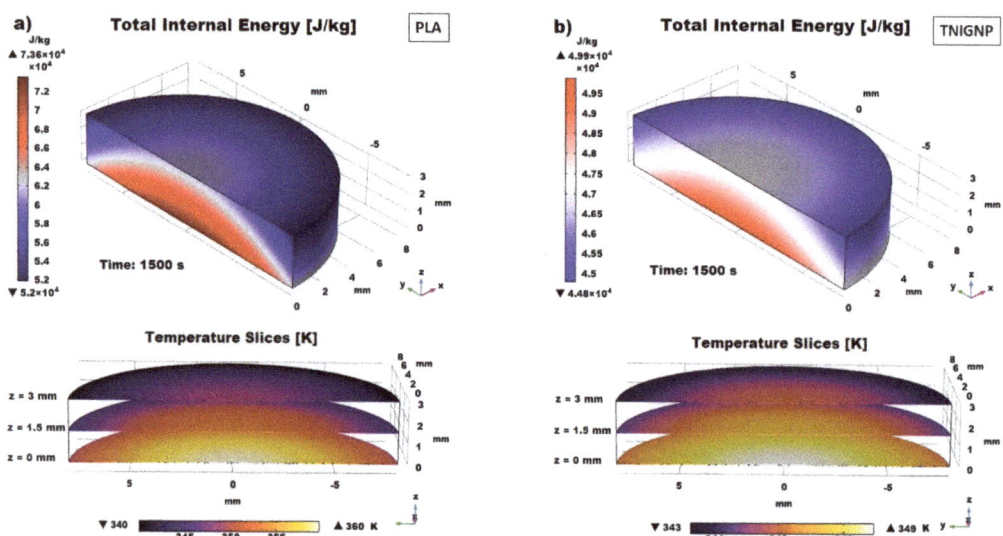

Figure 18. 3D views of the total internal energy due to the heating of the simulated discs when they act as a heat sink: PLA in (**a**) and TNIGNP in (**b**). Temperature slices (at z = 0 mm, z = 1.5 mm and z = 3 mm) established within solid disks are also reported.

3.6.3. Heat Fluxes (Conductive/Convective) Analysis

Figure 19a,b report the internal conductive heat flux trend (average values) and of the external one by natural convection as function of time, respectively. By observing them, stands out immediately that both fluxes, regardless the nature of heat sink if it is based on pure PLA or including TNIGNP, appear to be roughly equal-sided, already before steady-state condition. At the end of observing time of 1500 s a value of 784 (W/m^2) and 792 (W/m^2) are estimated for the conductive heat flux whereas 514.4 (W/m^2) and 514.1 (W/m^2) are the values for the convective flux of PLA and TNIGNP, respectively. On the other hand, as clearly highlighted in the magnification of the first 60 s depicted in the inset of Figure 19a, significant changes can be observed during the transient phase, especially for conductive heat flux. This difference is due to the graphene particles, which, as already explained, favor thermal transport when introduced in the polymer matrix by reducing the temperature gradient proportionally, by means of thermal conductivity. When thermal equilibrium is reached, the conductive heat flux will have been completely transferred to the surrounding environment, and there will be no further changes with respect to the thermal properties.

In light of this, the moment of time at 300 s, which falls approximately in the middle of the transient phase, is taken as a reference for the performance comparison with respect to the heat fluxes of the two simulated disks. At this selected time, a difference of about 42 (W/m^2) and 45 (W/m^2) is estimated for the conductive and convective flux, respectively. In steady-state condition (1500 s) the conductive and convective fluxes stabilize to a value of 701.29 (W/m^2) and 705.61 (W/m^2) or −514.04 (W/m^2) and −514.49 (W/m^2) for PLA and TNIGNP heat sink, respectively.

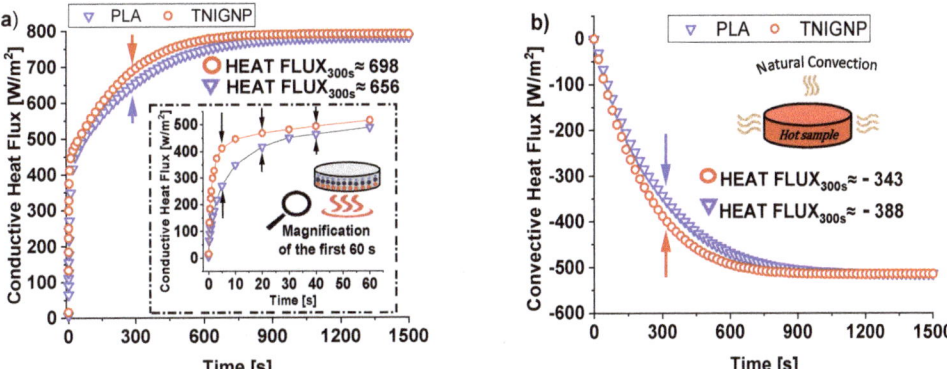

Figure 19. Conductive (**a**) and convective (**b**) heat flux evolution over time.

3D views of their distribution profiles within the solid are shown in Figure 20. More specifically, Figure 20a,b respectively report the conductive heat flux (z component) for PLA and TNIGNP along some selected equidistant cross-sections of the materials. Consulting such graphics with their relative color bars not only helps to quantify these fluxes, but also to discriminate the different rates (greater for PLA containing TNIGNP than for pure PLA). Finally, Figure 20c,d illustrate the profile of convective heat flux at the solid/air exchange surface for pure and filled polymer. It is worth noting that, in both cases, the convective flux is greater in the central part of the solids, due to both the greater exchange surface compared to the side walls as well as the higher temperature achieved in this region, in line with the thermal profiles previously discussed.

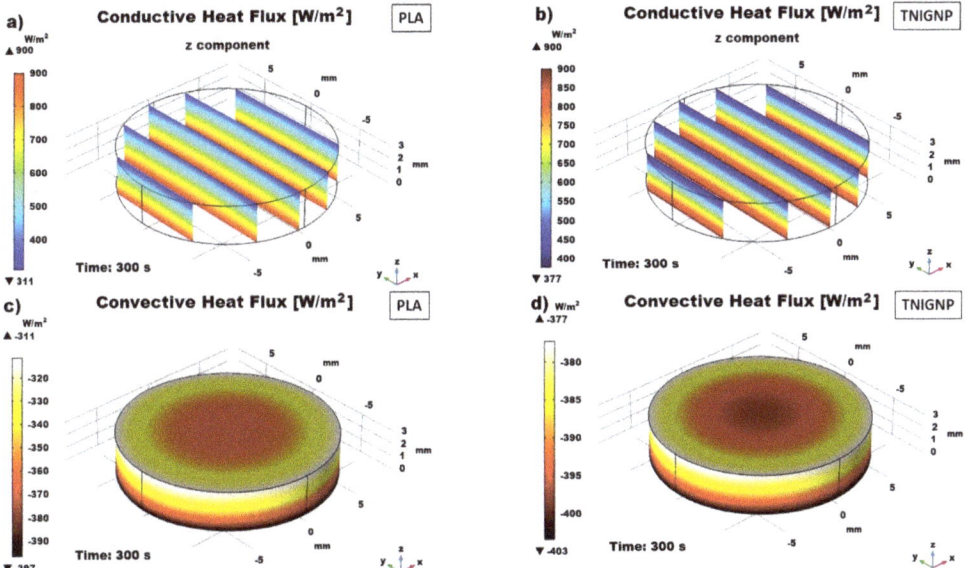

Figure 20. 3D views of the conductive heat flux in some cross-sections of PLA (**a**) and TNIGNP (**b**); their corresponding convective heat flux is presented in (**c**,**d**), respectively.

4. Discussion

The thermal behavior of poly(lactic acid) (PLA) filled with different carbonaceous fillers was experimentally, theoretical and numerically investigated.

Nanocomposites based on N7000-type CNTs, despite their small A.R. of 150, showed better thermal conductivity than those realized with TNIMH4. This result may be attributed to their shorter length, which leads to their easy and better dispersion, which in turn results in better interparticle contacts in the percolating network. The TNIMH4 nanotubes show an A.R. of 1000, which, combined with the longer length, most likely prevents good dispersion in the melt during the extrusion process. Of course, this different dispersion state conditions the thermal properties of the resulting composites.

The best thermal performances were revealed for composites including graphene-based particles, rather than carbon nanotubes. The explanation for this is to be found in the different interface interactions between the organic polymer and the carbonaceous fillers. For the 1-dimensional ones, like nanotubes, the Kapitza resistance (Rk) shows higher values due to the inner surfaces that are poorly wetted by the PLA in contrast to what happens for graphene nanoplatelets, where these surface impregnation phenomena are favored by their planar shapes. In this last case, a lowering of the differences of phonon density of the states between the constituent phases determines a more effective phononic heat flow.

Despite the high intrinsic thermal conductivities of both fillers (as reported in Table 1), the overall thermal performances of the resulting nanocomposites were decisively far from these values. The reason for this is the fact that, in percolated structures, macroscale properties are strongly conditioned by a great number of factors that occur at the micro- and nanoscale level, such as aggregations and dispersion states, functional group effects, crystallinity and surface tensions of the matrix, and so on. All these effects are not yet fully understood or predictable in the design stage, and therefore, further studies are needed in order to add knowledge in the field and to obtain new findings. Theoretical and simulation studies can support experimental research in order to achieve this objective. Design of Experiment (DoE), combined with Main Factor Plot (MFP) and the Response Surface Methodology (RSM) can be successfully applied in many experimental situations, and more recently, they have increasingly been used in the field of the design and development of materials. In this context, these statistical approaches appear to be particularly useful for experiments, especially when they are based on destructive tests, which lead to the depletion of resources, or when they are based on rare/expensive materials, as in the case of carbon-based fillers. These techniques improve experimental efficiency, as well as being able to assist the designer in comprehending factor interactions, since they provide detailed information with respect to process evolution. Additionally, their use allows us to achieve new insights into the behavior of advanced nanocomposites. The DoE approach employed in this study made it possible to quantify the influence of each of the investigated fillers, while RSM allowed us to derive a polynomial equation that was able to correlate their loading levels with the thermal conductivity. Given the reliability of the predictive models, whose results match very well with the experimental data, the theoretical and numerical studies are able to provide useful information during the design stage of nanocomposites by limiting the classical experimental activity based on onerous trial and error approaches, in which something is tested until one finds the most successful parameters. Multiphysics simulations make it possible to investigate the thermal behavior of polymer-based nanocomposites for their potential use in heat transfer applications. In the present study, by comparing the thermal performances of two heat sinks, based on unfilled PLA and reinforcement with 9 wt% TNIGNP, it was proved that the introduction of filler significantly improved the thermal properties. Therefore, simulation results also confirmed the key role of graphene in improving the thermal features of materials in which it is dispersed. Moreover, thanks to the thermal simulation studies, the working temperatures of the designed heat sinks can be tailored to be lower than the glass transition, thus avoiding the risk of material degradation in the case of their use in heat transfer applications.

5. Conclusions

This paper mainly addressed the study of the thermal behavior of poly(lactic acid) composites reinforced with different carbonaceous fillers produced via melt compounding. An experimental characterization in terms of morphology, electrical and thermal properties was carried out, and the results were discussed with reference to theoretical expectations. Carbon nanotubes, especially of type N7000, were found to be the best performing in terms of electrical properties, given their greater efficiency in creating percolation paths within the polymer. The percolation threshold was achieved with a filler concentration lower than 1.5 wt%, whereas a remarkable electrical conductivity of 2 S/m was measured at the highest investigated loading. For nanocomposites based on both types of graphene, the EPT fell in a wider range of [3÷6] wt%, and an electrical conductivity value of about 7×10^{-2} S/m was achieved in the best case (9 wt% TNIGNP). In contrast, industrial graphene nanoplatelets (TNIGNP) were found to be the best performing from a thermal point of view, since an increment of the thermal conductivity by 254% with respect to that of unfilled polymer was achieved in the best case (at concentration of 9 wt%). By using MWCNTs, the improvement in thermal conductivity with respect to the pure PLA was reduced to the still-interesting value of 112%, due to the conductivity value of 0.436 W/m·K measured for nanocomposites including 9 wt% N7000. Design of Experiment, including the response surface methodology, was performed to introduce predictive numerical models that would be useful for the design of such advanced materials. With reference to the best thermally performing nanocomposite (TNIGNP) and unfilled PLA, multiphysics simulations were carried out to numerically investigate and compare their thermal behavior when used as heat sinks for potential heat-transfer applications. A lower surface heating and a lower achieved temperature were observed at equilibrium (about 354 K and 344 K, for neat PLA and PLA with TNIGNPs), as well as a more efficient environmental heat exchange, especially during the transient phase, with a difference of 45 [W/m^2] at an observation time of 300 s for the two simulated heat sinks. At the same time, a greater heat convective flux (642 W/m^2) was exhibited by the TNIGNP-based heat sink with respect to that realized with pure PLA (613 W/m^2), which is an indicator of better thermal transport.

Author Contributions: Conceptualization G.S. and R.G.; Writing—original draft G.S.; Investigation of electrical and thermal properties G.S. and E.I.; Morphological analysis E.I.; Test samples E.I.; Multiphysics Simulation studies: R.G.; Writing—review R.K., V.R. and E.I. All authors have read and agreed to the published version of the manuscript.

Funding: This research was funded by Marie Skłodowska-Curie Actions (MSCA) Research and Innovation Staff Exchange (RISE) H2020-MSCA-RISE-2016, Project Acronym: Graphene 3D—Grant Number: 734164. This research was partially supported by the project BG05 M2OP001-1.001-0008 "National Center for Mechatronics and Clean Technologies", funded by the Operating Program "Science and Education for Intelligent Growth" of Republic of Bulgaria.

Institutional Review Board Statement: Not applicable.

Informed Consent Statement: Not applicable.

Data Availability Statement: Data are contained within the article.

Acknowledgments: The authors acknowledge support from H2020-SGS-FET-Graphene Flagship-881603-Graphene Core 3. The author would like to acknowledge the contribution of the COST Action CA15107, COST Action CA 19118 (Contract KP-06-COST-11 with the BNSF), and Bilateral collaboration between IMech, BAS and CNR, Napoli/Portici (2019–2021).

Conflicts of Interest: The authors declare no conflict of interest.

References

1. Namazi, H. Polymers in our daily life. *BioImpacts* **2017**, *7*, 73. [CrossRef] [PubMed]
2. Foti, D. Preliminary analysis of concrete reinforced with waste bottles PET fibers. *Constr. Build. Mater.* **2011**, *25*, 1906–1915. [CrossRef]

3. Horrocks, R.; Sitpalan, A.; Zhou, C.; Kandola, B.K. Flame retardant polyamide fibres: The challenge of minimising flame retardant additive contents with added nanoclays. *Polymers* **2016**, *8*, 288. [CrossRef]
4. Lunt, J. Large scale production, properties and commercial applications of polylactic acid polymers. *Polym. Degrad. Stab.* **1998**, *59*, 145–152. [CrossRef]
5. Gupta, B.; Revagade, N.; Hilborn, J. Poly(lactic) fiber: An overview. *Prog. Polym. Sci.* **2007**, *32*, 455–482. [CrossRef]
6. Le Phuong, H.A.; Izzati Ayob, N.A.; Blanford, C.F.; Mohammad Rawi, N.F.; Szekely, G. Nonwoven membrane supports from renewable resources: Bamboo fiber reinforced poly (lactic acid) composites. *ACS Sustain. Chem. Eng.* **2019**, *7*, 11885–11893. [CrossRef]
7. Bhagabati, P.; Bhasney, S.M.; Bose, D.; Remadevi, R.; Setty, M.; Rajkhowa, R.; Katiyar, V. Silk and Wool Protein Microparticle-Reinforced Crystalline Polylactic Acid Biocomposites with Improved Cell Interaction for Targeted Biomedical Applications. *ACS Appl. Polym. Mater.* **2020**, *2*, 4739–4751. [CrossRef]
8. Van den Eynde, M.; Van Puyvelde, P. 3D Printing of Poly(lactic acid). In *Industrial Applications of Poly(Lactic Acid)*; Di Lorenzo, M., Androsch, R., Eds.; Advances in Polymer Science; Springer: Cham, Germany, 2017; Volume 282, pp. 139–158.
9. Vyavahare, S.; Teraiya, S.; Panghal, D.; Kumar, S. Fused deposition modelling: A review. *Rapid Prototyp. J.* **2020**, *26*, 176–201. [CrossRef]
10. Mark, J.E. *Polymer Data Handbook*; Oxford University Press: London, UK, 2009; p. 1264.
11. Durakovic, B. Design for additive manufacturing: Benefits, trends and challenges. *Period. Eng. Nat. Sci.* **2018**, *6*, 179–191. [CrossRef]
12. Singamneni, S.; Yifan, L.V.; Hewitt, A.; Chalk, R.; Thomas, W.; Jordison, D. Additive Manufacturing for the Aircraft Industry: A Review. *J. Aeronaut. Aerosp. Eng.* **2019**, *8*, 1–13. [CrossRef]
13. Javaid, M.; Haleem, A. Additive manufacturing applications in medical cases: A literature based review. *Alex. J. Med.* **2018**, *54*, 411–422. [CrossRef]
14. Salmi, M. Additive Manufacturing Processes in Medical Applications. *Materials* **2021**, *14*, 191. [CrossRef] [PubMed]
15. Hoerber, J.; Glasschroeder, J.; Pfeer, M.; Schilp, J.; Zaeh, M.; Franke, J. Approaches for additive manufacturing of 3D electronic applications. *Procedia CIRP* **2014**, *17*, 806–811. [CrossRef]
16. Bourell, D.; Kruth, J.P.; Leu, M.; Levy, G.; Rosen, D.; Beese, A.M.; Clare, A. Materials for additive manufacturing. *CIRP Ann.* **2017**, *66*, 659–681. [CrossRef]
17. Kruth, J.P.; Leu, M.C.; Nakagawa, T. Progress in additive manufacturing and rapid prototyping. *CIRP Ann.* **1998**, *47*, 525–540. [CrossRef]
18. Chen, H.; Ginzburg, V.V.; Yang, J.; Yang, Y.; Liu, W.; Huang, Y.; Du, L.; Chen, B. Thermal conductivity of polymer-based composites: Fundamentals and applications. *Prog. Polym. Sci.* **2016**, *59*, 41–85. [CrossRef]
19. Zhidong, H.; Fina, A. Thermal conductivity of carbon nanotubes and their polymer nanocomposites: A review. *Progr. Polym. Sci.* **2011**, *36*, 914–944.
20. Guo, Y.; Ruan, K.; Shi, X.; Yang, X.; Gu, J. Factors affecting thermal conductivities of the polymers and polymer composites: A review. *Compos. Sci. Technol.* **2020**, *193*, 108134. [CrossRef]
21. Ramasubramaniam, R.; Chen, J.; Liu, H. Homogeneous carbon nanotube/polymer composites for electrical applications. *Appl. Phys. Lett.* **2003**, *83*, 2928–2930. [CrossRef]
22. Jin, J.; Lin, Y.; Song, M.; Gui, C.; Leesirisan, S. Enhancing the electrical conductivity of polymer composites. *Eur. Polym. J.* **2013**, *49*, 1066–1072. [CrossRef]
23. Huang, X.; Jiang, P.; Tanaka, T. A review of dielectric polymer composites with high thermal conductivity. *IEEE Electr. Insul. Mag.* **2011**, *27*, 8–16. [CrossRef]
24. Ngo, I.-L.; Jeon, S.; Byon, C. Thermal conductivity of transparent and flexible polymers containing fillers: A literature review. *Int. J. Heat Mass Trans.* **2016**, *98*, 219–226. [CrossRef]
25. Yang, Y. *Physical Properties of Polymer Handbook*; Springer Science & Business Media: New York, NY, USA, 2007; pp. 155–164.
26. Maldovan, M. Sound and heat revolutions in phononics. *Nature* **2013**, *503*, 209–217. [CrossRef]
27. Nan, C.W.; Shen, Y.; Ma, J. Physical properties of composites near percolation. *Annu. Rev. Mater. Res.* **2010**, *40*, 131–151. [CrossRef]
28. Pezzana, L.; Riccucci, G.; Spriano, S.; Battegazzore, D.; Sangermano, M.; Chiappone, A. 3D Printing of PDMS-Like Polymer Nanocomposites with Enhanced Thermal Conductivity: Boron Nitride Based Photocuring System. *Nanomaterials* **2021**, *11*, 373. [CrossRef]
29. Ji, J.; Chiang, S.W.; Liu, M.; Liang, X.; Li, J.; Gan, L.; Du, H. Enhanced thermal conductivity of alumina and carbon fibre filled composites by 3-D printing. *Thermochim. Acta* **2020**, *690*, 178649. [CrossRef]
30. Laureto, J.; Tomasi, J.; King, J.A.; Pearce, J.M. Thermal properties of 3-D printed polylactic acid-metal composites. *Prog. Addit. Manuf.* **2017**, *2*, 57–71. [CrossRef]
31. Huang, C.; Qian, X.; Yang, R. Thermal conductivity of polymers and polymer nanocomposites. *Mater. Sci. Eng. R Rep.* **2018**, *132*, 1–22. [CrossRef]
32. Kotsilkova, R.; Ivanov, E.; Georgiev, V.; Ivanova, R.; Menseidov, D.; Batakliev, T.; Angelov, V.; Xia, H.; Chen, Y.; Bychanok, D.; et al. Essential Nanostructure Parameters to Govern Reinforcement and Functionality of Poly (lactic) Acid Nanocomposites with Graphene and Carbon Nanotubes for 3D Printing Application. *Polymers* **2020**, *12*, 1208. [CrossRef] [PubMed]

33. Guadagno, L.; Raimondo, M.; Vertuccio, L.; Mauro, M.; Guerra, G.; Lafdi, K.; De Vivo, B.; Lamberti, P.; Spinelli, G.; Tucci, V. Optimization of graphene-based materials outperforming host matrices. *RSC Adv.* **2015**, *5*, 36969–36978. [CrossRef]
34. Spinelli, G.; Lamberti, P.; Tucci, V.; Kotsilkova, R.; Ivanov, E.; Menseidov, D.; Naddeo, C.; Romano, V.; Guadagno, L.; Adami, R.; et al. Nanocarbon/Poly(Lactic) Acid for 3D Printing: Effect of Fillers Content on Electromagnetic and Thermal Properties. *Materials* **2019**, *12*, 2369. [CrossRef] [PubMed]
35. ISO 22007-2:2015. *Plastic—Determination of Thermal Conductivity and Thermal Diffusivity—Part 2: Transient Plane Heat Source (Hot Disc) Method*; Swedish Standards Institute: Stockholm, Sweden, 2015.
36. Wagner, J.R.; Mount, E.M.; Giles, H.F. Design of experiments. In *Plastics Design Library, Extrusion*, 2nd ed.; William Andrew Publishing: Norwich, NY, USA, 2014; pp. 291–308.
37. Berk, J.; Berk, S. ANOVA, Taguchi, and Other Design of Experiments Techniques: Finding needles in haystacks. In *Quality Management for the Technology Sector*; Berk, J., Berk, S., Eds.; Butterworth-Heinemann: Oxford, UK, 2000; pp. 106–123.
38. Montgomery, D.C. *Design and Analysis of Experiments*, 8th ed.; John Wiley & Sons: Hoboken, NJ, USA, 2001.
39. Rosenberger, A.G.; Hardt, J.C.; Dragunski, D.C.; da Silva, F.F.; Bittencourt, P.R.S.; Bariccatti, R.A.; Caetano, J. Use of experimental design to obtain polymeric microfibers with carbon nanotubes. *Adv. Manuf. Polym. Compos. Sci.* **2020**, *6*, 115–126.
40. Hardian, R.; Liang, Z.; Zhang, X.; Szekely, G. Artificial intelligence: The silver bullet for sustainable materials development. *Green Chem.* **2020**, *22*, 7521–7528. [CrossRef]
41. Vicario, G. Computer experiments: Promising new frontiers in analysis and design of experiments. *Stat. Appl.* **2006**, *18*, 231–249.
42. Bird, R.; Stewart, W.; Lightfoot, E. *Transport Phenomena*, 2nd ed.; John Wiley and Sons: New York, NY, USA, 2002.
43. Mansor, M.R.; Fadzullah, S.H.S.M.; Masripan, N.A.B.; Omar, G.; Akop, M.Z. Comparison between functionalized graphene and carbon nanotubes: Effect of morphology and surface group on mechanical, electrical, and thermal properties of nanocomposites. In *Functionalized Graphene Nanocomposites and Their Derivatives*, 1st ed.; Jawaid, M., Bouhfid, R., el Kacem Qaiss, A., Eds.; Elsevier: Oxford, UK, 2019; pp. 177–204.
44. Mutiso, R.M.; Winey, K.I. Electrical Conductivity of Polymer Nanocomposites. In *Polymer Science: A Comprehensive Reference*, 1st ed.; Moeller, M., Matyjaszewski, K., Eds.; Elsevier: Oxford, UK, 2012; Volume 7, pp. 327–344.
45. Foygel, M.; Morris, R.D.; Anez, D.; French, S.; Sobolev, V.L. Theoretical and computational studies of carbon nanotube composites and suspensions: Electrical and thermal conductivity. *Phys. Rev. B* **2005**, *71*, 104201. [CrossRef]
46. Johner, N.; Grimaldi, C.; Balberg, I.; Ryser, P. Transport exponent in a three-dimensional continuum tunneling-percolation model. *Phys. Rev. B* **2008**, *77*, 174204. [CrossRef]
47. Li, C.; Thostenson, E.-T.; Chou, T.-W. Dominant role of tunneling resistance in the electrical conductivity of carbon nanotube–based composites. *Appl. Phys. Lett.* **2007**, *91*, 223114. [CrossRef]
48. De Vivo, B.; Lamberti, P.; Spinelli, G.; Tucci, V. Numerical investigation on the influence factors of the electrical properties of carbon nanotubes-filled composites. *J. Appl. Phys.* **2013**, *113*, 244301. [CrossRef]
49. Hone, J.; Whitney, M.; Piskoti, C.; Zettl, A. Thermal conductivity of single-walled carbon nanotubes. *Phys. Rev. B* **1999**, *59*, 2514–2516. [CrossRef]
50. Balandin, A.A. Thermal properties of graphene and nanostructured carbon materials. *Nat. Mater.* **2001**, *10*, 569–581. [CrossRef] [PubMed]
51. White, C.C.; Hunston, D.L.; Tan, K.T.; Filliben, J.J.; Pintar, A.; Schueneman, G. A Systematic Approach to the Study of Accelerated Weathering of Building Joint Sealants. *J. ASTM Int.* **2012**, *9*, 1–17. [CrossRef]
52. Croarkin, C.; Tobias, P.; Filliben, J.J.; Hembree, B.; Guthrie, W. *Handbook 151: NIST/SEMATECH e-Handbook of Statistical Methods*; NIST Interagency/Internal Report (NISTIR); National Institute of Standards and Technology: Gaithersburg, MD, USA, 2002. [CrossRef]
53. Kuehl, R.O. *Design of Experiment: Statistical Principles of Research Design and Analysis*, 2nd ed.; Duxbury Press: Pacific Grove, CA, USA, 1999.
54. Box, G.E.P.; Wilson, K.B. On the experimental attainment of optimum conditions. In *Breakthroughs in Statistics*; Springer: New York, NY, USA, 1992; pp. 270–310.
55. Khuri, A.I. Multiresponse surface methodology. In *Design and Analysis of Experiments. Handbook of Statistics 13*; Ghosh, S., Rao, C.R., Eds.; Elsevier Science: Amsterdam, The Netherlands, 1996; pp. 377–406.
56. Draper, N.R.; Lin, D.K.J. Response surface designs. In *Design and Analysis of Experiments. Handbook of Statistics 13*; Ghosh, S., Rao, C.R., Eds.; Elsevier Science: Amsterdam, The Netherlands, 1996; pp. 343–375.

Article

Hierarchical Microstructure of Tooth Enameloid in Two Lamniform Shark Species, *Carcharias taurus* and *Isurus oxyrinchus*

Jana Wilmers [1,*], Miranda Waldron [2] and Swantje Bargmann [1,3]

1   Chair of Solid Mechanics, University of Wuppertal, 42119 Wuppertal, Germany; bargmann@uni-wuppertal.de
2   Electron Microscope Unit, University of Cape Town, Cape Town 7701, South Africa; miranda.waldron@uct.ac.za
3   Wuppertal Center for Smart Materials, University of Wuppertal, 42119 Wuppertal, Germany
*   Correspondence: wilmers@uni-wuppertal.de; Tel.: +49-202-439-2086

Abstract: Shark tooth enameloid is a hard tissue made up of nanoscale fluorapatite crystallites arranged in a unique hierarchical pattern. This microstructural design results in a macroscopic material that is stiff, strong, and tough, despite consisting almost completely of brittle mineral. In this contribution, we characterize and compare the enameloid microstructure of two modern lamniform sharks, *Isurus oxyrinchus* (shortfin mako shark) and *Carcharias taurus* (spotted ragged-tooth shark), based on scanning electron microscopy images. The hierarchical microstructure of shark enameloid is discussed in comparison with amniote enamel. Striking similarities in the microstructures of the two hard tissues are found. Identical structural motifs have developed on different levels of the hierarchy in response to similar biomechanical requirements in enameloid and enamel. Analyzing these structural patterns allows the identification of general microstructural design principles and their biomechanical function, thus paving the way for the design of bioinspired composite materials with superior properties such as high strength combined with high fracture resistance.

Keywords: enameloid; microstructure; shark; teeth

1. Introduction

Shark teeth have always exerted a special kind of fascination on humans as they are the perfectly designed, highly efficient natural weapons of a deadly hunter. The teeth are arranged in multiple rows behind each other in the shark's jaws (Figure 1) and exhibit a variety of shapes and sizes among different species, ranging from flattened domes over needles to triangular cutting tools with sharp, serrated edges. Even between closely related species, a wide variety of tooth shapes can be found [1] and may in fact be used to identify species [2]. These morphological variations have been attributed to differences in feeding behavior and, thus, mechanical loads [3–5]. Optimal functionality is further guaranteed by regular shedding and replacement of the teeth. The replacement rate varies between species and with age and water temperature. For spotted ragged-tooth sharks (*Carcharias taurus*), an average tooth loss rate of 1.06 teeth per day was identified [6], which means an individual spotted ragged-tooth shark will shed over 13,500 teeth in a lifetime. For other species, even more rapid replacement can be found [7,8].

Based on mechanical studies [3,9], it has been proposed that the frequent tooth replacement of shark teeth is due to wear rather than failure by tooth fracture. Wear on a small scale would not impact the structural strength of the tooth but would reduce its efficiency as a cutting or piercing tool. Studies aimed at understanding the biomechanics of shark teeth have found remarkably similar stress patterns even for drastically different tooth morphologies under static loading [9]. Dynamic tests show the increased cutting efficiency of serrated edges compared to smoother teeth, as well as their drastically faster

Citation: Wilmers, J.; Waldron, M.; Bargmann, S. Hierarchical Microstructure of Tooth Enameloid in Two Lamniform Shark Species, *Carcharias taurus* and *Isurus oxyrinchus*. *Nanomaterials* **2021**, *11*, 969. https:// doi.org/10.3390/nano11040969

Academic Editor: Daniel Kiener

Received: 16 March 2021
Accepted: 6 April 2021
Published: 9 April 2021

Publisher's Note: MDPI stays neutral with regard to jurisdictional claims in published maps and institutional affiliations.

Copyright: © 2021 by the authors. Licensee MDPI, Basel, Switzerland. This article is an open access article distributed under the terms and conditions of the Creative Commons Attribution (CC BY) license (https:// creativecommons.org/licenses/by/ 4.0/).

mechanical wear [3]. In many shark species, however, tooth-on-prey contact leading to wear is comparatively rare as only large prey is manipulated with teeth [10].

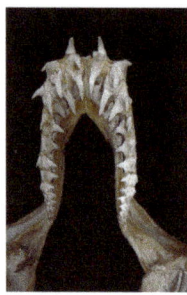

(**a**) jaw of *I. oxyrinchus*   (**b**) *C. taurus* at Two Oceans Aquarium, Cape Town

Figure 1. Shark teeth are arranged in series. Multiple rows of teeth grow behind each other and are continuously replaced over a lifetime. The primary function of the teeth of *I. oxyrinchus* (**a**,**b**) is to pierce (not cut) prey. The jaws of both shark species can be opened very wide to swallow prey in one piece.

It is well known that the mechanical properties of biological materials strongly depend on their complex microstructures. Hypermineralized tissues like shark tooth enameloid consist almost exclusively of nanoscale brittle mineral crystallites [4], yet exhibit considerable strength and exceptionally high fracture toughness way beyond that of the individual constituents. The relationship between the sophisticated microstructural hierarchy and the mechanical performance of other highly mineralized tissues such as amniote enamel or mollusc nacre has attracted considerable research interest, e.g., [11–15]. Shark enameloid, while similar in appearance and function to amniote enamel, is considerably less understood.

Enameloid consists of elongated fluorapatite ($Ca_2(PO_4)F$) crystallites with a roughly hexagonal cross-section visible in TEM images in different shark species [16,17]. The crystallites have a width of 50 nm to 80 nm and a length exceeding 1000 nm [17,18]. They are densely packed and arranged in a complex hierarchical structure. In all neoselachian sharks, the majority of the enameloid cover consists of bundled crystallite enameloid (BCE) and the tooth's outer surface is covered in single crystal enameloid(SCE) (Single crystal enameloid is also commonly referred to as 'shiny-layered enameloid' reflecting its optical appearance) [19–21]. In the BCE, fluorapatite nanocrystallites are arranged in bundles that may be oriented longitudinally, radially, or circumferentially within the tooth, with slight variations occurring between different shark species [20,22]. In modern sharks (*selachimorpha*), bundles close to the dentine-enameloid junction are three-dimensionally interwoven, a structure type referred to as tangled bundled enameloid (TBE). Further from the dentine, the bundle arrangement becomes more regular, with the bundles aligned parallel to each other and to the tooth's longitudinal axis, a structural motif referred to as parallel bundled enameloid (PBE).

This characteristic layered structure, is often referred to as a 'triple layered structure', with TBE, PBE and SCE each constituting one layer. However, the microstructure of shark enameloid has been found to be more complex than this description implies as it does not take into account, e.g., radial structural elements described in [21]. Instead, a nomenclature discerning the inner enameloid consisting of parallel bundled and tangled bundled enameloid, jointly referred to as the BCE unit, and an outer enameloid layer referred to as the ridge/cutting edge layer (RCEL) has been suggested in [19]. The RCEL itself consists of an external layer of single crystal enameloid and an internal layer of circumferential bundles. The inner and the outer enameloid are separated clearly as visible in micrographs, while transitions between the substructures within each layer are generally smooth. Elemental analysis of shark tooth enameloid shows that the crystallite composition varies slightly between the inner and the outer enameloid, with the outer enameloid being richer in the substituting ions magnesium and sodium than the inner, bundled enameloid [18].

The enameloid microstructures in other neoselachian species vary more drastically from the patterns identified in modern sharks. Batoids (rays and skates), for instance, exhibit a less complex microstructure, with some species even losing bundles completely [23,24].

This study characterizes and compares the enameloid microstructure of two modern lamniform sharks, *Isurus oxyrinchus* (shortfin mako shark) and *Carcharias taurus* (spotted ragged-tooth shark). A description of enameloid hierarchical microstructure in the style of the established description of amniote enamel [14,25,26] is proposed. As amniote enamel serves the same protective function as enameloid, structural similarities may be indicative of biomechanical function. Such similarities between enameloid structure and the enamel of different mammalian species are identified and discussed here. By focusing on structural motifs and patterns in shark enameloid, the foundation for microstructure design in synthetic composite materials is laid. As biological materials exhibit sophisticated structures on multiple length scales, they commonly combine highly desirable properties such as high fracture toughness and strength which are difficult to obtain in typical engineering materials [27]. Thus, mimicking some of nature's microstructural design principles may be the key in designing high performance composites.

2. Materials and Methods

Teeth of two species of the order *Lamniformes*, the spotted ragged-tooth or sand tiger shark (*Carcharias taurus*) and the shortfin mako shark (*Isurus oxyrinchus*), were studied. Both species feed mainly on bony fish and occasionally other elasmobranches or even sea turtles. Their teeth are used primarily to incapacitate prey before ingestion or, less often, to rip pieces from larger prey.

Samples of *I. oxyrinchus* teeth were acquired from Hout Bay, South Africa. The *C. taurus* teeth samples had been shed naturally and were kindly provided by Two Oceans Aquarium, Cape Town, South Africa. The teeth originate from one or more of the 5 adult females living in captivity (but born free) at the aquarium, weighing 80 kg to 170 kg. Teeth samples were stored at ambient conditions before specimen preparation and imaging. Teeth of both species were placed in a mold, covered in Spurrs resin and put in an oven at 60 °C for 12 h to allow the resin to set. The teeth were then ground in the longitudinal and transverse sections (Figure 2) and the exposed areas were polished with 0.2 µm aluminum powder. The samples were sonicated at 50 Hz in distilled water for 2 min and then etched with 10% hydrochloric acid for 2 min and rinsed in distilled water for 5 min. They were carbon coated in an evaporation coater and viewed with a Tescan MIRA SEM at 5 kV. The SEM images for both species at different magnifications are gathered in the Supplementary Material Figures S1–S15.

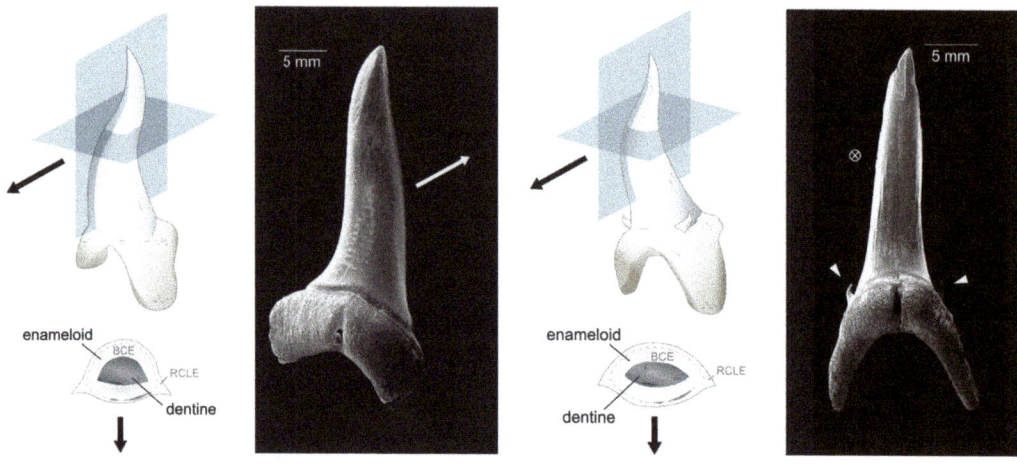

(a) tooth of *I. oxyrinchus*   (b) tooth of *C. taurus*

Figure 2. Tooth morphology of the two studied species. Both species have slender, dagger-like teeth without serrated edges. The teeth are curved slightly to improve the grip on prey. Sketches on the left show the tooth and its transversal cross-section. Arrows indicate the facial direction. The shaded planes correspond to the section planes used in imaging. The enameloid cover consists of inner and outer enameloid, marked in the cross-section with BCE and RCLE, respectively. *I. oxyrinchus* teeth (**a**) are monocuspid while *C. taurus* teeth (**b**) are tricuspid with a large main cusp and small outer cusplets indicated by white arrowheads. The *I. oxyrinchus* tooth (**a**) exhibits a small amount of wear on the lingual side of the tip. The *C. taurus* tooth depicted in (**b**) is fractured at the tip.

3. Results: Microstructure Description of Enameloid in *I. oxyrinchus* and *C. taurus*

The teeth of both studied species, *I. oxyrinchus* and *C. taurus*, have remarkably similar morphologies (Figure 2). The teeth are slender and needle-like and possess no serrated edges. This morphology is commonly associated with piercing of prey [5]. Both are curved in an S-shape with the tooth protruding forward and then curving backward into the oral cavity and only the tip curving forward again. The anterior side especially of *I. oxyrinchus* teeth is flattened. The most pronounced difference between teeth of the two species is the presence of small tricuspids in *C. taurus* not found in *I. oxyrinchus*.

The teeth imaged in this work were approximately 10 mm in length and had a lenticular cross-section with a width of 2.5 mm in the upper region where the teeth were sectioned. Other samples were larger, with lengths reaching 15 mm for *I. oxyrinchus* and 20 mm for *C. taurus*. The enameloid cover in the studied transversal sections had a thickness of ca. 0.5 mm for *I. oxyrinchus* and ca. 0.4 mm for *C. taurus*. The enameloid thickness varied slightly between different teeth; in the longitudinally sectioned tooth of *C. taurus*, the thickness on the anterior side of the tooth was 0.4 mm and on the posterior side it was 0.7 mm.

3.1. Enameloid Microstructure of Isurus oxyrinchus

The enameloid cover of *I. oxyrinchus* teeth exhibits the typical layered structure in longitudinal section as seen in Figure 3a,b. The outermost layer with a thickness of 30 μm to 35 μm consists of densely packed crystals (Figure 3c) in the upper region of the tooth. In the lower region, no single crystal enameloid is visible and the outer enameloid consists solely of a single layer of circumferentially oriented bundles as seen in Figure 3f.

The outer enamel covers a layer of parallel bundled enameloid (Figure 3d,e) with a thickness of ca. 170 μm. In this region, crystal bundles are aligned parallel to each other and oriented along the tooth's longitudinal axis. Closer to the dentine, this clear arrangement becomes less regular and transitions smoothly into entangled bundles. In the longitudinal section depicted in Figure 3, the transition from dentine in the tooth's core to the enameloid cover appears smooth (Figure 3b) due to only minor etching. In the transversal section Figure 4a, a sharp junction between enameloid and dentine is apparent.

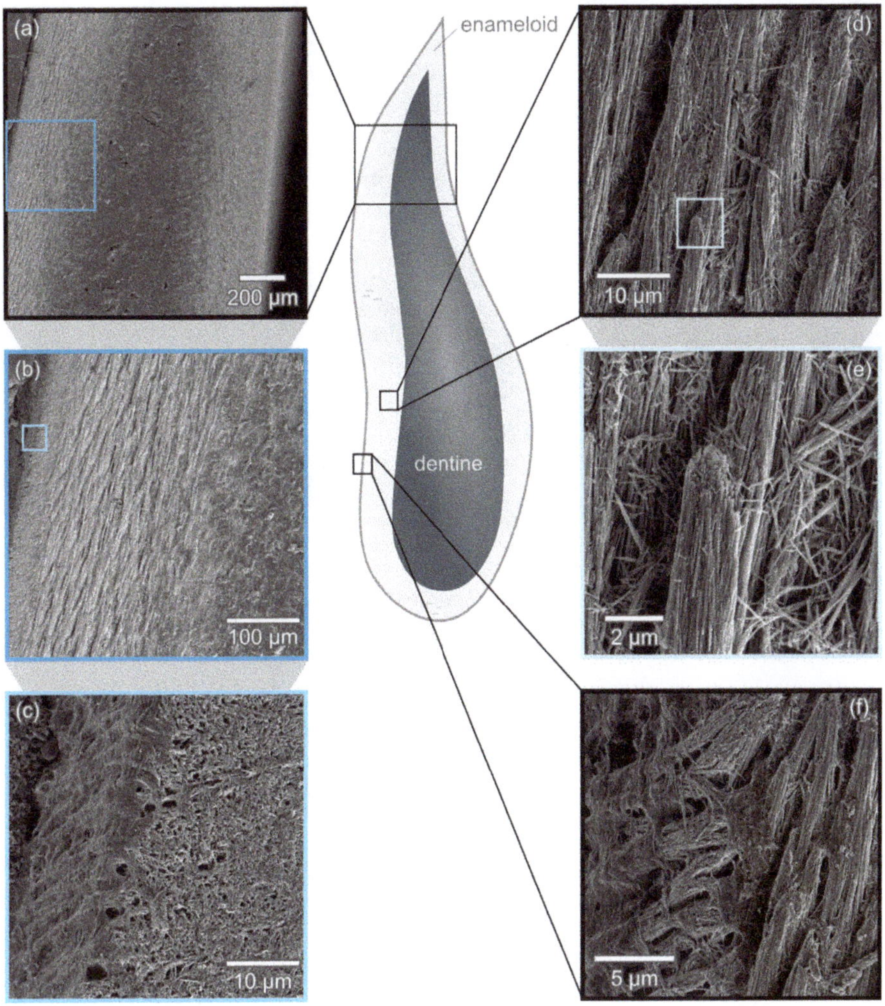

Figure 3. *I. oxyrinchus* tooth in longitudinal section. The enameloid cover is characteristically layered as seen in (**a**,**b**). Zooming into the outer enameloid, a densely packed outer layer is found in the top part of the tooth (**c**) while in the lower part, distinct circumferential bundles are apparent (**f**). (**d**) The outer enameloid surrounds a layer of parallel aligned crystal bundles. (**e**) Magnified view of the broken tip of a bundle from (**d**).

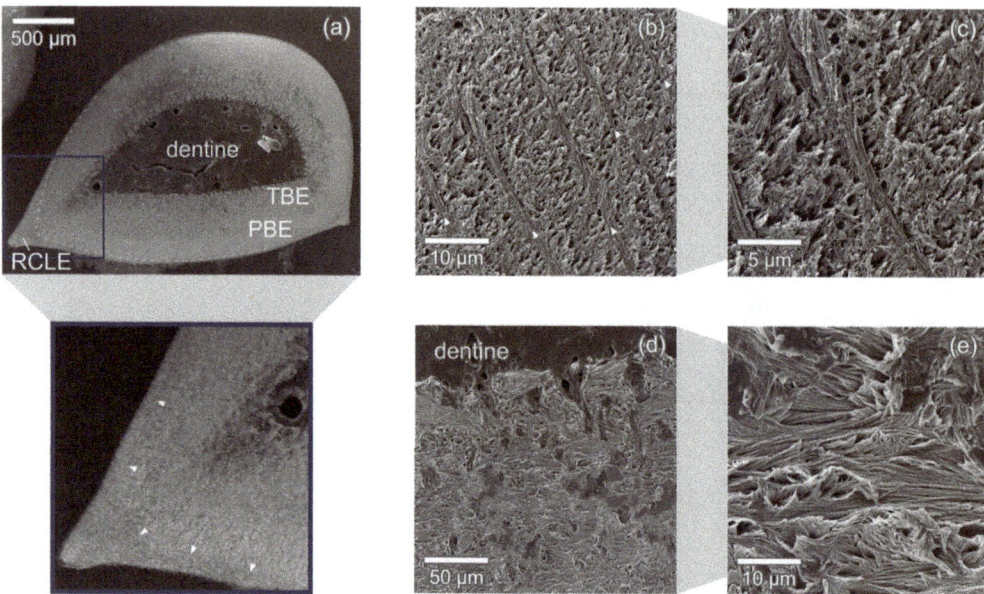

Figure 4. Transversal section of *I. oxyrinchus* tooth. The tooth's enameloid cover is built up in layers (**a**). The outer enameloid is designated RCLE and clearly separated (dashed line, arrow heads in zoomed-in view) from the inner enameloid consisting of PBE and TBE. (**b**,**c**) show the outer edge of the enameloid cover and radial elements (arrow heads) running towards the outer enameloid. In between the radial elements, bundles are oriented parallel to the tooth's longitudinal axis. (**d**,**e**) show three-dimensional entanglement of bundles close to the dentine-enameloid junction.

The bundles in the inner enameloid have a diameter of ca. 4 µm to 9 µm and consist of densely packed crystallites with an average width of 66 nm visible in Figure 5a. This size is in good agreement with the range of 50 nm to 80 nm found for *I. oxyrinchus* in [17,18]. The crystallites are several micrometers in length, but an exact measurement is impossible in the micrographs as the visible length exceeds the field of view. Within a bundle, the elongated crystallites are aligned parallel to each other without visible pores or gaps between them (Figure 5c). Neighboring crystallites remain in contact in the TBE when the bundle curves and changes direction (Figure 5b).

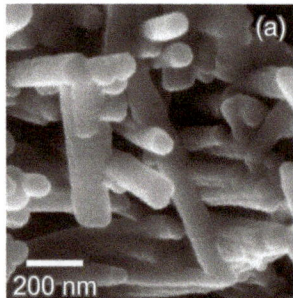  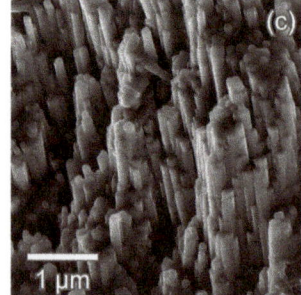

Figure 5. Fluorapatite crystallites in *I. oxyrinchus* tooth. (**a**) Fracture surface showing individual crystallites. (**b**) The thin, rod-like crystallites change orientation in the TBE (transversal section). Neighboring crystal rods remain parallel. (**c**) Densely packed crystallites within a crystallite bundle (transversal section). The crystallites are aligned parallel to each other and in close contact with each other.

Inner, bundled enameloid and outer enameloid are clearly separated with a distinct boundary clearly visible in the medial and distal tooth edges in transversal section Figure 3a. In these edges, the outer enameloid is thicker while in the middle of the lingual and labial faces, the outer enameloid's thickness is 30 µm to 35 µm as found in the longitudinal section.

In the transversal section, the different regions of the inner enameloid are more distinct due to stronger etching. Adjacent to the enameloid-dentine junction, crystallite bundles are three-dimensionally interwoven (Figure 4d,e). The TBE is approximately 270 µm to 290 µm thick. The surrounding PBE region is only marginally thinner. The inner enameloid, thus, appears to be made up of equal amounts of TBE and PBE. Exact measurements, however, are difficult as the transition between the regions is smooth.

The parallel bundles of the inner enameloid are interspersed with regularly arranged radial elements with a thickness of ca. 1.5 µm (Figure 4b,c). The distance between two radial elements ranges from 6 µm to 10 µm. This corresponds to the diameter of an individual bundle, thus, the radial elements appear to separate individual rows of parallel aligned bundles. The crystallites within the radial elements are oriented at almost a 90° angle to the bundle direction (i.e., parallel to the imaging plane) and run from the enameloid-dentine junction radially outwards to the outer tooth surface. As shown in Figure 4b, the radial elements merge into the outer enameloid layer which consists of densely packed, randomly oriented crystallites.

3.2. Enameloid Microstructure of Carcharias taurus

Figure 6 depicts the longitudinal section of a *C. taurus* tooth. In this section, no outer enameloid is visible and the PBE appears to extend right to the tooth edge (Figure 6a,d) with no outer enameloid visible. A thin layer of outer enameloid or remainders of one could possibly be obscured by the embedding material covering the tooth's edge. The PBE transitions smoothly into a region of TBE with three-dimensionally interwoven crystallite bundles as visible in Figure 6b,c. The enemaloid-dentine junction is sharp and clearly visible with no dentine reaching into the enameloid.

In the transversal section in Figure 7a, the typical layered structure of modern shark's enameloid is readily identifiable. The outer enameloid is clearly visible in the medial and distal tooth edges and narrows drastically towards the middle of the lingual and the labial faces. In the middle of these tooth faces, the outer enameloid appears to vanish completely (Figure 7b).

The transition from parallel aligned crystallite bundles (PBE) to the entangled, interwoven bundles (TBE) in the inner enameloid is smooth without any sharp boundaries. The PBE region has a thickness of about 115 µm measured in the transversal section Figure 7. The TBE with a thickness of ca. 215 µm is significantly thicker.

The crystal bundles in the inner enameloid have a diameter of 5 µm with the crystals' long axes running parallel to the bundle's long axis, as seen in Figure 6e,f. Figure 8 shows close-up micrographs of fluorapatite crystallites. The crystallites have an average thickness of 60 nm and are aligned along their elongated longitudinal axes. The crystallites are densely packed within a bundle and appear to be in direct contact along the whole crystallite length. This contact and the alignment are maintained even when the crystallites curve and change direction in the TBE (Figure 8b).

Close to the edge, radial elements intersecting the transversal plane at approximately 90° can be seen in Figure 7b,c. Each of these radial elements has a thickness of ca. 1.5 µm and the distance between two elements is approximately 5 µm or slightly higher (up to 7.5 µm) which correlates to one bundle diameter. The crystallites within the radial elements are aligned in parallel with each other and their long axes are oriented at almost a 90° angle to the long axes of the bundles as seen in Figure 9a which shows two radial elements with the crystallite bundles between them.

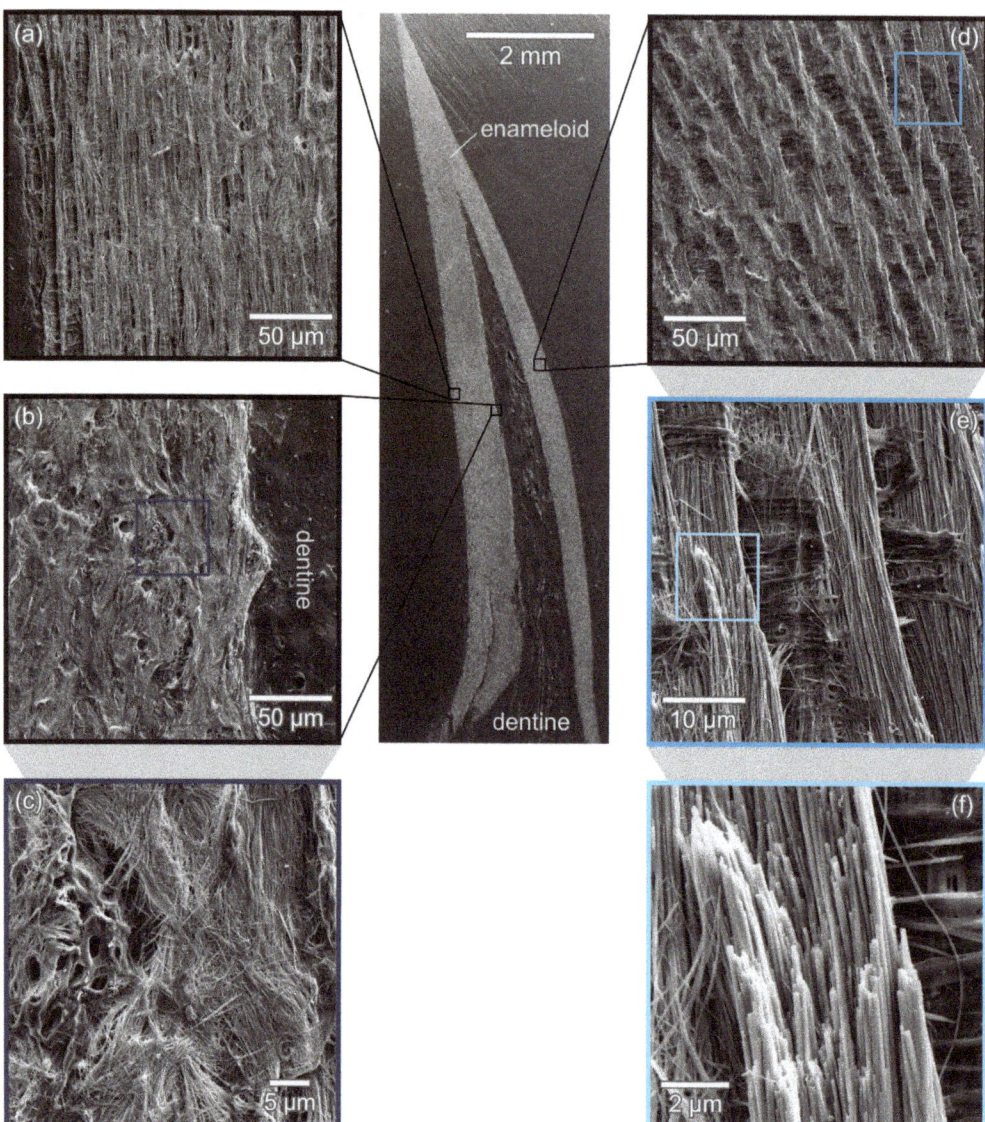

Figure 6. *C. taurus* tooth in longitudinal section. In the area imaged in (**a**), parallel bundle enameloid (PBE) reaches all the way to the outer surface which is obscured by embedding material. (**b**,**c**) show three-dimensionally entangled bundles belonging to the TBE close to the dentine. (**d**–**f**) show a sequence of images with increasing magnification into the PBE in the outer range of the inner enameloid. Between the bundles, remnants of layers between the bundles are visible. An individual bundle (**f**) consists of densely packed, parallel aligned crystallites.

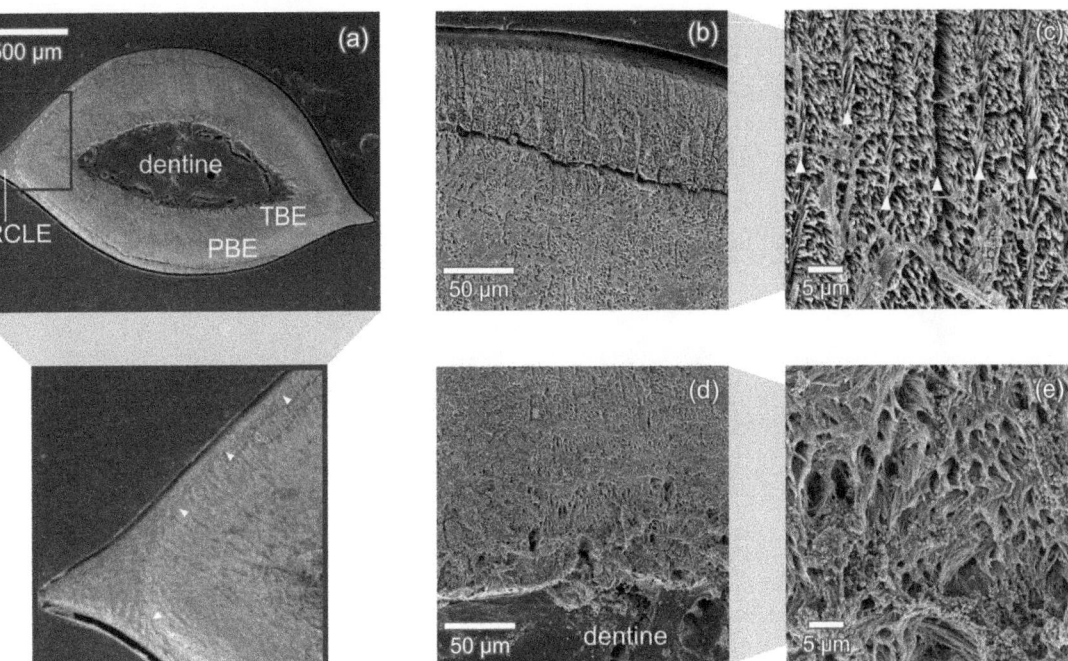

Figure 7. Transversal section of *C. taurus* tooth. (**a**) The enameloid cover is made up of an outer enameloid (RCLE) clearly separated (dashed line, arrow heads in zoomed-in view) from the inner enameloid which itself consists of TBE and PBE. (**b**,**c**) In the middle of the tooth edge, the outer enameloid appears to be worn away and parallel aligned crystal bundles (PBE) intersected by evenly spaced thin radial elements (arrow heads) are visible. (**d**,**e**) In the inner enameloid close to the dentine, the bundles are interwoven and change orientation frequently in all directions.
The crack apparent in (**a**,**b**) likely occurred during specimen preparation.

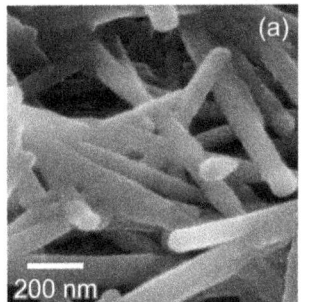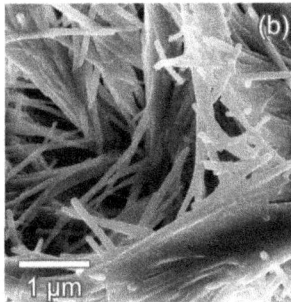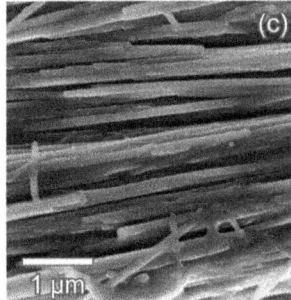

Figure 8. Fluorapatite crystallites in *C. taurus* tooth. (**a**) Individual crystallites at the polished sectioning plane. (**b**) Crystallites curve and change orientation in the TBE while remaining parallel to their neighbors (transversal section). (**c**) Within a bundle in the inner enameloid, crystallites are in close contact with each other and aligned along their long axis (transversal section).

 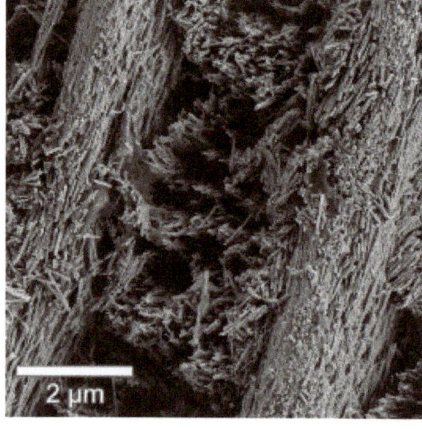

(a) enameloid of *C. taurus*   (b) enamel of *M. rufogriseus*

Figure 9. Transversal sections of (**a**) the outer enameloid of *C. taurus* and (**b**) the modified radial enamel in a tooth of *Macropus rufogriseus* (red-necked wallaby). Both species exhibit radial sheets of densely packed crystallites separating rows of bundled crystallites. The hydroxyapatite crystallites of the mammalian enamel are thinner than the fluorapatite of enameloid but the higher-level bundles and radial elements have the same size in both tissues. Despite the significant size difference between the two species, the imaged teeth are of similar size with a crown height of 10 mm in *C. taurus* and ca. 7 mm in *M. rufogriseus*.

4. Microstructural Features of Modern Shark Enameloid

4.1. Lamniform Shark Enameloid Structure

Isurus oxyrinchus and *Carcharias taurus* exhibit a remarkably similar enameloid microstructure corresponding to the typical layered structure known for modern sharks [19,20]. The enameloid of non-selachian species shows drastically varying levels of complexity with some exhibiting crystallite bundles [24] while the fossilized teeth of some *Ctenacanthiformes* are so highly mineralized that individual crystallites cannot be distinguished [28].

The microstructural complexity in modern sharks has been suggested to be a functional adaptation to the specialized feeding behavior in comparison to other chondrichthyes [28]. The enameloid microstructure of *I. oxyrinchus* and *C. taurus* enameloid described here exhibits a clear hierarchical set-up. It consists of thin, needle-like crystallites with a length exceeding a few micrometers. The crystallites of both species have an average width of ca. 60 nm to 65 nm, which falls perfectly into the range found in [18] for *I. oxyrinchus* and is in the same range as the 50 nm identified in [17]. Table 1 summarizes the sizes of all structural features in both species.

The majority of the enameloid cover is composed of these crystallites, arranged in densely packed crystallite bundles. The enameloid region adjacent to the dentine is in both species characterized by intertwining bundles, corresponding to the TBE. In *C. taurus*, the boundary between dentine and TBE is well-defined while in *I. oxyrinchus*, especially in longitudinal section (Figure 3), the transition is smooth and some dentine extends into the enameloid cover. A similarly smooth transition has been found for *I. oxyrinchus* in [4] and for *C. taurus* in [29].

The TBE is covered by PBE in which the crystallite bundles are aligned parallel to each other and the tooth's longitudinal axis in both studied species. In *I. oxyrinchus*, TBE and PBE regions are of similar thickness while in *C. taurus*, the TBE layer makes up approximately two thirds of the bundled enameloid unit (Table 1).

Table 1. Size of structural features in modern shark enameloid and mammalian enamel.

	I. oxyrinchus Enameloid	*C. taurus* Enameloid	Wallaby Enamel	Human Enamel
thickness TBE [µm]	270–290	215	-	-
thickness PBE [µm]	250	115	-	-
thickness outer enameloid [µm]	30–35	10–30	-	-
crystal width [nm]	50–75	55–70	45–50	50–70 [30,31]
bundle diameter [µm]	4–9	5	3–4	4–8 [32]
thickness of radial elements [µm]	1.5	1.5	1–1.5	-
distance radial elements [µm]	6–10	5	4	-

The outermost layer of a modern shark's tooth typically consists of single crystal enameloid in which crystals are oriented randomly. In the studied teeth of *I. oxyrinchus*, this outer layer is visible and in the lower regions of the tooth, a single layer of circumferential bundles as also found in [18] can be identified. The outer enameloid in *I. oxyrinchus* here is with ca. 30 µm thickness significantly thicker than the one described in [18] for the same species.

For *C. taurus*, no outer enameloid layer could be found in longitudinal section. The outer enameloid is thick in the tooth's medial and distal edges and thins over the tooth's circumference until it vanishes in the middle of the lingual and labial faces (Figure 7). The *C. taurus* tooth studied in [29] exhibits an outer enameloid layer of approximately uniform thickness. This suggests that a thin outer layer might have been worn away, either during normal function, after shedding due to environmental conditions [10] or during etching. The outer enameloid of *C. taurus* does not exhibit circumferential bundles as are found in *I. oxyrinchus*. This is the most significant difference in the enameloid microstructures of the two species.

From the outer enameloid, thin radial elements reach into the inner enameloid where they intersperse individual rows of parallel bundles. The presence of such radial elements has been documented for *I. oxyrinchus* [18], *C. taurus* [29] and numerous other lamniform shark species [19,20,33].

4.2. Radial Elements in Enameloid

Generally, the radial elements found in the outer enameloid and PBE are referred to as 'radial bundles' [19,20]. Their geometry is described as 'ribbon-like' bundles in *I. oxyrinchus* [18]. FIB-SEM tomography of the enameloid of two carcharhiniform sharks [33] shows the three-dimensional arrangement of radial elements close to the SCE. The crystallites are arranged within a thin layer between the bundles with frequent gaps in the layer and transversal connections between adjacent layers apparent.

In the studied section of *I. oxyrinchus*, the crystals within the radial elements are aligned parallel to the transversal section (Figure 4) while in the transversal section of *C. taurus* (Figure 7), the crystallites in the radial elements intersect the imaging plane at an angle, compare also Figure 9a. From these micrographs, the radial element layers appear to be continuous thin sheets with two directions (the radial and the axial one) being much larger than the thickness and have no gaps readily apparent. The sheets separate rows of crystal bundles in which the crystals are oriented at an angle to the sheet crystals. In both species, this angle appears to be close to 90°.

This structural motif of thin sheets separating rows of crystal bundles from each other can also be found in the dental enamel of some herbivorous mammals [34,35] and is there referred to as *modified radial enamel*. Figure 9 shows the radial elements in *C. taurus* enameloid and the modified radial enamel in a molar of *Macropus rufogriseus*, the red-necked wallaby, showing the striking similarity between the structures. The crystallites in the marsupial's enamel are hydroxyapatite and slightly thinner than the fluorapatite crystals of *C. taurus*'s enameloid. Similarly, the bundles are thinner than in the enameloid, but the sheets have the same thickness of 1.5 µm (Table 1). The similar length scale of both structures is remarkable when considering the large size difference between the two species. The investigated teeth are of similar size with a crown height of 10 mm in *C. taurus* and ca. 7 mm in *M. rufogriseus*. The diameter of the shark tooth is with 2.5 mm of the same order as an individual cusp of *M. rufogriseus*'s molar. For mammalian enamel, the size of microstructural elements is independent of the size of a single tooth or the animal [14].

Modified radial enamel in mammals is generally interpreted as an adaptation to large axial stresses arising due to horizontal mastication movements that are especially common in grazing species. These axial stresses favor crack propagation in radial direction, i.e., cracks travelling perpendicular to the tooth's longitudinal axis from the surface towards the softer dentine. Modified radial enamel introduces vertical decussation planes, i.e., planes of abrupt crystal orientation discontinuities, within the crack path. This results in deflection of the crack and twisting of the crack surface, thus, effectively slowing or arresting cracks before reaching the dentine and protecting the tooth from catastrophic fracture [14]. Slender shark teeth like the ones studied here are loaded in bending during holding and shaking of prey [9]. This loading case results in dominant axial tensile stresses which would cause radial crack growth. The decussation planes introduced by the presence of radial sheets, thus, can distort the crack path. The crystallite bundles are oriented axially and, thus, are stiff under bending and do not exhibit significant weak propagation paths for radial cracks. The radial elements, therefore, provide toughening against chipping of the tooth, i.e., crack propagation parallel to the tooth surface, without introducing additional weakness against crack propagation under bending.

5. Structural Hierarchy in Dental Materials: Similarities between Enameloid and Enamel

Shark enameloid and amniote enamel consist of different minerals and have been shown to differ in evolutionary origin [19,36] but both dental tissues serve the same function as the outermost layer of the tooth that protects the softer dentine. Thus, identification of common microstructure patterns—such as radial elements—can give valuable insight into biomechanical function, as seen in Section 4.2, and may form the basis for the design of bioinspired composite materials. Therefore, the hierarchical microstructure of shark enameloid is in the following described in comparison with amniote enamel for which a well-established terminology exists ([26], see also [14]). Remarkably, both tissues exhibit the same five levels of hierarchy, depicted in Figure 10.

The smallest building block in each tissue is the individual nanocrystal which is designated as Level 0 of the hierarchy. Level I describes the local arrangement of crystallites. In the shark enameloid studied here, individual crystallites are either aligned in parallel or—in the shiny layer enameloid—fully randomly arranged. The parallel aligned enameloid crystals are densely packed into crystallite bundles on Level II which in the enamel nomenclature is also referred to as the *module* level. In dental enamel, different modules can be identified [14] but the most prominent are the 'rods' or 'prisms' typical for mammalian enamel. These generally have a rounded cross-section with diameters of 2 µm to 10 µm and, within the rods, enamel crystallites are predominantly arranged parallel to each other and to the rod's length axis. This arrangement is identical to the bundles in shark enameloid.

The assembly of these modules forms Level III of the hierarchy. Different assemblies such as TBE and PBE can be referred to as enameloid types, in analogy to 'enamel types'. Much of the functional adaptation of mammalian enamel occurs on this third level of the hierarchy that is characterized by locally varying arrangements of modules and single crystallites as is the case in the discussed parallel bundled enameloid interspersed with radial elements. Over the thickness of the enameloid cap, different enameloid types occur, and some enameloid types may only be found in certain regions of the tooth. For *I. oxyrinchus* and *C. taurus*, the enameloid consists of an innermost layer of TBE, surrounded by PBE which in turn is covered by the outer enameloid. This varying pattern constitutes Level IV of the structural hierarchy and is the equivalent to the *schmelzmuster* (from the German 'Zahnschmelz' for tooth enamel) of enamel.

In many evolved mammalian species, the enamel consists of an inner layer of decussated enamel types in which rods are interwoven or arranged at sharp angles to each other [14]. This inner layer is commonly covered by radial enamel in which the rods aligned parallel and the whole tooth is covered in a thin layer of "prism-less", i.e., single crystallite enamel. This typical schmelzmuster is interpreted to be an adaptation for fracture resistance, as cracks can easily travel along the boundaries between parallel aligned bundles but the presence of decussation deflects and bifurcates the crack path, resulting in toughening and arrest of the crack before critical failure. The layered structure of a shark tooth is a clear match for this set-up, suggesting that the parallel bundled enameloid increases stiffness of the tooth necessary for piercing while the tangled bundled enameloid increases fracture resistance. Strikingly, the relative thickness of the TBE layer found in *I. oxyrinchus* and *C. taurus* in this study (Table 1) corresponds exactly to the 50% to 65% of decussated enamel found in mammalian enamel [14].

Remark: The structural hierarchy of *Isurus oxyrinchus* teeth is also discussed in [18]. A six-level hierarchy description is proposed in which the fluorapatite unit cell constitutes Level 1 and, thus, the smallest building block and Level 6 describes the whole tooth. The higher-level structural features identified in [18] correspond to Levels II–IV described above.

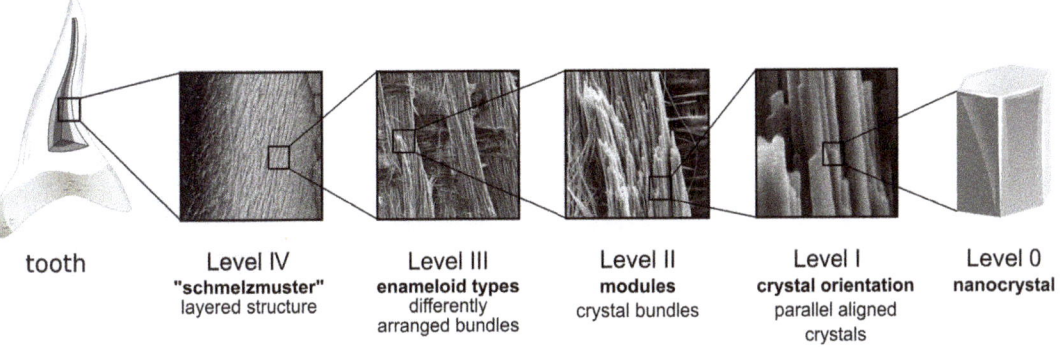

Figure 10. Structural hierarchy of shark teeth enameloid. The enameloid cover of an individual shark tooth exhibits a five-level hierarchical microstructure. On Level IV, the *schmelzmuster*, different structural patterns are combined to improve the biomechanical function of the tooth. These Level III patterns are the different enameloid types and consist of the Level II enameloid modules. In shark enameloid, the most common module is a bundle of parallel aligned crystallites (Level I). The lowest level of the hierarchy, Level 0, is the individual nanoscale crystallite.

6. Conclusions

Hard tissues such as shark enameloid are biological composites of nanoscale mineral crystals arranged in intricate hierarchical patterns interspersed with only minor amounts of remnant protein. The microstructural design in these tissues results in a macroscopic material that is stiff, strong and tough despite consisting almost completely of brittle mineral.

Analysis of micrographs of two lamniform shark species, the shortfin mako (*Isurus oxyrinchus*) and the spotted ragged-tooth shark (*Carcharias taurus*), reveals the hierarchical structure of enameloid. Fluorapatite nanocrystallites (Level 0) are arranged in bundles (Level II) that themselves are arranged in a layered pattern over the enameloid cover (Level IV). The microstructural arrangements found in both species are remarkably similar, both contain parallel aligned bundles (PBE) interspersed with radial elements and an inner layer of three-dimensionally interwoven TBE. In this work, the hierarchy of enameloid microstructure is discussed in comparison to the well-established nomenclature used to describe the microstructure of amniote enamel [26]. The microstructures of both tissues exhibit the same five hierarchical levels, despite their different evolutionary origin.

Remarkably, on different hierarchical levels, direct structural analogues for shark enameloid microstructure patterns could be identified in mammalian enamel. Shark enameloid contains radial elements separating rows of crystal bundles on Level III, a structural motif strikingly similar to the mammalian *modified radial enamel*. Another Level III enameloid type, the tangled bundle enameloid, is structurally identical to the decussated *irregular enamel* of marsupials and proboscideans [35,37]. On Level IV, these different enameloid types are arranged in a layered structure that corresponds to the typical schmelzmusters of mammalian enamel.

These striking structural similarities allow one to draw conclusions on enameloid microstructure function based on the wealth of information available for enamel [14]. Modified radial enamel, for instance, has been found to increase fracture resistance while maintaining high stiffness. Likewise, the radial elements in enameloid are likely to increase fracture toughness by providing 'easy' crack propagation paths along the discontinuities in crystallite orientation. These deliberately introduced crack paths result in energy dissipation through crack deflection and can guide fracture away from sensitive parts of the tooth. Similarly, the characteristic layered structure of lamniform shark enameloid is functionally identical to mammalian enamel with an inner decussated enamel providing fracture toughness and protection of the soft dentine from cracks.

By transferring such biological design principles to synthetic materials, unique property combinations such as high strength combined with high fracture resistance may be achieved. Mimicking the full hierarchical structure of a biological tissue over all involved length scales may still be unachievable with modern fabrication techniques, but adopting individual structural motifs has been used successfully in the development of 'self-sharpening' knives inspired by rat teeth [38] or impact resistant glass based on nacre [39]. As the described structural motifs of radial elements and the layered structure have developed independently in the different tissues, they should be investigated further with regards to the toughening mechanisms they provide and to derive design principles to be applied in bioinspired composites and metamaterials.

Supplementary Materials: The SEM micrographs of the studied teeth are available online at https://www.mdpi.com/2079-4991/11/4/969/s1, Figures S1–S15.

Author Contributions: Conceptualization, J.W. and S.B.; SEM imaging, M.W.; formal analysis, J.W. and S.B.; investigation, J.W.; resources, S.B.; writing—original draft preparation, J.W.; writing—review and editing, J.W., S.B. and M.W.; visualization, J.W. All authors have read and agreed to the published version of the manuscript.

Funding: This research received no external funding.

Informed Consent Statement: Not applicable.

Acknowledgments: Specimen of *Carcharias taurus* (spotted ragged-tooth shark) kindly provided by Two Oceans Aquarium, South Africa. Specimen of *Macropus rufogriseus* (red-necked wallaby, Figure 9b) kindly provided by Zoo Duisburg, Germany. Micrograph Figure 9b courtesy of M. Wurmshuber, Chair of Materials Physics, Montanuniversität Leoben, Austria.

Conflicts of Interest: The authors declare no conflict of interest.

References

1. Shimada, K. Dental homologies in lamniform sharks (Chondrichthyes: Elasmobranchii). *J. Morphol.* **2002**, *251*, 38–72. [CrossRef]
2. Whitenack, L.B.; Gottfried, M.D. A morphometric approach for addressing tooth-based species delimitation in fossil mako sharks, *Isurus* (Elasmobranchii: Lamniformes). *J. Vertebr. Paleontol.* **2010**, *30*, 17–25. [CrossRef]
3. Corn, K.A.; Farina, S.C.; Brash, J.; Summers, A.P. Modelling tooth–prey interactions in sharks: The importance of dynamic testing. *R. Soc. Open Sci.* **2016**, *3*, 160141. [CrossRef]
4. Enax, J.; Prymak, O.; Raabe, D.; Epple, M. Structure, composition, and mechanical properties of shark teeth. *J. Struct. Biol.* **2012**, *178*, 290–299. [CrossRef]
5. Frazzetta, T. The mechanics of cutting and the form of shark teeth (Chondrichthyes, Elasmobranchii). *Zoomorphology* **1988**, *108*, 93–107. [CrossRef]
6. Correia, J.P. Tooth loss rate from two captive sandtiger sharks (*Carcharias Taurus*). *Zoo Biol.* **1999**, *18*, 313–317. [CrossRef]

7. Luer, C.A.; Blum, P.C.; Gilbert, P.W. Rate of Tooth Replacement in the Nurse Shark, *Ginglymostoma Cirratum*. *Copeia* **1990**, *1*, 182–191. [CrossRef]
8. Botella, H.; Valenzuela-Ríos, J.I.; Martínez-Pérez, C. Tooth replacement rates in early chondrichthyans: A qualitative approach. *Lethaia* **2009**, *42*, 365–376. [CrossRef]
9. Whitenack, L.B.; Simkins, D.C.; Motta, P.J. Biology meets engineering: The structural mechanics of fossil and extant shark teeth. *J. Morphol.* **2010**, *272*, 169–179. [CrossRef]
10. Weber, K.; Winkler, D.E.; Kaiser, T.M.; Žigaitė, Ž.; Tütken, T. Dental microwear texture analysis on extant and extinct sharks: Ante- or post-mortem tooth wear? *Palaeogeogr. Palaeoclimatol. Palaeoecol.* **2021**, *562*, 110147. [CrossRef]
11. Barthelat, F. Architectured materials in engineering and biology: Fabrication, structure, mechanics and performance. *Int. Mater. Rev.* **2015**, *60*, 413–430. [CrossRef]
12. Gao, H.; Ji, B.; Jäger, I.L.; Arzt, E.; Fratzl, P. Materials become insensitive to flaws at nanoscale: Lessons from nature. *Proc. Natl. Acad. Sci. USA* **2003**, *100*, 5597–5600. [CrossRef] [PubMed]
13. Liu, Z.; Zhang, Z.; Ritchie, R.O. On the Materials Science of Nature's Arms Race. *Adv. Mater.* **2018**, *30*, 1705220. [CrossRef] [PubMed]
14. Wilmers, J.; Bargmann, S. Nature's design solutions in dental enamel: Uniting high strength and extreme damage resistance. *Acta Biomater.* **2020**, *107*, 1–24. [CrossRef] [PubMed]
15. Wang, Y.; Naleway, S.E.; Wang, B. Biological and bioinspired materials: Structure leading to functional and mechanical performance. *Bioact. Mater.* **2020**, *5*, 745–757. [CrossRef]
16. Daculsi, G.; Kerebel, L. Ultrastructural study and comparative analysis of fluoride content of enameloid in sea-water and fresh-water sharks. *Arch. Oral Biol.* **1980**, *25*, 145–151. [CrossRef]
17. Chen, C.; Wang, Z.; Saito, M.; Tohei, T.; Takano, Y.; Ikuhara, Y. Fluorine in Shark Teeth: Its Direct Atomic-Resolution Imaging and Strengthening Function. *Angew. Chem. Int. Ed.* **2014**, *53*, 1543–1547. [CrossRef]
18. Enax, J.; Janus, A.M.; Raabe, D.; Epple, M.; Fabritius, H.O. Ultrastructural organization and micromechanical properties of shark tooth enameloid. *Acta Biomater.* **2014**, *10*, 3959–3968. [CrossRef]
19. Cuny, C.; Risnes, S. The enameloid microstructure of the teeth of synechodontiform sharks (Chondrichthyes: Neoselachii). *Vertebr. Paleontol.* **2005**, *3*, 8–19.
20. Guinot, G.; Cappetta, H. Enameloid microstructure of some Cretaceous Hexanchiformes and Synechodontiformes (Chondrichthyes, Neoselachii): New structures and systematic implications. *Microsc. Res. Tech.* **2011**, *74*, 196–205. [CrossRef]
21. Enault, S.; Guinot, G.; Koot, M.B.; Cuny, G. Chondrichthyan tooth enameloid: Past, present, and future. *Zool. J. Linn. Soc.* **2015**, *174*, 549–570. [CrossRef]
22. Lübke, A.; Enax, J.; Loza, K.; Prymak, O.; Gaengler, P.; Fabritius, H.O.; Raabe, D.; Epple, M. Dental lessons from past to present: Ultrastructure and composition of teeth from plesiosaurs, dinosaurs, extinct and recent sharks. *RSC Adv.* **2015**, *5*, 61612–61622. [CrossRef]
23. Enault, S.; Cappetta, H.; Adnet, S. Simplification of the enameloid microstructure of large stingrays (Chondrichthyes: Myliobatiformes): A functional approach. *Zool. J. Linn. Soc.* **2013**, *169*, 144–155. [CrossRef]
24. Manzanares, E.; Botella, H.; Delsate, D. On the enameloid microstructure of Archaeobatidae (Neoselachii, Chondrichthyes). *J. Iber. Geol.* **2018**, *44*, 67–74. [CrossRef]
25. von Koenigswald, W. Levels of complexity in the microstructure of mammalian enamel and their application in studies of systematics. *Scanning Microsc.* **1992**, *6*, 195–217.
26. von Koenigswald, W.; Sander, P.M. Glossary of terms used for enamel microstructures. In *Tooth Enamel Microstructure*; von Koenigswald, W., Sander, P., Eds.; Balkema: Rotterdam, The Netherlands, 1997; pp. 267–280.
27. Ritchie, R.O. The conflicts between strength and toughness. *Nat. Mater.* **2011**, *10*, 817–822. [CrossRef]
28. Gillis, J.A.; Donoghue, P.C. The homology and phylogeny of chondrichthyan tooth enameloid. *J. Morphol.* **2007**, *268*, 33–49. [CrossRef] [PubMed]
29. van Vuuren, L.J.; Loch, C.; Kieser, J.A.; Gordon, K.C.; Fraser, S.J. Structure and mechanical properties of normal and anomalous teeth in the sand tiger shark *Carcharias Taurus*. *J. Zoo Aquar. Res.* **2015**, *3*, 29–36.
30. Beniash, E.; Stifler, C.A.; Sun, C.Y.; Jung, G.S.; Qin, Z.; Buehler, M.J.; Gilbert, P.U.P.A. The hidden structure of human enamel. *Nat. Commun.* **2019**, *10*, 4383. [CrossRef]
31. Zheng, J.; Huang, H.; Shi, M.; Zheng, L.; Qian, L.; Zhou, Z. In vitro study on the wear behaviour of human tooth enamel in citric acid solution. *Wear* **2011**, *271*, 2313–2321. [CrossRef]
32. Boushell, L.W.; Sturdevant, J.R. Clinical Significance of Dental Anatomy, Histology, Physiology, and Occlusion. In *Sturdevant's Art & Science of Operative Dentistry*, 7th ed.; Ritter, A., Ed.; Elsevier Health Sciences: Amsterdam, The Netherlands, 2017.
33. Fellah, C.; Douillard, T.; Maire, E.; Meille, S.; Reynard, B.; Cuny, G. 3D microstructural study of selachimorph enameloid evolution. *J. Struct. Biol.* **2021**, *213*, 107664. [CrossRef] [PubMed]
34. Pfretzschner, H. Enamel microstructure and hypsodonty in large mammals. In *Structure, Function and Evolution of Teeth*; Smith, P., Tchernov, E., Eds.; Freund Publishing House Ltd.: Tel Aviv, Israel, 1992.
35. von Koenigswald, W. Two different strategies in enamel differentiation: Marsupialia versus Eutheria. In *Development, Function and Evolution of Teeth*; Teaford, M.F., Smith, M.M., Ferguson, M.W.J., Eds.; Cambridge University Press: Cambridge, UK, 2000; pp. 107–118.

36. Kemp, N.E. *Sharks, Skates, and Rays: The Biology of Elasmobranch Fishes*; Chapter Integumentary System and Teeth; The Johns Hopkins University Press: Baltimore, MD, USA, 1999; pp. 43–68.
37. Ferretti, M.P. Enamel Structure of *Cuvieronius Hyodon* (Proboscidea, Gomphotheriidae) discussion on enamel evolution in elephantoids. *J. Mamm. Evol.* **2008**, *15*, 37–58. [CrossRef]
38. Meyers, M.A.; Lin, A.Y.M.; Lin, Y.S.; Olevsky, E.A.; Georgalis, S. The cutting edge: Sharp biological materials. *JOM* **2008**, *60*, 19–24. [CrossRef]
39. Yin, Z.; Hannard, F.; Barthelat, F. Impact-resistant nacre-like transparent materials. *Science* **2019**, *364*, 1260–1263. [CrossRef] [PubMed]

Article

Preparation of Nanoscale Urushiol/PAN Films to Evaluate Their Acid Resistance and Protection of Functional PVP Films

Kunlin Wu [1,2], Bing-Chiuan Shiu [1], Ding Zhang [1,3], Zhenhao Shen [1], Minghua Liu [2] and Qi Lin [1,*]

1. Fujian Engineering and Research Center of New Chinese Lacquer Materials, Ocean College, Minjiang University, Fuzhou 350108, China; Kunlinwu2020@163.com (K.W.); bcshiu@mju.edu.cn (B.-C.S.); dingzhang1874@126.com (D.Z.); s951623874@126.com (Z.S.)
2. College of Environment and Resources, Fuzhou University, Fuzhou 350108, China; mhliu2000@fzu.edu.cn
3. College of Materials Science and Engineering, Fuzhou University, Fuzhou 350108, China
* Correspondence: linqi@mju.edu.cn

Abstract: Different amounts of urushiol were added to a fixed amount of polyacrylonitrile (PAN) to make nanoscale urushiol/PAN films by the electrospinning method. Electrospinning solutions were prepared by using dimethylformamide (DMF) as the solvent. Nanoscale urushiol/PAN films and conductive Poly(3,4-ethylenedioxythiophene):poly(styrenesulfonate)(PEDOT:PSS)/polyvinyl pyrrolidone (PVP) films were prepared by electrospinning. In order to prepare an electrospun sandwich nanoscale film, urushiol/PAN films were deposited as both the top and bottom layers and PEDOT:PSS/PVP film as the inner layer. When the PAN to urushiol ratio was 7:5, the fiber diameter ranged between 150 nm and 200 nm. The single-layer urushiol/PAN film could not be etched after being immersed into 60%, 80%, and 100% sulfuric acid (H_2SO_4) for 30 min, which indicated the improved acid resistance of the PAN film. The urushiol/PAN film was used to fabricate the sandwich nanoscale films. When the sandwich film was immersed into 80% and 100% H_2SO_4 solutions for 30 min, the structure remained intact, and the conductive PVP film retained its original properties. Thus, the working environment tolerability of the functional PVP film was increased.

Keywords: urushiol; polyacrylonitrile (PAN); electrospinning; acid membrane; nanofibers

1. Introduction

Raw lacquer (RL) is a milky white gel-like liquid harvested from *oxicodendron vernicifluum*. It is a natural polymer-based, environmentally friendly composite material and mainly consists of urushiol (60–65%), water (20–30%), colloidal substances (5–7%), glycoprotein composition (2%), and laccase (0.2%) [1,2]. Urushiol is generally extracted from raw lacquer using ethanol and acetone as solvents. Urushiol is a derivative of catechol with long side chains of different unsaturation degrees [3,4]. Because of its special structure, urushiol has shown many excellent properties, including chemical corrosion resistance, high gloss, stable heat resistance, etc. [5–7]. Raw lacquer is a kind of original coating material in China, which has a history of more than 8000 years. Some ancient lacquerware is still well preserved, which shows that urushiol has excellent durability and corrosion resistance. Many researchers have done a lot of research on the performance optimization of urushiol. Deng et al. [8] prepared an excellent corrosion-resisting graphene/raw lacquer composite coating by modifying waterborne graphene with lignin tripolymer (LT; acted as an aqueous stabilizer) and subsequently adding it to RL. Zheng et al. [9] proposed a facile one-pot synthesis method for silver nanoparticles (AgNPs) encapsulated in polymeric urushiol (PUL). Silver nitrate catalyzed the polymerization of urushiol in PUL, and the antibacterial rate of the 0.1% AgNPs coating was 100% in laboratory experiments. In addition, polyamidoamine (PAMAM) is also used to improve the alkali resistance of raw lacquer. Zhang et al. [10] reported that when an RL/PAMAM film was immersed in 15% NaOH for seven days, PAMAM molecules and urushiol were cross-linked to form a dense

Citation: Wu, K.; Shiu, B.-C.; Zhang, D.; Shen, Z.; Liu, M.; Lin, Q. Preparation of Nanoscale Urushiol/PAN Films to Evaluate Their Acid Resistance and Protection of Functional PVP Films. *Nanomaterials* **2021**, *11*, 957. https://doi.org/10.3390/nano11040957

Academic Editor: Jürgen Eckert

Received: 5 March 2021
Accepted: 6 April 2021
Published: 9 April 2021

Publisher's Note: MDPI stays neutral with regard to jurisdictional claims in published maps and institutional affiliations.

Copyright: © 2021 by the authors. Licensee MDPI, Basel, Switzerland. This article is an open access article distributed under the terms and conditions of the Creative Commons Attribution (CC BY) license (https://creativecommons.org/licenses/by/4.0/).

structure, and this significantly improved the alkali resistance of the RL/PAMAM film. Jeong et al. [11] prepared urushiol powders with different amounts of 3-(trimethoxysilyl) propyl methacrylate (TPM). The as-prepared powders manifested excellent antibacterial activity, good antioxidant activity, and very high thermal stability. Cheng et al. [12] proposed an efficient and green approach to simultaneously reduce and functionalize graphene with urushiol. Urushiol with unsaturated long alkyl chains was modified on the graphene surface to disperse graphene in the organic solvent and the polymer matrix, resulting in enhanced interfacial interaction between graphene and the polymer matrix.

Polyacrylonitrile (PAN) is generally obtained from acrylonitrile extracted from petroleum through free radical polymerization. Acrylonitrile units in a macromolecular chain are connected by a head–tail method. The strength of polyacrylonitrile fibers is not high, and their abrasion resistance and fatigue resistance are poor. In addition, PAN has good weather resistance and solar resistance [13,14]. PAN nanoscale films are often used as proton exchange membranes in fuel cells [15], industrial wastewater heavy metal adsorption filtration membranes [16], and oil-water filtration membranes in marine and river environments [17]. Furthermore, PAN nanoscale films are also used in combat wear, sports, and biomedical wearable applications [13]. In order to ensure the stability of PAN nanoscale films in the aforesaid application conditions, the chemical stability of PAN is crucial.

Polyvinyl pyrrolidone (PVP) is a non-ionic polymer and is obtained through bulk polymerization, solution polymerization, and other polymerization methods using vinyl pyrrolidone (NVP) as the raw material. PVP is the most typical N-vinylamide-based polymer. As a synthetic water-soluble polymer, PVP has the characteristics of water-soluble polymeric materials, such as excellent film-forming and adhesive properties, good hygroscopicity, superb solubilization, and cohesion [18]. Lee et al. [19], developed a silver nanowire-based electrode on PVP-coated PET with low resistance and high conductivity. Zhang et al. [20] prepared precursor solutions by dissolving polymer mixtures of PAN and PVP with different weight ratios into a constant volume of dimethylformamide (DMF) through electrospinning. PVP nanoscale films are widely used in supercapacitors [21], CO_2 adsorption [22], and lithium-ion batteries [23]. Poly(3,4-ethylenedioxythiophene) polystyrene sulfonate (PEDOT:PSS)-doped PVP is extensively used in super electrodes [24,25] and gas sensors [26] due to its better conductive effect. However, in application, PVP is not acid-resistant and is soluble in water, resulting in a short film life. Wu et al. [27] added raw lacquer to PVP and prepared nanofiber films by electrospinning, which improved the mechanical properties and acid resistance of the PVP films. However, adding raw lacquer would change some of the unique properties of PVP itself. Therefore, a coating is needed that can, not only improve the service life of the PVP film, but also maintain the PVP characteristic performance.

In the present work, urushiol was extracted from a lacquer of *oxicodendron vernicifluum* by the rotary evaporation method. Using dimethylformamide (DMF) as the solvent, different amounts of urushiol were added into a certain amount of polyacrylonitrile (PAN), and urushiol/PAN nanofilms were prepared by the electrospinning method. The addition of urushiol can improve the shortcomings of low mechanical strength and chemical stability of PAN films. The improvement of the performance of PAN films is of great significance to its application. In addition, to further explore the use of urushiol/PAN nanoscale films as protective materials, a sandwich structure of urushiol/PAN-coated PVP nanoscale film was prepared by electrospinning. This sandwich nanoscale film was immersed in sulfuric acid and pure water to investigate its corrosion resistance.

2. Experiment

2.1. Materials

Polyacrylonitrile (PAN; molecular weight = 150,000), polyvinylpyrrolidone (PVP; molecular weight = 360,000), and PEDOT:PSS were obtained from Sigma, St. Louis, MO, USA. Chinese raw lacquer (RL; Shanxi, China) was filtered with gauze. Sodium hydroxide (NaOH) and sulfuric acid (H_2SO_4) (Shanghai Chenghai Chemical Industry Co. Ltd.,

Shanghai, China) and absolute ethanol (Xilong Science Co. Ltd., Guangdong, China) were all analytically pure.

2.2. Preparation of Urushiol, Urushiol/PAN Mixture, and PEDOT:PSS/PVP Mixture

Preparation of urushiol: Twenty grams of anhydrous ethanol were added to 5 g of raw lacquer in a beaker and sonicated for 2 h. Impurities in the solution were filtered by a filter paper, and anhydrous ethanol in the solution was removed by the rotary evaporation extraction method at 60 °C to obtain urushiol [28].

Preparation of urushiol/PAN mixture: The preparation method of the 5:5 urushiol/PAN mixture was taken as an example. Ten grams of DMF were first added to 0.5 g of urushiol in a beaker and evenly stirred. Subsequently, 0.5 g of PAN was added to the resultant and stirred with a magnetic stirrer at 450 r/min for 8 h to obtain a mixture of urushiol/PAN. Table 1 presents different urushiol/PAN mixture ratios used in the current experiment.

Table 1. Voltage values of urushiol/polyacrylonitrile (PAN) and Poly(3,4-ethylenedioxythiophene): poly(styrenesulfonate)(PEDOT:PSS)/polyvinyl pyrrolidone (PVP) films prepared with different urushiol/PAN mixing ratios.

Urushiol/PAN (g)	Voltage 1 (kV)	Voltage 2 (kV)	Voltage 3 (kV)	The Average (kV)
5% PAN	29.54	27.32	27.00	27.95
3:5	23.61	27.08	26.08	25.59
4:5	23.65	23.53	25.02	24.06
5:5	23.31	23.85	23.44	23.53
6:5	23.05	23.25	22.96	23.09
7:5	22.31	22.36	22.79	22.49
PEDOT:PSS/PVP (g)	Voltage 1 (kV)	Voltage 2 (kV)	Voltage 3 (kV)	The Average(kV)
4:1	27.51	28.07	27.68	27.75

Preparation of PEDOT:PSS/PVP mixture: Ten grams of DMF were added to 8 g of PEDOT:PSSS in a beaker and evenly stirred. Subsequently, 2 g of PVP was added to the resultant mixture and stirred with a magnetic stirrer at 450 r/min for 8 h to prepare a mixture of PEDOT:PSS/PVP.

2.3. Characterization

The IR spectra of the membranes were detected by an FT-IR spectrometer (MPIR8400S, Shimadzu, Kyoto, Japan) based on the ATR method. Sixteen scans were conducted for each film at a resolution of 4 cm^{-1}. The SEM images of the as-prepared nanofilms were captured by a field-emission SEM (Nova Nano SEM 230, FEI, Hillsboro, OR, USA) at an acceleration voltage of 2 kV. The fiber size distribution (fiber diameter) was evaluated in Image-Pro Plus 6.0 software. For tensile strength (σ) measurements, each film was cut into a width (b) of 10 mm and a length of 100 mm and tested on a universal material testing machine (Instron 1185, Instron, Chicago, IL, USA) at a speed of 2 mm/min to record the peak load (P) of film fracture. The tensile strength (σ) was determined by the following formula.

$$\sigma = \frac{P}{bd} \tag{1}$$

2.4. Electrospinning Process

Preparation of urushiol/PAN and PEDOT:PSS/PVP nanofilm was completed using electrospinning equipment (JDF05, Changsha Nayi Instrument Technology Co. Ltd., Changsha, China). The mixtures were infused into a syringe. The syringe needle type was 23 G, and its inner diameter was 0.34 mm. Ten grams of dimethylformamide were used to dissolve the urushiol and PAN of different qualities, and repeated tests were conducted on different samples. Table 1 presents the average voltage values obtained from different samples. The metallic syringe needle served as the anode, while the collection board acted as the cathode, and they interacted with each other under an externally applied high voltage (Figure 1). The temperature and humidity during the experiment were strictly controlled. The humidity in the air was ≤60% at room temperature. At the feed flow rate of 0.55 mL/h, the horizontal and vertical distances between the needle tip and the collecting reel center were 14 cm and 11 cm, respectively.

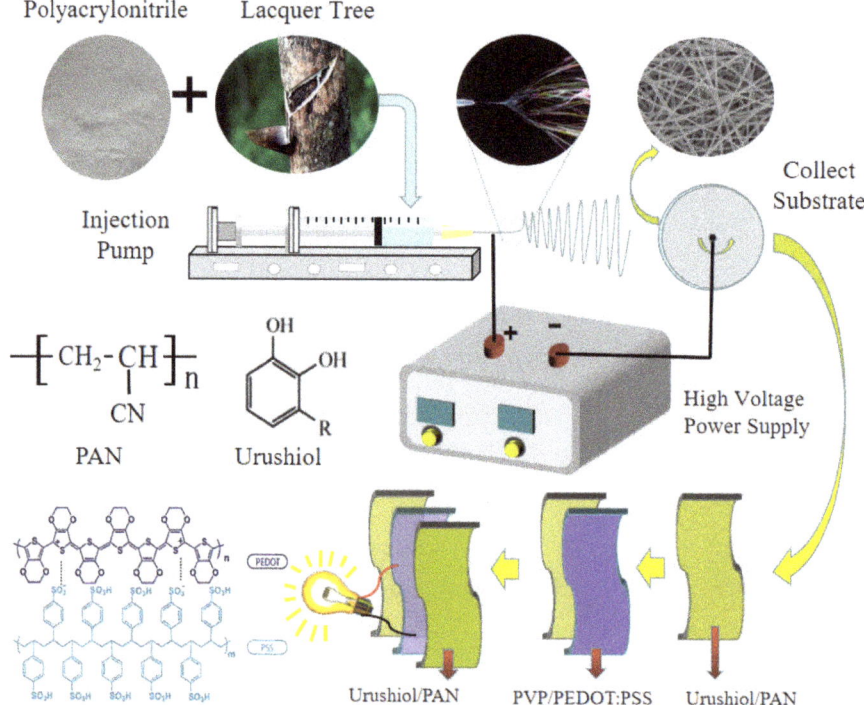

Figure 1. Schematic diagrams of electrospun urushiol/PAN and PEDOT:PSS/PVP nanofilms.

3. Results and Discussion

3.1. Effects of Urushiol/PAN Ratio on Fiber Morphology

Figure 2a reveals that the nanoscale PAN film prepared by electrospinning had good fiber morphology with an average fiber diameter of 40–70 nm. Figure 2b expresses that when the urushiol/PAN mixing ratio was 3:5, the fiber diameter ranged between 100 nm and 180 nm. As the viscosity and surface tension of the mixed solution increased with the addition of urushiol, the electrospinning voltage decreased accordingly, resulting in a good fiber morphology [29]. Table 1 presents the voltage values for different urushiol/PAN mixing ratios. When the urushiol/PAN mixing ratio was 6:5, the fiber diameter ranged between 140 nm and 240 nm. When the urushiol/PAN mixing ratio was 7:5 and the voltage was about 22.5 kV, the fiber diameter ranged between 100 nm and 250 nm, resulting in the optimal fiber morphology. With a further increase in the urushiol amount, voltage adjustment could not yield a good fiber morphology. When the mixture reaches the collecting substrate, the solvent evaporates rapidly, and the PAN solidifies into a fiber. Due to the long drying time of urushiol, when the urushiol content increases, it is easy to cause the undried fibers to stick together.

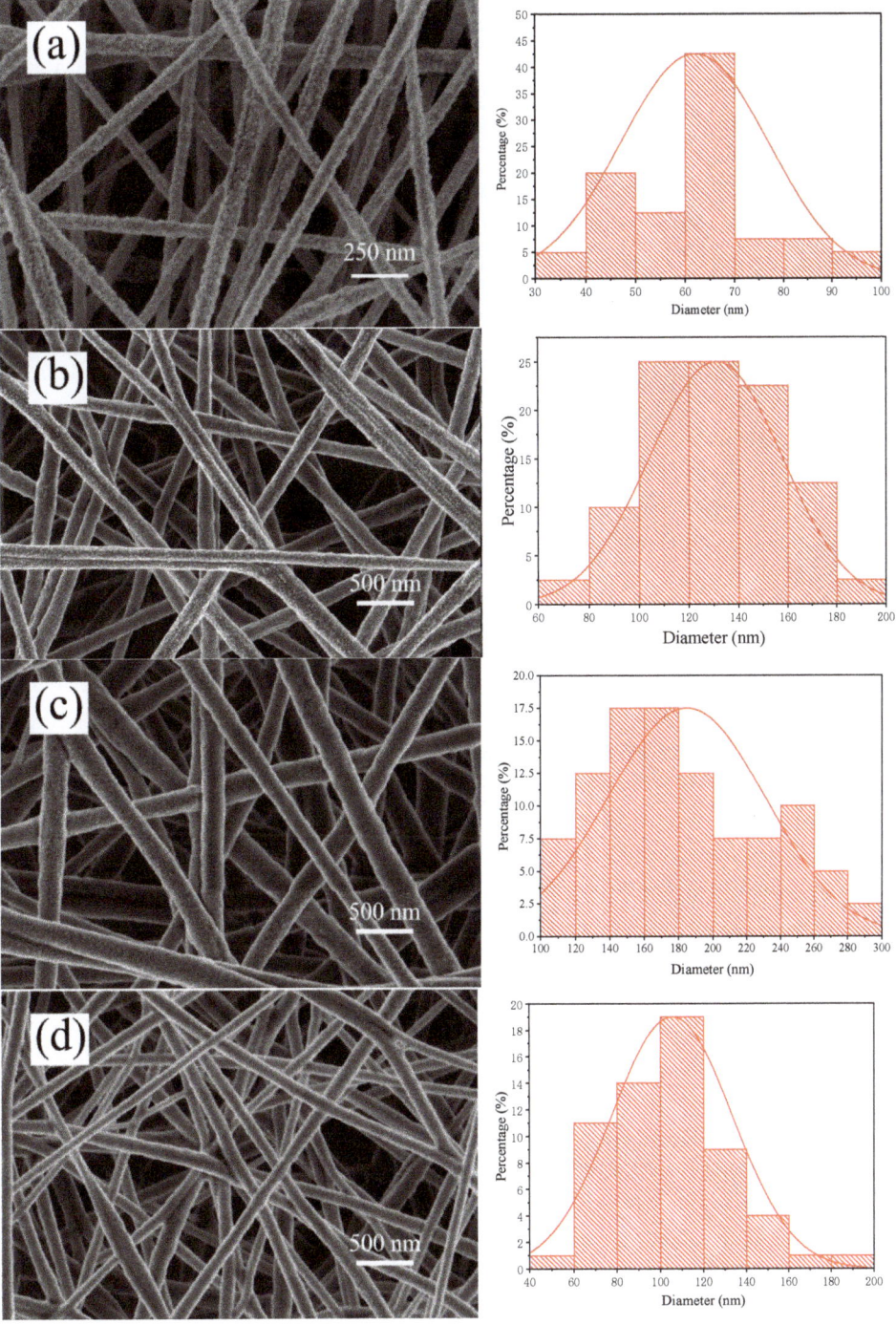

Figure 2. Cont.

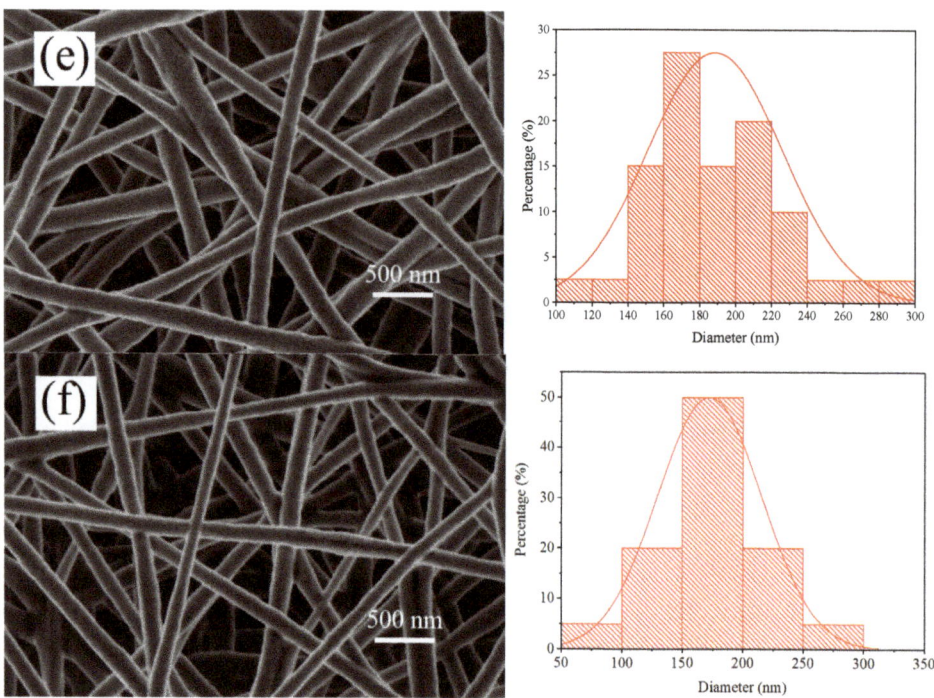

Figure 2. SEM images of urushiol/PAN nanofilm surfaces for different mass ratios: (**a**) Without urushiol (5% PAN), (**b**) 3:5, (**c**) 4:5, (**d**) 5:5, (**e**) 6:5, (**f**) 7:5.

In order to further observe changes in the morphology of the nanoscale PAN film after urushiol addition, the cross-sectional view of the nanoscale urushiol/PAN film was observed by SEM. It is clear from Figure 3a that without urushiol, the gap between PAN film layers was large, which can be attributed to the smaller fiber diameter of the nanoscale PAN nanofilm without urushiol. As the amount of urushiol was increased, the structure of the nanoscale film became denser, manifesting a better acid-resisting performance at the same rotating speed of the collecting drum (Figure 3b–e). When the film was cut, its fibers experienced a curling phenomenon due to the brittle fracture of liquid nitrogen; however, this phenomenon was not observed when urushiol was added in a low proportion (Figure 3e,f). When the urushiol/PAN mixing ratio was 6:5 or 7:5, the fiber curling phenomenon occurred.

3.2. FT-IR Spectra Analysis

Figure 4 displays the FT-IR spectra of the nanoscale films prepared with different urushiol/PAN mixing ratios. Urushiol is a mixture of several unsaturated branched derivatives of catechol. It contains two adjacent phenolic hydroxyl groups with the characteristics of phenol. It also has unsaturated aliphatic hydrocarbons on the benzene ring with the characteristics of unsaturated bonds. The absorption peaks of urushiol C-O-H groups at 3439 cm^{-1} and 1150 cm^{-1} were O-H stretching vibration and C-O stretching vibration [30]. The peak at 1363 cm^{-1} corresponds to the ring stretching of the aromatic carbons near the phenolic O-H, and the one at 1275 cm most probably to C-H or O-H bending. With the addition of PAN, the intensity of these peaks did not decrease, indicating that the number of O-H groups on the urushiol aromatic ring did not decrease. The vibration peaks of urushiol phenyl were observed at 1621 cm^{-1} and 1597 cm^{-1} [28], indicating that the urushiol/PAN membrane retained the structural characteristics of urushiol after adding

PAN. The PAN nanofibers displayed characteristic vibrations at 2243 cm^{-1} for ν(C≡N), 2929 cm^{-1} for ν(CH$_2$), and 1452 cm^{-1} for δ(CH$_2$) [31]. The characteristic peak of PAN still existed after the addition of urushiol, further proving that the absorption peak and basic element of PAN were not damaged by the presence of urushiol.

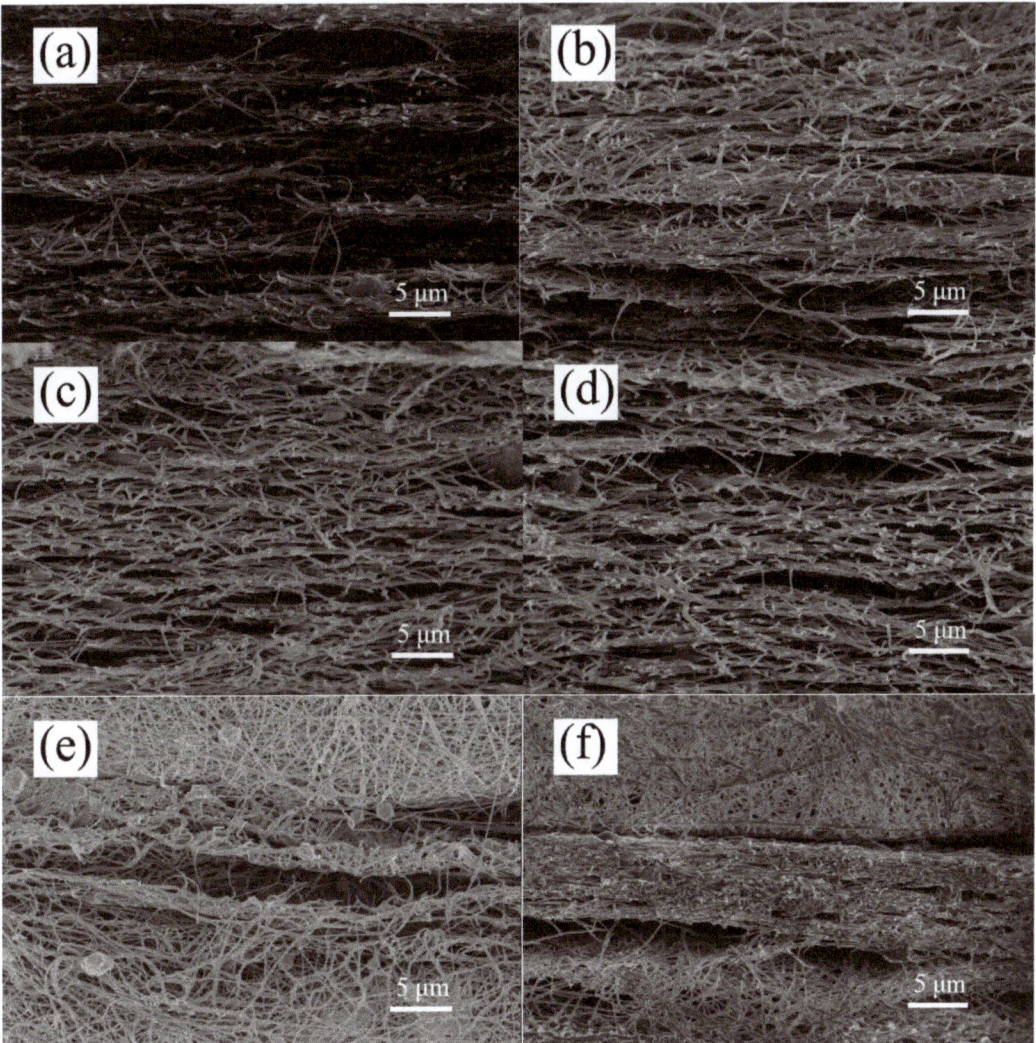

Figure 3. SEM images of the cross-sections of urushiol/PAN nanofilms for different mass ratios: (**a**) Without urushiol (5% PAN), (**b**) 3:5, (**c**) 4:5, (**d**) 5:5, (**e**) 6:5, (**f**) 7:5.

3.3. Acid Resistance of Nanofilms

Figure 5a displays the changes in the pure nanoscale PAN film immersed in 60% (a$_1$), 80% (a$_2$), and 100% (a$_3$) sulfuric acid for 30 min. It is evident that the pure PAN film retained a good film-forming performance in the 60% sulfuric acid solution; however, it easily dissolved in the 80% and 100% sulfuric acid solutions. As the amount of urushiol increased, the overall film morphology in 80% and 100% sulfuric acid solutions was incomplete; however, the film morphology remained after 30 min. This is because urushiol has excellent acid resistance and the structure of urushiol will not be destroyed by electrospinning technology. The FT-IR spectra of urushiol/PAN also showed that the urushiol specific groups still existed, indicating that the urushiol/ PAN prepared by electrospinning retained the original excellent performance of urushiol. After PAN was mixed with urushiol, regardless of the mixing ratio, the film formation and morphological integrity after immersion in the 100% sulfuric acid solution for 30 min were superior to those in the 80% sulfuric acid solution. When the urushiol/PAN mixing ratio reached 7:5, the nanoscale urushiol/PAN film manifested good structural integrity in the 80% sulfuric acid solution. In order to further investigate this phenomenon, elemental analyses, and electron microscopic observations were performed on the film soaked in different sulfuric acid solutions.

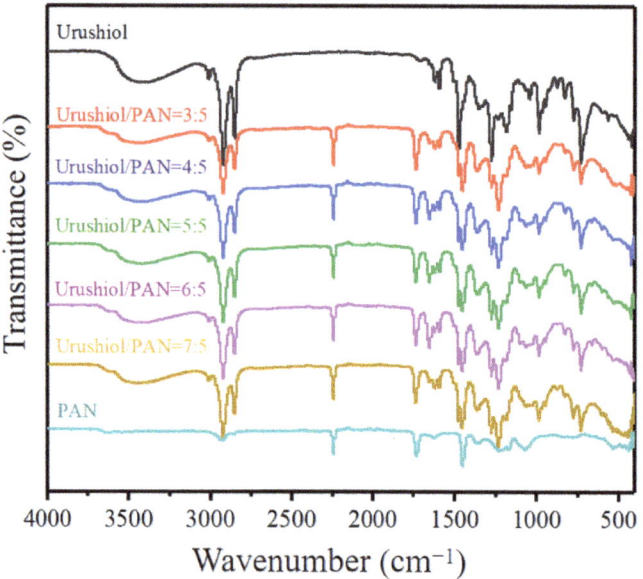

Figure 4. FT-IR spectra of urushiol/PAN nanofilms with different urushiol/ PAN mixing ratios.

3.4. EDS Analysis

When the ratio of urushiol to PAN was less than 5:7, the film was damaged to varying degrees in 80% sulfuric acid solution. Figure 6 displays the SEM and EDS analysis results of the 7:5 nanoscale urushiol/PAN film immersed in different sulfuric acid solutions for 30 min and dried for 24 h. The acid resistance of the nanoscale PAN film after urushiol addition was greatly improved. The fiber morphology remained intact when the film was immersed in 60% and 80% sulfuric acid solutions. Table 2 presents the weight percentage changes of oxygen, nitrogen, and carbon. When the ratio of urushiol to PAN was lower than 5:7, the nanoscale urushiol/PAN film could be etched and dissolved in 80% sulfuric acid; however, urushiol/PAN film remained a complete shape in the 100% sulfuric acid solution without rupture or dissolution. The EDS analysis reveals that the nanoscale urushiol/PAN film was quickly oxidized in the strong acid solution and formed a dense oxide layer on the surface [32]. Figure 7 shows the FT-IR spectra of the urushiol/PAN film treated with 100% sulfuric acid. The characteristic absorption peaks of PAN at 2243 cm^{-1}, 2929 cm^{-1}, and 1640 cm^{-1} disappear. The disappearance of characteristic absorption peaks at 2243 cm^{-1}, may be due to the hydrolysis of C≡N to carboxylic acid under acidic conditions. Because C=C has an electrophilic addition reaction with sulfuric acid, resulting in the disappearance of characteristic absorption peaks at 2929 cm^{-1} and 1640 cm^{-1}.

Table 2. EDS results of the 7:5 nanoscale urushiol/PAN film immersed in different concentrations of sulfuric acid.

Percentage by Weight	C K	N K	O K
60%	66.89	23.11	10.00
80%	71.10	20.20	8.70
100%	76.74	0	23.26

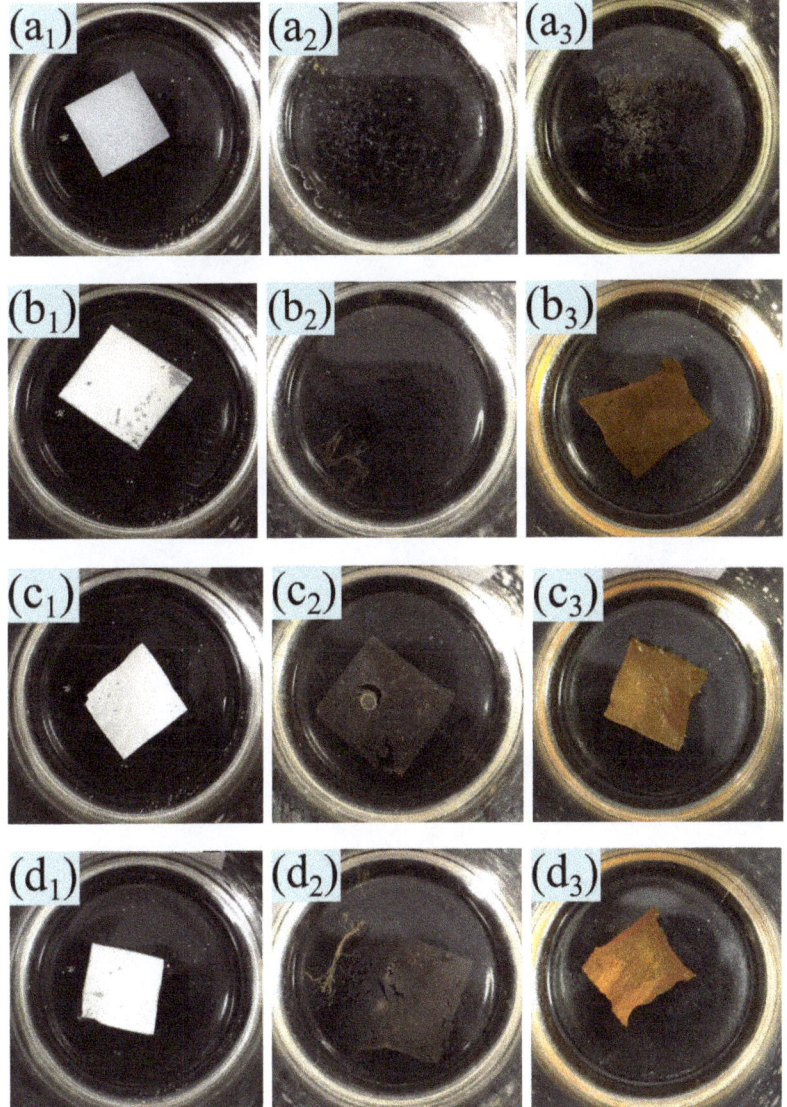

Figure 5. *Cont.*

Figure 5. Morphologies of nanoscale urushiol/PAN films immersed in (a_1) 60%, (a_2) 80%, and (a_3) 100% sulfuric acid solutions for 30 min: (**a**) Without urushiol (5% PAN), (**b**) 3:5, (**c**) 4:5, (**d**) 5:5, (**e**) 6:5, (**f**) 7:5.

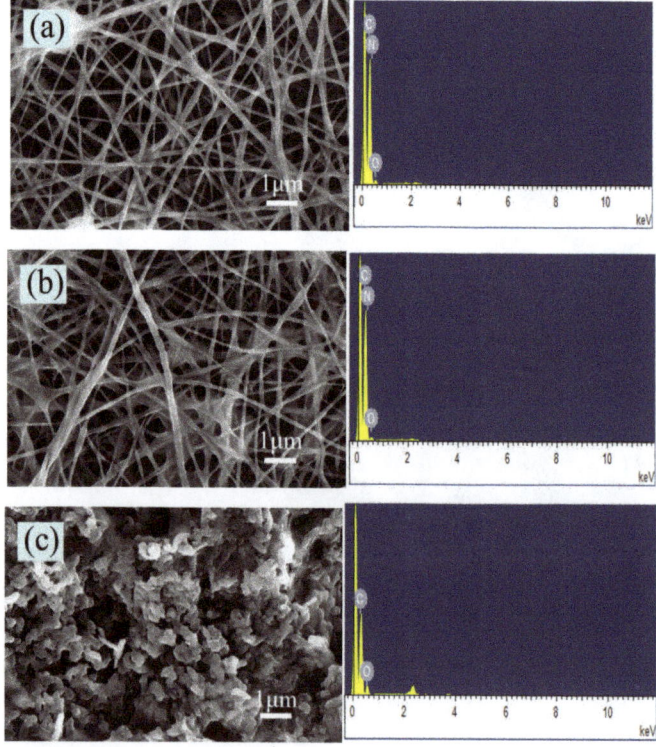

Figure 6. SEM images and EDS results of the 7:5 nanoscale urushiol/PAN film immersed in (**a**) 60%, (**b**) 80%, and (**c**) 100% sulfuric acid solutions.

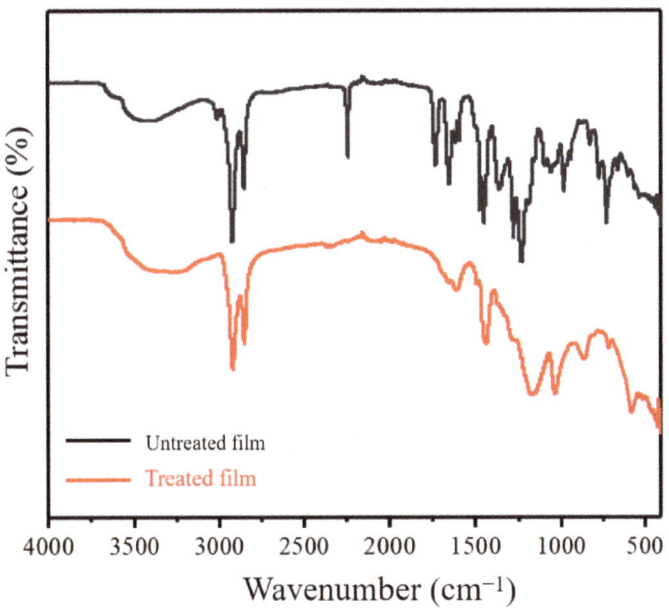

Figure 7. FT-IR spectra of the 7:5 nanoscale urushiol/PAN film immersed in 100% sulfuric acid solutions.

3.5. Water Contact Angle and Tensile Strength

Electrospun nanoscale films have the advantages of small fiber diameter and large specific surface area. PAN has good weather resistance and poor wear resistance. Moreover, water can easily penetrate a nanoscale PAN film. Figure 8 presents the water contact angle test results for the nanoscale PAN film. The contact angle was 0° after about 15 s, suggesting hydrophilic properties of the nanoscale PAN film. In contrast, the water contact angle of the nanoscale 7:5 urushiol/PAN film was about 80° after 15 s; therefore, the nanoscale PAN film mixed with urushiol had better water resistance. This is because urushiol is a hydrophobic material, and the addition of urushiol also showed the hydrophobicity of urushiol/PAN nanofilms. At the same time, the internal structure of the urushiol/PAN nanofilm was relatively dense after adding urushiol, which slowed down the infiltration of water molecules into the inner membrane.

The tensile strength of the nanoscale urushiol/PAN film is presented in Table 3. With the addition of urushiol, the tensile strength of the nanoscale PAN film was significantly enhanced. This is because of the high hardness of urushiol, adding urushiol can improve the mechanical properties of the film. At the same time, with the increase in urushiol proportion, the internal structure of urushiol/ PAN nanofilms became denser and denser, which was equivalent to the increase in the number of fibers per unit area.

Table 3. Tensile strength results of urushiol/PAN nanofilms.

Urushiol/PAN (g)	Wide (mm)	Thickness (mm)	Peak Load (N)	Tensile Strength (MPa)
5% PAN	9	2.817×10^{-1}	1.751	0.691
3:5	9	9.254×10^{-2}	0.850	0.856
4:5	9	1.371×10^{-1}	2.551	2.067
5:5	9	8.058×10^{-2}	1.950	2.689
6:5	9	9.106×10^{-2}	2.651	3.235
7:5	9	4.506×10^{-2}	1.751	4.318

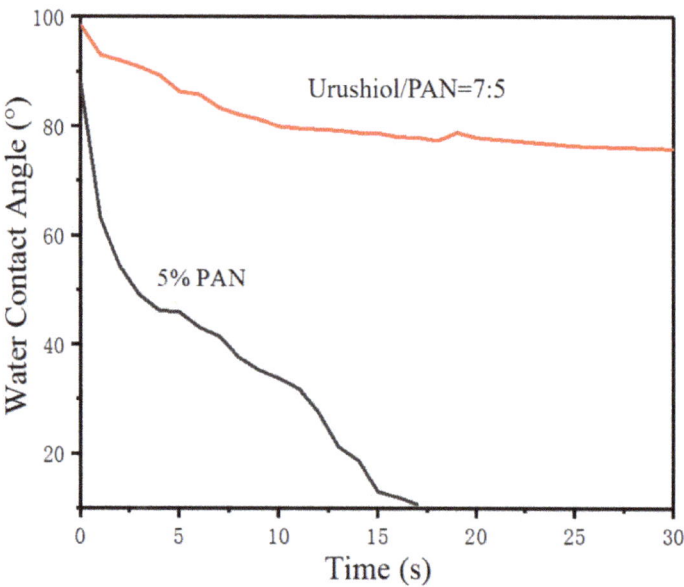

Figure 8. Water contact angle test results of nanoscale urushiol/PAN and PAN films.

3.6. A Test for Acid Resistance of Sandwich Structure

Figure 9 presents the test results of the sandwich nanoscale urushiol/PAN film. Both top and bottom layers were electrospun PAN films, and the middle inner layer was electrospun PVP film (Figure 9a). PVP in the middle layer was immediately dissolved when the sandwich nanoscale film was immersed in water; thus, the nanoscale PVP film disappeared, resulting in a separation phenomenon, which limits the application of many functional nanoscale PVP films due to their hydrolysis and intolerance to acids. The SEM images of the electrospun urushiol/PAN: PEDOT:PSS/PVP:urushiol/PAN film (urushiol/PAN mixing ratio = 7:5; urushiol/PAN films were deposited at the top and bottom of a PEDOT:PSS/PVP film (inner layer)) immersed in 80% and 100% sulfuric acid solutions for 30 min are displayed in Figure 9c,d, respectively. No layer separation was observed in this case, and the overall film morphology and structure remained intact. The EDS analysis result is displayed in Figure 9b. The characteristic sulfur peak in the conductive PEDOT:PSS polymer still existed after the sandwich nanoscale film was immersed in the 100% sulfuric acid solution. The surface of the urushiol/PAN film is oxidized to form an oxide layer, preventing the acid from corroding the inner structure and hindering the etching reaction. Therefore, the electrospun 7:5 nanoscale urushiol/PAN film had good acid resistance.

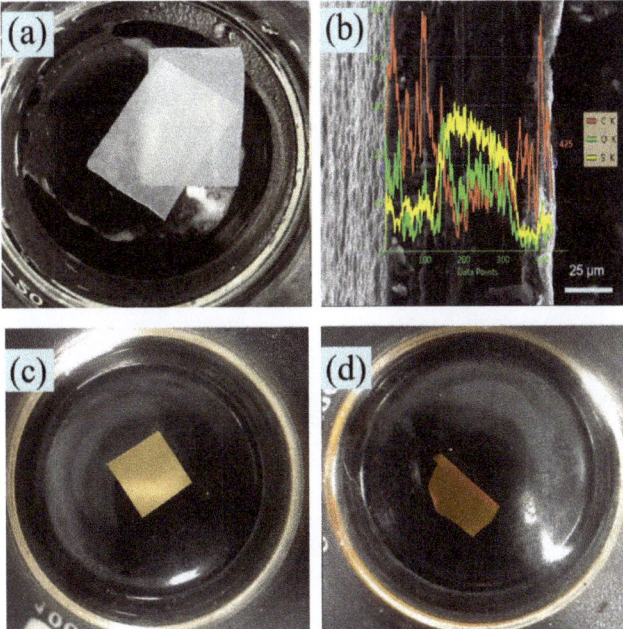

Figure 9. (a) Picture of sandwich electrospun PAN:PVP:PAN film, (b) EDS result of the sandwich urushiol/PAN:PEDOT:PSS/PVP:urushiol/PAN film, (c) picture of the urushiol/PAN:PEDOT:PSS/PVP:urushiol/PAN film immersed in the 80% sulfuric acid solution for 30 min, (d) picture of the urushiol/PAN:PEDOT:PSS/PVP:urushiol/PAN film immersed in the 100% sulfuric acid solution for 30 min.

4. Conclusions

In the current study, different amounts of urushiol were added to a fixed amount of PAN to prepare a nanoscale PAN film with good acid-resisting properties. When the urushiol and PAN mixing ratio reached 3:5, the nanoscale PAN film exhibited the same anti-etching effect as metal substances. The urushiol/PAN film could not be dissolved in the 100% sulfuric acid solution because a dense oxidized layer appeared on the film surface, preventing the acidic etching reaction with the inner layer. However, the urushiol/PAN film could be dissolved in the 80% sulfuric acid solution. As 80% sulfuric acid is less oxidizing than 100% sulfuric acid, the acidic solution could penetrate the film for the acidic etching reaction. When the urushiol and PAN mixing ratio was 7:5, a superior acid resistance effect was noticed, and the film remained intact in either 80% or 100% sulfuric acid solutions. Therefore, the urushiol/PAN mixing ratio of 7:5 was used to protect the nanoscale PVP film. In order to prepare an electrospun sandwich nanoscale film, urushiol/PAN films were deposited as top and bottom layers on a PEDOT:PSS/PVP film (inner layer). After the immersion of the nanoscale PEDOT:PSS/PVP film in 80% and 100% sulfuric acid solutions, the sulfur peak of PEDOT:PSS was still noticed, demonstrating good acid resistance of the urushiol/PAN film. When urushiol was added in a low proportion, PAN yielded an acidic reaction between 80% and 100% sulfuric acid solutions. Therefore, urushiol/PAN films can be used as unique materials with strong acid resistance and weak acid dissolution properties in the future.

Author Contributions: Methodology, K.W. and B.-C.S.; formal analysis, K.W., B.-C.S and Q.L.; data curation, K.W., D.Z. and Z.S.; resources, Q.L. and M.L.; writing—original draft preparation, K.W. and B.-C.S.; writing—review and editing, K.W. and B.-C.S.; supervision, Q.L. and M.L.; All authors have read and agreed to the published version of the manuscript.

Funding: This work was financially supported by the National Innovation Program for College Students (202010395016) and Fuzhou Municipal Science Foundation (2020-GX-2).

Conflicts of Interest: The authors declare no conflict of interest.

References

1. Lu, R.; Harigaya, S.; Ishimura, T.; Nagase, K.; Miyakoshi, T. Development of a fast drying lacquer based on raw lacquer sap. *Prog. Org. Coat.* **2004**, *51*, 238–243. [CrossRef]
2. Yang, J.; Deng, J.; Zhang, Q.; Shen, Q.; Li, D.; Xiao, Z. Effects of polysaccharides on the properties of Chinese lacquer sap. *Prog. Org. Coat.* **2015**, *78*, 176–182. [CrossRef]
3. Xia, J.; Chen, B.; Chen, Q. Synthesis of anisotropic Janus composite particles based on poly(urushiol borate). *Polym. Eng.* **2018**, *58*, 2304–2310. [CrossRef]
4. Lu, R.; Yoshida, T.; Miyakoshi, T. Oriental Lacquer: A Natural Polymer. *Polym. Rev.* **2013**, *53*, 153–191. [CrossRef]
5. Kim, D.; Jeon, S.L.; Seo, J. The preparation and characterization of urushiol powders (YPUOH) based on urushiol. *Prog. Org. Coat.* **2013**, *76*, 1465–1470. [CrossRef]
6. Zhou, C.; Hu, Y.; Yang, Z.; Yuan, T.; Huang, J.; Li, P.; Liu, Y.; Zhang, S.; Yang, Z. Facile synthesis and characterization of urushiol analogues from tung oil via ultraviolet photocatalysis. *Prog. Org. Coat.* **2018**, *120*, 240–251. [CrossRef]
7. Kim, H.S.; Yeum, J.H.; Choi, S.W. Urushiol/polyurethanenalogues from tung oil via ultraviolet photocatalysis. *Prog. Coat.* **2009**, *65*, 341–347. [CrossRef]
8. Deng, Y.; Bai, W.; Chen, J.; Zhang, X.; Wang, S.; Lin, J.; Xu, Y. Bio-inspired electrochemical corrosion coatings derived from graphene/natural lacquer composites. *RSC Adv.* **2017**, *7*, 45034–45044. [CrossRef]
9. Zheng, L.; Lin, Y.; Wang, D.; Chen, J.; Xu, Y. Facile one-pot synthesis of silver nanoparticles encapsulated in natural polymeric urushiol for marine antifouling. *RSC Adv.* **2020**, *10*, 13936–13943. [CrossRef]
10. Zhang, D.; Xia, J.; Xue, H.; Zhang, Y.; Lin, Q. Improvement on properties of Chinese lacquer by polyamidoamine. *Polym. Eng.* **2020**, *60*, 25371–25379. [CrossRef]
11. Jeong, S.; Kim, D.; Seo, J. Preparation and the antioxidant and antibacterial activities of urushiol powders (YPUOH). *Prog. Org. Coat.* **2014**, *77*, 981–987. [CrossRef]
12. Cheng, H.; Lin, J.; Su, Y.; Chen, D.; Zheng, X.; Zhu, H. Green synthesis of soluble graphene in organic solvent via simultaneous functionalization and reduction of graphene oxide with urushiol. *Mater. Today Commun.* **2020**, *23*, 100938–100944. [CrossRef]
13. Sahay, R.; Agarwal, K.; Anbazhagan, S.; Raghavan, N.; Baji, A. Helicoidally Arranged Polyacrylonitrile Fiber-Reinforced Strong and Impact-Resistant Thin Polyvinyl Alcohol Film Enabled by Electrospinning-Based Additive Manufacturing. *Polymers* **2020**, *12*, 2376. [CrossRef] [PubMed]
14. Karbownik, I.; Rac-Rumijowska, O.; Fiedot-Toboa, M.; Rybicki, T.; Helena, T. The Preparation and Characterization of Polyacrylonitrile-Polyaniline (PAN/PANI) Fibers. *Materials* **2019**, *12*, 664. [CrossRef]
15. Zeng, Y.; He, Z.; Hua, Q.; Xu, Q.; Min, Y. Polyacrylonitrile Infused in a Modified Honeycomb Aluminum Alloy Bipolar Plate and Its Acid Corrosion Resistance. *ACS Omega* **2020**, *5*, 16976–16985. [CrossRef]
16. Jamshidifard, S.; Koushkbaghi, S.; Hosseini, S.; Rezaei, S.; Karamipour, A.; Irani, M. Incorporation of UiO-66-NH2 MOF into the PAN/chitosan nanofibers for adsorption and membrane filtration of Pb(II), Cd(II) and Cr(VI) ions from aqueous solutions. *Hazard. Mater.* **2019**, *368*, 10–20. [CrossRef]
17. Naseeb, N.; Mohammed, A.; Laoui, T.; Khan, Z. A Novel PAN-GO-SiO$_2$ Hybrid Membrane for Separating Oil and Water from Emulsified Mixture. *Materials* **2019**, *12*, 212. [CrossRef]
18. Koczkur, K.M.; Mourdikoudis, S.; Polavarapu, L.; Skrabalak, S.E. Polyvinylpyrrolidone (PVP) in nanoparticle synthesis. *Dalton Trans.* **2015**, *44*, 17883–17905. [CrossRef] [PubMed]
19. Lee, S.H.; Lim, S.; Kim, H. Smooth-surface silver nanowire electrode with high conductivity and transparency on functional layer coated flexible film. *Thin Solid Films* **2015**, *589*, 403–407. [CrossRef]
20. Zhang, Z.; Li, X.; Wang, C. Polyacrylonitrile and Carbon Nanofibers with Controllable Nanoporous Structures by Electrospinning. *Macromol. Mater. Eng.* **2010**, *294*, 673–677. [CrossRef]
21. Tan, S.; Kraus, T.J.; Li-Oakey, K.D. Understanding the supercapacitor properties of electrospun carbon nanofibers from Powder River Basin coal. *Fuel* **2019**, *245*, 148–159. [CrossRef]
22. Zainab, G.; Babar, A.A.; Ali, N.; Aboalhassan, A.A.; Ding, B. Electrospun Carbon Nanofibers with Multi-Aperture/Opening Porous Hierarchical Structure for Efficient CO$_2$ Adsorption. *Colloid Interface Sci.* **2019**, *561*, 659–667. [CrossRef]
23. Li, Z.; Yin, Q.; Hu, W.; Zhang, J.; Guo, J.; Chen, J.; Sun, T.; Du, C.; Shu, J.; Yu, L. Tin/tin antimonide alloy nanoparticles embedded in electrospun porous carbon fibers as anode materials for lithium-ion batteries. *Mater. Sci.* **2019**, *54*, 9025–9033. [CrossRef]
24. Cárdenas-Martínez, J.; España-Sánchez, B.L.; Esparza, R.; Vila-Nio, J.A. Flexible and transparent supercapacitors using electrospun PEDOT:PSS electrodes. *Synth. Met.* **2020**, *267*, 116436–116442. [CrossRef]
25. Han, W.; Wang, Y.; Su, J.; Xin, X.; Guo, Y.; Long, Y.Z.; Ramakrishna, S. Fabrication of nanofibrous sensors by electrospinning. *Sci. China* **2019**, *62*, 886. [CrossRef]
26. Zhang, Q.; Wang, X.; Fu, J.; Liu, R.; He, H.; Ma, J.; Miao, Y.; Ramakrishna, S.; Long, Y. Electrospinning of Ultrafine Conducting Polymer Composite Nanofibers with Diameter Less than 70 nm as High Sensitive Gas Sensor. *Materials* **2018**, *11*, 1744. [CrossRef]

27. Wu, K.; Zhang, D.; Liu, M.; Lin, Q.; Shiu, B.C. A Study on the Improvement of Using Raw Lacquer and Electrospinning on Properties of PVP Nanofilms. *Nanomaterials* **2020**, *10*, 1723. [CrossRef]
28. Je, H.; Won, J. Natural urushiol as a novel under-water adhesive. *Chem. Eng. J.* **2021**, *404*, 126424–126431. [CrossRef]
29. Aijaz, M.O.; Karim, M.R.; Alharbi, H.F.; Alharthi, N.H. Novel optimised highly aligned electrospun PEI-PAN nanofibre mats with excellent wettability. *Polymer* **2019**, *180*, 121665–121677. [CrossRef]
30. Gao, R.; Wang, L.; Lin, Q. Effect of hexamethylenetetramine on the property of Chinese lacquer film. *Prog. Org. Coat.* **2019**, *133*, 16–173. [CrossRef]
31. Wu, M.; Wang, Q.; Li, K.; Wu, Y.; Liu, H. Optimization of stabilization conditions for electrospun polyacrylonitrile nanofibers. *Polym. Degrad. Stab.* **2012**, *97*, 1511–1519. [CrossRef]
32. Das, T.N. Oxidation of phenol in aqueous acid: Characterization and reactions of radical cations vis-à-vis the phenoxyl radical. *J. Phys. Chem. A* **2005**, *109*, 3344–3351. [CrossRef] [PubMed]

Article

Fabricating Femtosecond Laser-Induced Periodic Surface Structures with Electrophysical Anisotropy on Amorphous Silicon

Dmitrii Shuleiko [1,2,*], Mikhail Martyshov [1], Dmitrii Amasev [3], Denis Presnov [1,4,5], Stanislav Zabotnov [1,2,6], Leonid Golovan [1,2], Andrei Kazanskii [1] and Pavel Kashkarov [1,6]

1. Faculty of Physics, Lomonosov Moscow State University, 1/2 Leninskie Gory, 119991 Moscow, Russia; martyshov@physics.msu.ru (M.M.); denis.presnov@phys.msu.ru (D.P.); zabotnov@physics.msu.ru (S.Z.); golovan@physics.msu.ru (L.G.); kazanski@phys.msu.ru (A.K.); kashkarov@physics.msu.ru (P.K.)
2. Big Data Storage and Analysis Center, Lomonosov Moscow State University, Lomonosovsky Avenue 27/1, 119192 Moscow, Russia
3. Prokhorov General Physics Institute of the Russian Academy of Sciences, 38 Vavilova st., 119991 Moscow, Russia; amasev.dmitriy@physics.msu.ru
4. Skobeltsyn Institute of Nuclear Physics, Lomonosov Moscow State University, 1/2 Leninskie Gory, 119991 Moscow, Russia
5. Quantum Technology Centre, Lomonosov Moscow State University, 1/35 Leninskie Gory, 119991 Moscow, Russia
6. National Research Centre "Kurchatov Institute", 1 Akademika Kurchatova sq., 123182 Moscow, Russia
* Correspondence: shuleyko.dmitriy@physics.msu.ru

Citation: Shuleiko, D.; Martyshov, M.; Amasev, D.; Presnov, D.; Zabotnov, S.; Golovan, L.; Kazanskii, A.; Kashkarov, P. Fabricating Femtosecond Laser-Induced Periodic Surface Structures with Electrophysical Anisotropy on Amorphous Silicon. *Nanomaterials* **2021**, *11*, 42. https://dx.doi.org/10.3390/nano11010042

Received: 24 November 2020
Accepted: 23 December 2020
Published: 26 December 2020

Publisher's Note: MDPI stays neutral with regard to jurisdictional claims in published maps and institutional affiliations.

Copyright: © 2020 by the authors. Licensee MDPI, Basel, Switzerland. This article is an open access article distributed under the terms and conditions of the Creative Commons Attribution (CC BY) license (https://creativecommons.org/licenses/by/4.0/).

Abstract: One-dimensional periodic surface structures were formed by femtosecond laser irradiation of amorphous hydrogenated silicon (a-Si:H) films. The a-Si:H laser processing conditions influence on the periodic relief formation as well as correlation of irradiated surfaces structural properties with their electrophysical properties were investigated. The surface structures with the period of 0.88 and 1.12 µm were fabricated at the laser wavelength of 1.25 µm and laser pulse number of 30 and 750, respectively. The orientation of the surface structure is defined by the laser polarization and depends on the concentration of nonequilibrium carriers excited by the femtosecond laser pulses in the near-surface region of the film, which affects a mode of the excited surface electromagnetic wave which is responsible for the periodic relief formation. Femtosecond laser irradiation increases the a-Si:H films conductivity by 3 to 4 orders of magnitude, up to 1.2×10^{-5} S·cm, due to formation of Si nanocrystalline phase with the volume fraction from 17 to 28%. Dark conductivity and photoconductivity anisotropy, observed in the irradiated a-Si:H films is explained by a depolarizing effect inside periodic microscale relief, nonuniform crystalline Si phase distribution, as well as different carrier mobility and lifetime in plane of the studied samples along and perpendicular to the laser-induced periodic surface structures orientation, that was confirmed by the measured photoconductivity and absorption coefficient spectra.

Keywords: laser-induced periodic surface structures; silicon nanocrystals; surface plasmon-polaritons; femtosecond laser pulses; amorphous silicon; electrophysical measurements; Raman spectroscopy

1. Introduction

Laser structuring of amorphous hydrogenated silicon (a-Si:H) surfaces has been attracting the attention of scientists for a long time, as it is a promising method for increasing the efficiency of solar cells based on this material. Such processing induces formation of surface inhomogeneities with characteristic dimensions comparable to the visible and near infrared radiation wavelength, which leads to an increase in the optical absorption of the film [1–3] due to enhanced scattering of the incident light on the structured surface. Laser modification also induces formation of a crystalline Si phase within a-Si:H films, that sufficiently retards photoelectric properties degradation under the sunlight action to these structures [1,4]. In this case, the irradiated a-Si:H film represents a nanocomposite in the form of silicon nanocrystals embedded into an amorphous matrix [4]. Laser pulses of various duration can be used for a-Si:H films modification, including nanoseconds [5]. However, only irradiation by ultrashort laser pulses enables uniform and accurate structuring of the a-Si:H film, owing to emergence of nonthermal melting mechanism during laser processing [6–8].

An additional feature observed during femtosecond laser irradiation of a-Si:H films is the formation of anisotropic laser-induced periodic surface structures (LIPSS or "ripples"). The characteristic scale of such ripples formed in air environment is comparable to the wavelength of the incident radiation [9]. In generally, similar technologies allow fabricating LIPPS with periods, which are significantly smaller than the laser wavelength. For example, femtosecond laser irradiation of semiconductor surfaces in various liquid environments might lead to formation of the LIPSS with subwavelength periods from 70 to 400 nm [10]. It is also possible to produce one-dimensional surface structures with a characteristic transverse size about 10 nm using two femtosecond laser beams with orthogonal polarizations and spatial overlapping [11]. Such structures may be interesting from the point of view of following studying dependencies of their electrical properties on the period. Nonetheless, in the presented work we will pay our attention to classical and easy to produce interference LIPSS with the micron-scale period which is comparable with the wavelength of the used femtosecond radiation.

Formation of such LIPSS is described by a mechanism which consists in interference of surface plasmon-polaritons generated by high-power ultrashort laser pulses with incident light [12,13]. Such LIPSS can serve as a diffraction grating, which is known to increase the efficiency of amorphous/crystalline Si heterojunction solar cells due to localization of incident light at a certain depth of the structure [14]. The formation of one-dimensional surface structures on the a-Si:H film surface also attracts interest for application in polarization optoelectronics, since such modified films demonstrate birefringence, dichroism [15,16], and anisotropy of electrical properties [12,15,17]. In general, for a wider set of materials one of the promising LIPSS applications is polarization-sensitive optical memory fabrication [16] including multilevel and highly stable coding [18]. In such a system, three bits of information or more can be written [19] in a single micron-scaled cell with the LIPSS formed within it. The storage of additional information in the same cell is achieved by variation of the formed LIPSS orientation, as well as dichroism and retardance values of these structures [20]. Finally, femtosecond laser-irradiated thin a-Si:H films containing both silicon nanocrystals and ordered laser-induced surface structures can be used to create hydrophobic coatings, sensors, and thin-film transistors for flat panel displays [9].

It should be noted that a lot of the listed above papers about LIPSS based on a-Si:H [9,15–17] have no information about influence of processing conditions on the orientation and period of the formed structures. Though, in the experiments [12] a rotation of the LIPSS direction at the a-Si:H film surface was demonstrated when the number of modifying laser pulses was changed, the observed effect requires theoretical substantiation taking into account the influence of nonequilibrium charge carriers excited in a-Si:H by high-power femtosecond laser pulses. At the same time, laser processing of an a-Si:H film in scanning mode can lead to inhomogeneous crystallization within the film surface plane due the scan lines formation in a case when the laser spot moves continuously in one direction and with a certain step in the orthogonal one [12]. Thus, when analyzing the electrophysical anisotropy of the femtosecond laser-irradiated a-Si:H films, it is necessary to take into account the simultaneous formation of the LIPSS and scan lines, the influence of which must be separated.

Within the above, the aims of our work are the following: to find a correlation between the formation of various LIPSS types and photoexcitation of appropriate surface electromagnetic wave modes taking into account the density of carriers photoinduced in a-Si:H film by femtosecond laser radiation; to analyze a correlation of the modified films structure with their electrical and photoelectrical properties, excluding the possible large-scale scan lines contribution.

2. Materials and Methods

Initial a-Si:H films with 600 nm thickness were fabricated by plasma-enhanced chemical vapor deposition [4] on glass substrates. Then the films were irradiated with a laser system Avesta based on a Cr:forsterite crystal by femtosecond laser pulses with the fundamental wavelength λ = 1250 nm. The laser pulse duration τ = 125 fs and the repetition rate f = 10 Hz, respectively. The irradiation was carried out in air at the normal pressure.

The films were processed by femtosecond laser pulses in a scanning mode, which was realized by moving the samples in the horizontal plane (XY) perpendicular to the laser beam. A system of two automated mechanical translators Standa was controlled by a personal computer and used to move the samples. The laser polarization vector was directed along the X axis. The setup scheme and its photo are given in Figure 1a,b, respectively.

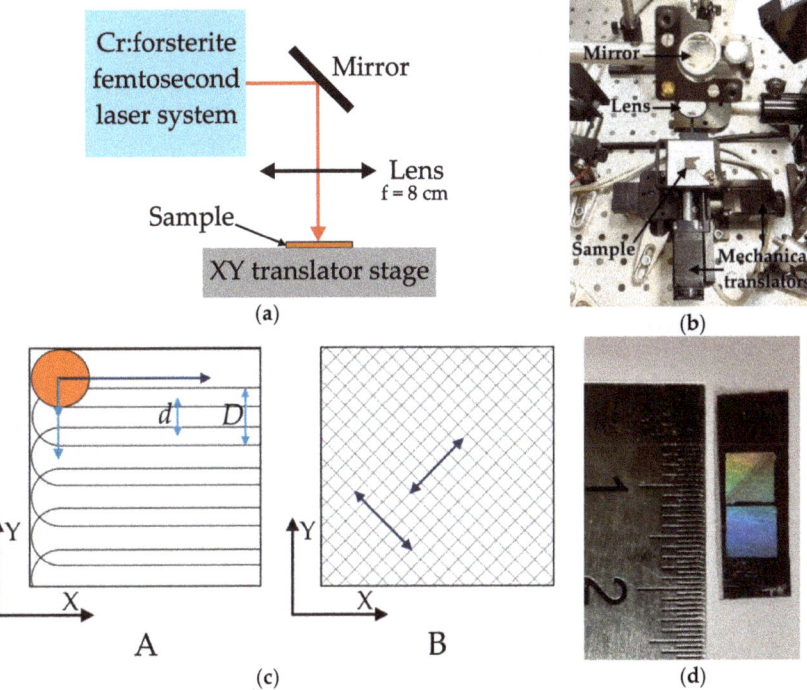

Figure 1. (a) Scheme of experimental setup used for processing the a-Si:H films by femtosecond laser pulses. (b) Photo of experimental setup. (c) Scanning mode templates used during a-Si:H femtosecond laser modification. (d) Photo of the areas at a-Si:H film surface, laser-modified in scanning mode B.

Two different scanning modes were used for sample fabrication. The first mode, indicated by the letter A in Figure 1c, was implemented by moving the a-Si:H film in the horizontal plane continuously with a given speed V along X axis, and discretely with a given step $d \leq D$ along the Y axis, where D is the laser spot diameter. As a result, so-called scan lines were formed on the treated area along X axis without unmodified gaps between them. Due to continuous movement along the X axis, the spots on a-Si:H surface from consecutive laser pulses were overlapping, which resulted into a total laser pulses number per the same area n = 30 (Table 1).

Table 1. Sample processing parameters.

Sample	Scanning Mode (Figure 1c)	Laser Pulse Fluence Q (J/cm^2)	Laser Spot Diameter D (μm)	Number of Laser Pulses n
1	A	0.5	150	30
2	B			750

The second mode, indicated as B in Figure 1c, was realized by simultaneous smooth movement of the a-Si:H film along both axes at the same speed. As a result, scan lines were formed diagonally, at an angle of 45° in relation to the sides of the treated area. At first the entire area of the sample was scanned along the diagonal lying in the 1 and 3 quadrants of the coordinate plane, and then the same area was scanned along the diagonal lying in the 2 and 4 quadrants. However, in this case the scanning was performed at lower speed V which resulted into a high number of the laser pulses n per the same area of the sample. In this case better uniformity of the treated surface was achieved in comparison with the mode A, and a possible contribution to the conductivity anisotropy related to scan lines presence was excluded.

Several square 4.5 × 4.5 mm^2 areas at the surfaces of a-Si:H films were irradiated with each template to provide enough treated surface for scanning electron microscopy (SEM) investigation as well as the electrophysical measurements. An example of irradiated areas at a-Si:H film surface is given in Figure 1d. The areas obtained using the scanning mode A were indicated as sample 1, while the areas irradiated by scanning mode B as sample 2, according to Table 1.

Images of the modified surface of the films were obtained by scanning electron microscopy (SEM) using a Carl Zeiss Supra 40 microscope. The phase composition analysis of both initial a-Si:H film and irradiated samples was carried out by Raman spectroscopy using a Horiba JobinYvon HR800 Raman microscope with excitation by He-Ne laser at the wavelength of 633 nm.

The electrical properties of initial and femtosecond laser-irradiated a-Si:H films were studied using a Keithly 6487 picoammeter and an optical nitrogen cryostat in vacuum of 10^{-3} Pa. The electrical measurements were carried out at direct current applied in the directions which are coplanar to the film surface. For this purpose, each time 4 square aluminum electric contacts were deposited by thermal resistive sputtering at the surface of unmodified a-Si:H and each modified area. The electric contact edges were directed both along and orthogonal to the fabricated LIPPS. Thus, the electrical measurements were conducted in two mutually orthogonal directions in surface plane.

To remove water from the surface, all samples were annealed at temperature of 433 K (~160 °C) in vacuum for 5 min before the electrical measurements. After that dark conductivity temperature dependencies were measured while the samples were cooling down to 273 K with a step of 2 K. For photoelectrical measurements, the samples were illuminated in the visible and near-infrared spectral range from 1 to 2.2 eV using a halogen incandescent lamp and a Lot Oriel monochromator. Absorption coefficient spectra of both initial a-Si:H film and irradiated samples were measured using a constant photocurrent method (CPM).

3. Theoretical Modeling of the LIPSS Formation

The formation of a periodic relief on metal, dielectric, and semiconductor surfaces irradiated by femtosecond laser pulses is often associated with the surface electromagnetic waves (SEW) generation [21]. For the normal LIPSS formation, the conditions for a certain SEW mode excitation—a surface plasmon-polariton—must be satisfied at the interface between two media [13]:

$$\text{Re}\varepsilon_2 < 0, \ |\text{Re}\varepsilon_2| > \text{Re}\varepsilon_1 > 0, \tag{1}$$

where $\varepsilon_1 = 1$ is the dielectric permittivity corresponding to air, and ε_2 corresponds to the irradiated material. Therefore, in the case of semiconductors and dielectrics, including a-Si:H, it is necessary to achieve the negative sign of Re ε_2 for surface plasmon-polariton excitation [12,13,21]. The conditions (1) can arise in a-Si:H under irradiation with high-power femtosecond laser pulses. Due to intense photoexcitation by incident radiation a high concentration of nonequilibrium charge carriers is generated, which leads to metallization of irradiated surface. According to the Drude model, relations (1) are satisfied when the nonequilibrium charge carriers density reaches the threshold value N_0, which corresponds to the Drude plasma resonance [13]:

$$N_0 = \left(\omega^2 + \gamma^2\right)(\varepsilon + 1)m^*/4\pi e_0^2, \qquad (2)$$

where ω is the incident laser radiation frequency, $\gamma = e_0/(m^*\mu)$ is the collision frequency, of the material, m^* and e_0 are the effective mass and charge of an electron, respectively, ε is the static dielectric permittivity for non-excited a-Si:H and μ is the electron mobility for the irradiated material.

On the other hand, the maximum value of nonequilibrium charge carrier's density N achieved in the a-Si:H film upon femtosecond laser irradiation can be estimated by the differential equation:

$$\frac{dN}{dt} = \frac{(1-R)I(t)}{\hbar\omega}\alpha + \frac{(1-R)^2 I^2(t)}{2\hbar\omega}\beta, \qquad (3)$$

where α, β are the one-photon and two-photon absorption coefficients, respectively; R is the reflection coefficient and $I(t)$ is the incident laser radiation intensity. Temporal distribution of the laser radiation intensity within the pulse for the Equation (3) was described as follows [22]:

$$I(t) = I_0\, t \exp(-4t/\tau), \quad \int_0^\infty I(t)dt = Q \qquad (4)$$

where $Q = 0.5$ J/cm^2 is the fluence of a laser pulse.

The following values of the absorption coefficients for a-Si:H were used in the Equation (3): $\alpha = 10$ cm^{-1} [23], $\beta = 37$ cm/GW [24]. The reflection coefficient R is given by the formula:

$$R = \left|\frac{\sqrt{\tilde{\varepsilon}}-1}{\sqrt{\tilde{\varepsilon}}+1}\right|^2, \qquad (5)$$

where $\tilde{\varepsilon}$ is the complex dielectric permittivity of irradiated material, which also depends on N, according to the Drude model [13].

In order to analyze the formation of LIPSS with different periods and orientations, depending on the nonequilibrium charge carriers concentration in the irradiated material, the model proposed by J Sipe et al. [25] and modified by J. Bonse et al. [26–28] was used. This so-called Sipe-Drude model utilizes so-called efficacy factor $\eta(\kappa_x, \kappa_y)$, which indicates the probability of LIPSS formation with period Λ under the action of incident laser radiation and in the presence of SEW excitation. The values of κ_x and κ_y are the surface lattice wave vector κ components normalized by the incident laser radiation wavelength:

$$|\kappa| = \lambda/\Lambda \qquad (6)$$

The model allows for relating the density of nonequilibrium charge carriers N excited by a laser pulse, the dielectric permittivity of the irradiated material, and the period of the formed LIPSS. The laser radiation parameters, such as wavelength, polarization, and angle of incidence, as well as the roughness of the surface are also taken into account.

The numerical calculation of $\eta(\kappa_x, \kappa_y)$, was carried out using the system of equations for the s-polarized radiation with a wavelength of 1.25 µm at normal incidence ($\Theta = 0$) [27]. The shape and filling factors determining the surface roughness were taken equal to 0.4

and 0.1, respectively, in accordance with [25]. The direction of the polarization vector was taken parallel to the x axis.

4. Experimental Results and Discussion

4.1. LIPSS Formation on the a-Si:H Film Surface

By means of SEM various one-dimensional periodic structures on the surface of the irradiated a-Si:H films were revealed. The orientation of these LIPSS changes depending on the number of femtosecond laser pulses n. Formation of surface gratings with ridges oriented orthogonally to the polarization vector of the modifying laser radiation was observed for the sample 1 at $n = 30$. The corresponding wave vector κ of these structures is parallel to the laser polarization. Such so-called "normal LIPSS" [25] are presented in Figure 2a. The period Λ of the lattice is 0.88 ± 0.03 μm.

Figure 2. Images of femtosecond laser-modified surfaces of (**a**) sample 1 and (**b**) sample 2, obtained by SEM at 5000× magnification. The arrows indicate the polarization of the femtosecond laser pulses. SEM images of femtosecond laser-modified surfaces of (**c**) sample 1 and (**d**) sample 2, obtained at 70× magnification.

At the surface of sample 2, modified with higher number of laser pulses $n = 750$, the formation of LIPSS with ridges orientation along the laser polarization is observed (Figure 2b). Such one-dimensional structures are called "anomalous LIPSS" [9], their wave vector κ is directed orthogonally to the laser polarization, and their period is $\Lambda = 1.12 \pm 0.02$ μm. Additionally, the scan lines with the width $d = 150$ μm are formed during irradiation of sample 1 as can be seen in Figure 2c. In this case the laser spot diameter is

D = 200 µm, thus no unmodified gaps between the scan lines are formed. It also can be seen in Figure 2d that for the sample 2 the scan lines are directed at 45° to the edges of the modified area.

It should be noted that the LIPSS of both types have periods close to the wavelength λ of used femtosecond laser pulses, but smaller than λ. The decrease of the LIPSS period compared to incident laser radiation varies from few percent for the second structure type and up to 30% for the first structure. Additionally, the orientation of all types of the structures is determined only by the direction of the femtosecond laser pulses polarization vector and does not depend on the laser beam scanning direction.

4.2. Raman Spectra Analysis for Irradiated and Unirradiated a-Si:H Films

To determine the phase composition of the films before the electrical measurements, the Raman spectra were obtained for both initial a-Si:H and irradiated samples. The presence of a wide band with a maximum at ω_A = 480 cm^{-1}, which is characteristic to amorphous silicon, can be seen in Raman spectra of all the samples in Figure 3a. Along with it, a narrow line near ω_C = 520 cm^{-1}, which corresponds to crystalline (nanocrystalline) silicon (nc-Si) [4,29], is present only in the irradiated samples.

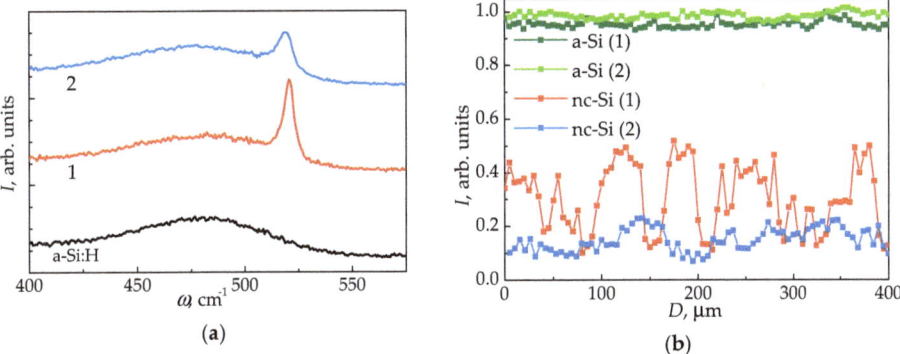

Figure 3. (**a**) Raman spectra for unirradiated amorphous silicon, as well as for samples 1 and 2 normalized by maximum intensity at ω_A. For sample 1, the spectrum corresponds to the middle of the scan line. (**b**) Integrated TO phonon modes intensities corresponding to a-Si and nc-Si for the samples 1 and 2, mapped in the direction orthogonal to the scan lines, and normalized by the maximum integrated a-Si:H TO phonon mode intensity.

The crystalline silicon (c-Si) phase volume fraction f_C in the irradiated films was calculated using the following expression [29]:

$$f_C = \frac{I_C}{\sigma_0 I_A + I_C}, \qquad (7)$$

where I_A and I_C are the integrated intensities of TO phonon modes corresponding to the frequencies ω_A and ω_C, σ_0 is the empirical ratio of the Raman scattering integrated cross sections for the crystalline and amorphous silicon phases, which is determined by the formed silicon nanocrystals size [30]. The size of nanocrystals estimated from the shift of the Raman line at 520 cm^{-1} [30] is ≈3.6 nm for sample 2 and ≈5 nm for sample 1. Corresponding σ_0 values varied from 0.9 to 0.6, while for bulk c-Si this value equals 0.1 [30]. Obtained according to (7) values of the nc-Si phase volume fraction for sample 1 and 2 are 28 ± 16% and 17 ± 4%, respectively. As can be seen in Figure 3b, the nc-Si phase distribution is significantly less homogeneous within the sample 1, compared to sample 2. This can be explained by the difference in scan lines formation on the irradiated surface when using different scanning modes. As a result, fluence is more nonuniformly distributed within the treated area cross section at applying the one-pass scanning mode A than in case of multi-pass mode B (Figure 1c).

The estimation of volume fractions of the c-Si phase within the samples is important for given below electrical study as well as calculations of the nonequilibrium electrons concentration generated into the conduction band during LIPSS formation; as such, simulation requires the values of the electron mobility and effective mass for the irradiated material.

4.3. Modeling of the LIPSS Formation on a-Si:H Surfaces

The formation of LIPSS with a period close to the incident radiation wavelength on the a-Si:H film requires the excitation of a lot of nonequilibrium electrons into the conduction band, within the surface layer of the film. According to expression (2), in case of a-Si:H surface irradiation by femtosecond laser pulses with 1.25 µm wavelength, the threshold value of the nonequilibrium carriers density required for plasmon-polariton generation is $N_0 = 8.2 \cdot 10^{21}$ cm^{-3}. The dielectric permittivity value $\varepsilon = 9.7$ was used in calculations, according to [12]. It should be noted that in the irradiated samples the formation of nc-Si phase was observed, as was shown in Section 4.2. Therefore, the value of the electron mobility μ for modified a-Si:H films was taken equal to 30 cm^2/V·s, which corresponds to microcrystalline silicon, according to Bergren et al. [31]. Note that the μ value used in calculation is almost an order of magnitude higher than the charge carrier mobility in a-Si:H (≈ 4.5 cm^2/V·s), but smaller than carrier mobility μ for bulk c-Si ≈ 1300 cm^2/V·s [32].

The maximum value of the nonequilibrium charge carriers density N is estimated from the Equation (3) as $1.2 \cdot 10^{22}$ cm^{-3} at the laser pulse fluence $Q = 0.5$ J/cm^2 and duration $\tau = 125$ fs. The obtained value for N is higher than the threshold value N_0. Thus, femtosecond laser pulses used in the experiment allow for achieving of high carrier density within the surface layer of a-Si:H film, that is necessary for LIPSS formation.

The LIPSS modeling according to the Sipe-Drude theory [25–28] was conducted by varying the N values until corresponding $\eta(\kappa_x, \kappa_y)$ two-dimension distributions (Figure 4a,b) match Fourier transform images (Figure 4c,d) of the real ripples shown in Figure 2a,b. As a result, the nonequilibrium electrons concentrations N and complex dielectric permittivities ε, which correspond to the observed ripples formation conditions, were determined. Obtained N and ε values are given in the Table 2.

As can be seen, the nonequilibrium electrons concentration exceeds the threshold value N_0 for the normal LIPSS, which corresponds to plasmon-polariton excitation conditions. On the other hand, for the sample 2 containing anomalous ripples, which are parallel to the laser polarization, the obtained $N < N_0$. This result is consistent with the mechanism of anomalous ripples formation proposed in [12,13].

According to this mechanism, the decrease of the nonequilibrium electrons concentration in near-surface layer of the film is caused by growth of electron thermal emission at increase of the laser pulses number. This effect is associated with intensified heating of the a-Si:H film surface due to the presence of a periodic surface relief formed by previous laser pulses [12]. Namely, a positive feedback emerges between the surface relief formation by femtosecond laser pulses and an increase in the absorption coefficient of the film in course of time. The increased optical absorption of the film is associated with more efficient scattering of incident radiation by the surface inhomogeneities, formed by previous laser pulses. Such inhomogenities enhance scattering of the incident light with wavelength close to their characteristic dimensions [1] according to Mie theory framework. In our case the characteristic dimension of scatterers is set by LIPSS period, which is close to the wavelength of the laser radiation used. Further, the decreased concentration of nonequilibrium electrons, generated by the laser pulse, affects the sign of the dielectric permittivity real part Re ε. As can be seen in Table 2, the value of Re ε, corresponding to the obtained N values, is negative for the normal LIPSS that are directed orthogonally to the femtosecond laser pulses polarization. Contrary, in case of anomalous LIPSS formation, when the ripples are directed along the polarization of incident femtosecond laser pulses, corresponding Re ε, is positive [13]. Such alteration of the Re ε sign in the near-surface region of the film from negative to positive enables excitation of a different SEW mode (TE instead of TM), and consequently the direction of the LIPSS changes. Thus, the period and orientation of both LIPSS types observed in the experiment are in a good agreement with the obtained values for the complex dielectric permittivity ε and the listed above mechanisms of their fabrication.

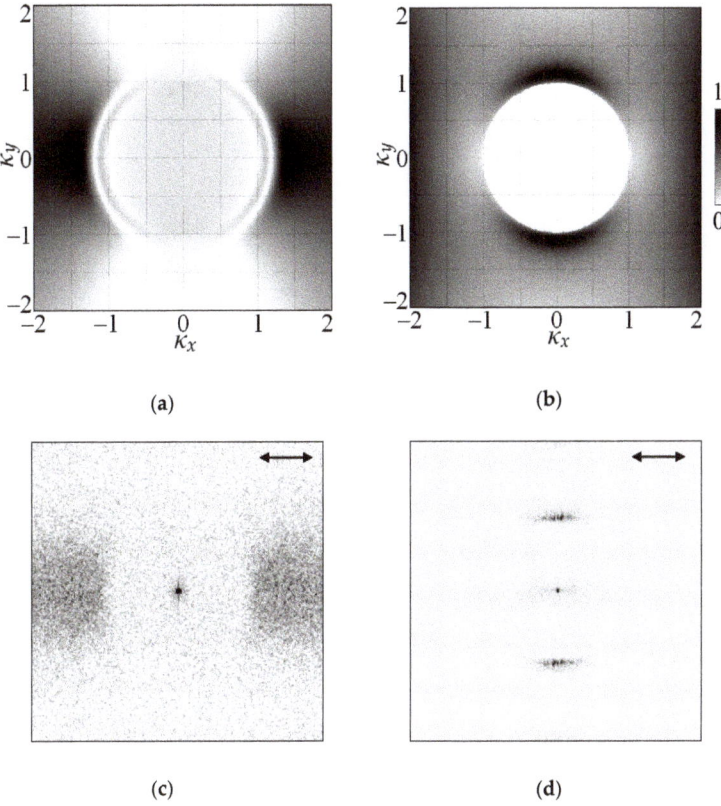

Figure 4. Results of the efficacy factor $\eta(\kappa_x, \kappa_y)$ modeling for the samples (**a**) 1 and (**b**) 2, and their comparison with the Fourier-transformed SEM images of the LIPSS experimentally observed on the surface of corresponding samples (**c,d**). The arrows indicate the orientation of the laser radiation polarization vector.

Table 2. LIPSS period and orientation, nonequilibrium electrons concentration N and the dielectric permittivity ε at the surface of a-Si:H films during irradiation with femtosecond laser pulses.

Sample	LIPSS Period (µm)	LIPSS Direction Relative to Polarization	N (cm^{-3})	ε
1	0.88 ± 0.03	⊥	9.3 × 10^{21}	−2.3 + 0.4i
2	1.12 ± 0.02	∥	6.5 × 10^{21}	1.3 + 0.3i

In addition, it should be noted that the periods of both normal and anomalous surface lattices formed by femtosecond laser pulses at the a-Si:H surface in the present work are smaller than those obtained by our group in the previous experiments using a lower laser fluence [12,21]. According to modeling within Sipe theory such decrease of the LIPSS period is attributed to a lower nonequilibrium electrons density achieved during laser processing. An explanation of the observed effect may be similarly given by increased intensity of the thermal emission of electrons from the surface of the a-Si:H due to a stronger heating of the film surface when femtosecond pulses with a higher fluence are applied [33,34]. An additional effect on a decrease in the period of the formed LIPSS can be provided by an increase in the contrast of the relief formed by laser pulses with a higher fluence. According to [35], as the depth of the ablated grooves of the LIPSS increases, the resonant wavelength of the surface electromagnetic wave shifts to smaller values.

4.4. Electrical and Photoelectrical Properties of the Modified a-Si:H Surface

The dark conductivity temperature dependencies of the unmodified a-Si:H and samples 1, 2 are given in Figure 5a. Each curve corresponds to different applied electric field vectors E relative to scan lines and formed LIPSS, according to Figure 5b. It is worth to mention that the scheme for sample 2 in Figure 5b do not show the scan lines, since they are directed at ±45° angles to the applied E accordingly the scanning mode B (Figure 1c), and therefore do not affect the conductivity.

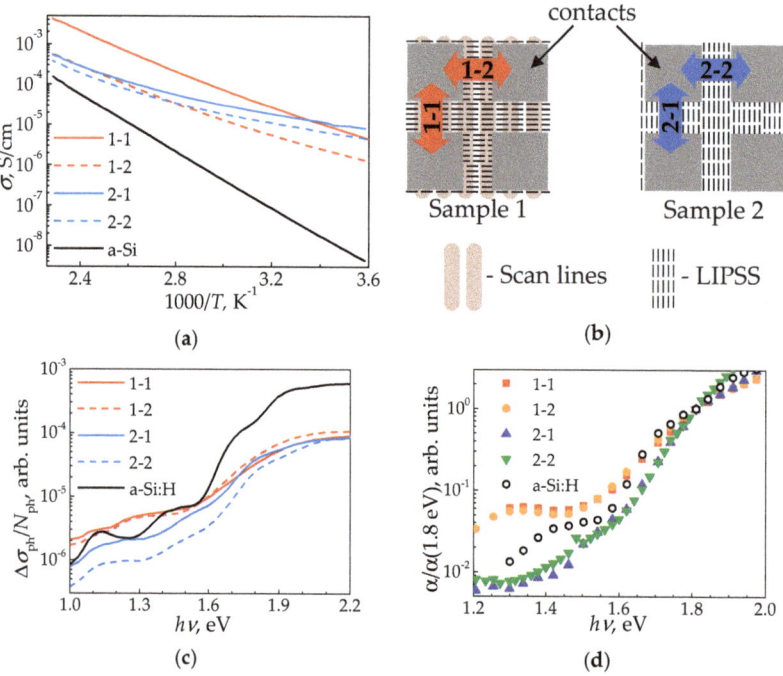

Figure 5. (a) Temperature dependencies of dark conductivity σ. (b) Electric contacts geometry relative to scan lines and LIPSS. Directions of the electric field vector E are marked as 1-1, 1-2, 2-1, and 2-2. (c) Spectra of photoconductivity σ_{ph}, normalized on the number of photons N_{ph}. (d) Spectra of absorption coefficient α, normalized on its value at the mobility gap with for a-Si:H (1.8 eV).

As can be seen from Figure 5a, the dark conductivity curves for all the irradiated samples lie higher and have a lesser slope than for the initial a-Si:H, which is explained by the nc-Si phase contribution to the conductivity of these samples. Additionally, the slope of conductivity temperature dependencies $\sigma(T)$ for all the irradiated samples demonstrate increase with temperature growth (corresponds to $1/T$ abscissa decrease in Figure 5a). It indicates an increase of the activation energy, which can be calculated from the temperature curve slope.

Since, in contrast to the initial a-Si:H, the modified regions of the film had a nonmonotonic dark conductivity temperature dependencies, the activation energies were calculated by approximating the obtained experimental curves by an exponent in a certain temperature ranges. In this case, the dark conductivity temperature dependence is described as follows:

$$\sigma = \sigma_0 exp\left(-\frac{E_A}{kT}\right), \tag{8}$$

where σ_0 is a constant, E_A is the activation energy, T is the temperature, k is the Boltzmann constant. To calculate the activation energy at temperatures of 300 and 400 K, the approximation temperature ranges from 273 to 333 K and from 373 to 433 K were taken, respectively. The values obtained in two mutually orthogonal directions for each sample were the same, taking into account the error of up to 0.05 eV. Obtained activation energies vary from 0.35 to 0.5 eV for sample 1, and from 0.21 eV to 0.47 eV for sample 2 with increasing the temperature from 300 to 400 K. Contrary, the activation energy of the initial a-Si:H is constant and equals 0.7 eV. The activation energy increase can be explained by simultaneous coexistence of both amorphous and nc-Si phases in the modified samples, when the amorphous phase contributes noticeably to dark conductivity only at higher temperatures.

The electrical measurements revealed that specific dark conductivity of a-Si:H films increased by 2 to 3 orders of magnitude as a result of femtosecond laser treatment (Table 3). On the other hand, the photoconductivity, calculated as the difference between the conductivity of illuminated sample and the dark conductivity, decreased by 4 times and more for irradiated samples. It should be clarified that observed dark conductivity increase is mainly influenced by the formation of more conductive nc-Si phase within a-Si:H under the action of high-power femtosecond laser pulses, which was confirmed by Raman spectroscopy in Section 4.2. Despite the nc-Si phase volume fraction that is almost 2 times smaller in sample 2 than in sample 1, as was calculated above, the dark conductivities of both samples were close. Such result can be explained by a more uniform nc-Si phase distribution within the sample 2, processed according to the template B, compared to the sample 1, modified using the template A with the scan lines formation. In the latter case, using template A leads to formation of the regions with significantly reduced nc-Si phase volume fraction f_C, within the sample along the scan lines edges, as can be seen in Figure 3b. In these regions, the f_C value can be lower than the percolation threshold for nc-Si, which for a homogeneous and isotropic system of silicon nanocrystals randomly distributed in an amorphous Si matrix corresponds to f_C = 16% [36]. Thus, the conductivity of such regions within the modified film will be determined mainly by the poorly conducting amorphous phase, which leads to reduction of the resulting dark conductivity of the sample 1, especially in the direction which is perpendicular to the scan lines.

Table 3. Dark conductivity and photoconductivity for the samples 1, 2 for different directions of the applied electric field **E** relative to LIPSS and scan lines, and for the unmodified a-Si:H film. The data is obtained at room temperature (293 K).

Sample	Direction of E	Dark Conductivity (10^{-6} S/cm)	Photoconductivity (10^{-6} S/cm)
1	⊥ LIPSS, ∥ scan lines	11 ± 2	0.8 ± 0.2
	∥ LIPSS, ⊥ scan lines	2.7 ± 0.1	1.0 ± 0.1
2	∥ LIPSS	12 ± 2	0.96 ± 0.05
	⊥ LIPSS	6.59 ± 0.06	0.6 ± 0.1
a-Si:H	–	0.018 ± 0.001	3.7 ± 0.1

Additionally, in-plane anisotropy of conductivity was revealed for the irradiated samples. The dark conductivity values differ up to 4 times in mutually orthogonal directions for sample 1 and up to 1.8 times–for sample 2, as shown in Table 3, while the dark conductivity of initial a-Si:H film is isotropic.

The dark conductivity anisotropy observed for both irradiated samples might be caused both by the form anisotropy of LIPSS [37] and the nonuniform nc-Si phase distribution within scan lines [12]. However, according to Figure 5b and Table 3 the dark conductivity is higher along the scan lines direction (geometry 1-1) than along the LIPSS (geometry 1-2) at room temperature. Thus, the scan lines presence has a stronger effect on the dark conductivity anisotropy. Contrarily, for the sample 2 the dark conductivity is higher along the LIPSS direction. In this case the scan lines were formed at the same angle of 45° to all the electric contacts edges, according to the template B in Figure 1c, and their influence on conductivity was minimized. In other words, conductivity anisotropy of the sample 2 is explained by the pronounced orientation-dependent depolarization of an external electric field inside the micron-scale LIPSS [12].

An additional contribution to the observed effect may as well be caused by anisotropic crystalline and amorphous Si phase distribution within LIPSS, according to [9,38]. In presence of the LIPSS with high amplitude on the surface of a-Si:H film, the charge transport along the LIPSS direction occurs in the highly conductive nc-Si phase which is mainly distributed along the LIPSS ridges [38]. At the same time, in the orthogonal direction, the charge transport paths are crossed by the ablated valleys of the surface relief, which contain mainly an amorphous phase with significantly lower conductivity than the nc-Si. In this case the magnitude of the conductivity anisotropy can be influenced by the depth of the surface relief. However, the period of such relief should not affect the conductivity: though the number of the ablated valleys per unit area is higher in case of small period of LIPSS, the amount of nc-Si phase in the single LIPSS ridge is also proportionally smaller.

It is worth to note that the electrical properties of the a-Si:H films modified by femtosecond laser pulses in air, including the conductivity anisotropy, can be affected by the surface oxidation, which was demonstrated in several works [9,17]. However, strong oxidation is accompanied by a change in a Raman spectrum, when the wide band at 480 cm^{-1} corresponding to a-Si:H broadens [39] or has significantly reduced intensity [17,40]. Such oxidized a-Si:H films can demonstrate extremely low conductivity, even despite the large c-Si volume fraction of 70% [17]. Additionally, Si surfaces passivated by a SiO_2 layer of 15 nm thickness demonstrate nonlinear conductivity [41]. However, in our case, no similar changes in the Raman spectra (Figure 3a), as well as in the conductivity dependencies (Figure 5a,c), were observed. It indicates that the probable silicon oxide phase formation does not significantly affect the conductivity of irradiated a-Si:H films.

Figure 5c as well as Table 3 also demonstrates pronounced anisotropy of the photoconductivity for sample 2. It might be related either to anisotropy of the electron and hole (marked as n and p in subscript indices, respectively) lifetime $\tau_{n,p}$ and mobility $\mu_{n,p}$, or by the features of the film absorption coefficient α:

$$\Delta\sigma_{ph} = e_0(\mu_n\tau_n + \mu_p\tau_p)G(\alpha), \tag{9}$$

where $G(\alpha)$ is the charge carrier generation rate. However, the spectral dependencies of the absorption coefficient α given in Figure 5d demonstrate isotropic behavior for sample 2. Thus, observed photoconductivity anisotropy is mostly affected by the $\tau_{n,p}$ and $\mu_{n,p}$ anisotropy in relation to the directions along and orthogonal to the LIPSS. This anisotropy may be caused by the different a-Si and nc-Si phase distributions inside the LIPSS. In this consideration along the LIPSS the carriers transport occurs mainly through nc-Si phase, due to its appropriate distribution within the surface relief according to [9], while in the orthogonal direction there is an alternation of a-Si and nc-Si phases. At the same time, the photoconductivity of sample 1 is almost isotropic (Figure 5c), considering the error; the photoconductivity of unirradiated a-Si:H is also isotropic.

In the 1.1–1.5 eV spectral range the photoconductivity spectral dependencies for sample 1 lie higher or near, than one for the initial a-Si:H, while for sample 2 these dependencies are lower, as can be seen in Figure 5c. This indicates a high optical absorption of sample 1 in this range, and low–for sample 2 compared to the initial film, which is confirmed by the absorption spectra obtained by CPM and given in Figure 5d. On the other hand, the photoconductivity above 1.6 eV spectral range for both irradiated samples are smaller compared to a-Si:H one.

In general, the spectral dependence of absorption coefficient is determined by the concentration of defects within the mobility gap, such as dangling bonds. In the 1.1–1.5 eV range the higher optical absorption of sample 1 compared to the initial film, indicates an increase of dangling bonds concentration within the sample 1, which may be caused by femtosecond laser-induced dehydrogenation of a-Si:H. The absorption spectrum in the range of 1.5–1.7 eV also allows to determine the disordering degree of the material. Optical transitions in this range correspond to the absorption edge and are described by the dependence:

$$\alpha(h\nu) = \alpha_0 \exp(-h\nu/E_U), \qquad (10)$$

where E_U is the Urbach energy, which describes the band tails width for the disordered material and α_0 is an absorption normalization constant, which depends on the material structure. Approximated from Figure 5d according to Equation (10) $E_U = 65 \pm 4$ and 62 ± 6 eV for the unmodified film and sample 2, respectively: however for sample 1 $E_U = 100 \pm 4$ eV. Obtained value is almost twice higher than for the initial a-Si:H, indicating wider band tails and, therefore, greater lattice disorder within the sample 1. Thus, the increased absorption of sample 1 can be explained by the nonuniform femtosecond laser-induced crystallization using the scanning mode A that is accompanied by effective dehydrogenation and dangling bonds formation.

Optical absorption of a femtosecond laser-irradiated a-Si:H film in the 1.1–1.5 eV range can be additionally increased compared to initial a-Si:H due to formation of surface relief with the typical subwavelength dimensions (the periods) according to Mie theory [1], as was mentioned in Section 4.3.

Decreased optical absorption in the 1.1–1.5 eV range for the sample 2, compared to unmodified a-Si:H and sample 1, most likely, can be attributed to excessive material ablation at n = 750 pulses from surface in comparison to sample 1 fabricated at 25 times less exposure of n = 30 pulses. As a result, the film thickness decreases, and the number of absorbed photons in the smaller volume becomes lower too. Decreased photoconductivity above 1.6 eV spectral range for both irradiated samples compared to a-Si:H can be attributed to a decrease of a-Si:H phase volume fraction due to laser-induced crystallization while optical absorption of c-Si in this spectral diapason is lower compared to a-Si:H [42,43].

5. Conclusions

In summary, the possibility of the subwavelength periodical relief formation at a-Si:H surface by femtosecond laser irradiation in different scanning modes by the laser beam was demonstrated. According to theoretical simulation the orientation of the formed surface structures depends on the density of nonequilibrium electrons photoexcited into conduction band by the femtosecond laser pulse during processing. The increase in the number of laser pulses leads to a stronger heating of the a-Si:H film owing to optical absorption enhance by the LIPSS which formed on previous irradiation steps. The heating, in turn, induces thermal emission of electrons from the irradiated surface and leads to change the dielectric permittivity value in the near surface layer during irradiation step by step. As a result, the photoexcited SEW changes its type from TM to TE and, as a consequence, the direction of the formed surface lattice is changed.

The formation of nc-Si phase with a volume fraction from 17 to 28% within the modified a-Si:H films increased their dark conductivity by 2 to 3 orders of magnitude compared to the nonirradiated film. Simultaneously observed pronounced electrophysical anisotropy of femtosecond laser-irradiated films can be attributed both to scan lines formation and LIPSS existence. The dark conductivity values for two orthogonal directions in surface plane may differ up to 4 times due to LIPSS form anisotropy and the nonuniform nc-Si phase distribution within the scan lines. The photoconductivity of the samples after irradiation decreased by 4 times and more due to amorphous silicon phase volume fraction reduction caused by material ablation and crystallization. Furthermore, the anisotropy of the photoconductivity is reached and explained by different charge carrier lifetime and mobility along and perpendicular to the LIPSS direction that is confirmed by the measured photoconductivity and absorption coefficient spectra.

Author Contributions: Conceptualization, S.Z., L.G., A.K. and P.K.; methodology, S.Z., L.G. and A.K.; software, D.S.; validation, D.S., M.M., D.A. and D.P.; formal analysis, D.S.; investigation, D.S., M.M., D.A. and D.P.; resources, D.A.; data curation, D.S.; writing—original draft preparation, D.S.; writing—review and editing, M.M., S.Z. and L.G.; visualization, D.S.; supervision, P.K.; project administration, M.M. All authors have read and agreed to the published version of the manuscript.

Funding: Experiments on femtosecond laser irradiation, SEM and calculations of the LIPSS formation were supported by the Ministry of Science and Higher Education of the Russian Federation in the framework of an Contract № 7/1251/2019 on 15.08.2019 on provision of a grant for state support of centers of the National Technology Initiative on the basis of educational organizations of higher education and scientific organizations. Raman spectroscopy investigations, as well as electrophysical measurements and calculations were supported by a Joint grant № 19-32-70026 from the Moscow Government and the Russian Foundation for Basic Research.

Data Availability Statement: Data sharing is not applicable to this article.

Conflicts of Interest: The authors declare no conflict of interest.

References

1. Hong, L.; Wang, X.C.; Zheng, H.Y.; He, L.; Wang, H.; Yu, H.Y. Femtosecond laser induced nanocone structure and simultaneous crystallization of 1.6 μm amorphous silicon thin film for photovoltaic application. *J. Phys. D Appl. Phys.* **2013**, *46*, 195109. [CrossRef]
2. Differt, D.; Soleymanzadeh, B.; Lükermann, F.; Strüber, C.; Pfeiffer, W.; Stiebig, H. Enhanced light absorption in nanotextured amorphous thin-film silicon caused by femtosecond-laser materials processing. *Sol. Energy Mater. Sol. Cells* **2015**, *135*, 72–77. [CrossRef]
3. Guo, A.; Ilyas, N.; Song, Y.; Lei, R.; Zhong, H.; Li, D.; Li, W. Irradiation Effect on Ag-Dispersed Amorphous Silicon Thin Films by Femtosecond Laser. *Proc. SPIE* **2019**, *11046*, 1104629. [CrossRef]
4. Emelyanov, A.; Kazanskii, A.; Kashkarov, P.; Konkov, O.; Terukov, E.; Forsh, P.; Khenkin, M.; Kukin, A.; Beresna, M.; Kazansky, P. Effect of the femtosecond laser treatment of hydrogenated amorphous silicon films on their structural, optical, and photoelectric properties. *Semiconductors* **2012**, *46*, 749–754. [CrossRef]
5. Kang, M.J.A.; Park, T.S.; Kim, M.; Hwang, E.S.; Kim, S.H.; Shin, S.T.; Cheong, B.-H. Periodic surface texturing of amorphous-Si thin film irradiated by UV nanosecond laser. *Opt. Mater. Express* **2019**, *9*, 4247–4255. [CrossRef]
6. Zhan, X.-P.; Hou, M.-Y.; Ma, F.-S.; Su, Y.; Chen, J.-Z.; Xu, H.-L. Room temperature crystallization of amorphous silicon film by ultrashort femtosecond laser pulses. *Opt. Laser. Technol.* **2019**, *112*, 363–367. [CrossRef]
7. Ionin, A.A.; Kudryashov, S.I.; Samokhin, A.A. Material surface ablation produced by ultrashort laser pulses. *Phys. Uspekhi* **2017**, *60*, 149–160. [CrossRef]
8. Huang, J.; Jiang, L.; Li, X.; Wang, A.; Wang, Z.; Wang, Q.; Hu, J.; Qu, L.; Cui, T.; Lu, Y. Fabrication of highly homogeneous and controllable nanogratings on silicon via chemical etching-assisted femtosecond laser modification. *Nanophotonics* **2019**, *8*, 869–878. [CrossRef]
9. Dostovalov, A.; Bronnikov, K.; Korolkov, V.; Babin, S.; Mitsai, E.; Mironenko, A.; Tutov, M.; Zhang, D.; Sugioka, K.; Maksimovic, J.; et al. Hierarchical anti-reflective laser-induced periodic surface structures (LIPSS) on amorphous Si films for sensing applications. *Nanoscale* **2020**, *12*, 13431–13441. [CrossRef]
10. Zhang, D.S.; Sugioka, K. Hierarchical microstructures with high spatial frequency laser induced periodic surface structures possessing different orientations created by femtosecond laser ablation of silicon in liquids. *Opto Electron. Adv.* **2019**, *2*, 190002. [CrossRef]
11. Lin, Z.; Liu, H.; Ji, L.; Lin, W.; Hong, M. Realization of ~10 nm features on semiconductor surfaces via femtosecond laser direct patterning in far field and in ambient air. *Nano Lett.* **2020**, *20*, 4947–4952. [CrossRef] [PubMed]
12. Shuleiko, D.V.; Potemkin, F.V.; Romanov, I.A.; Parhomenko, I.N.; Pavlikov, A.V.; Presnov, D.E.; Zabotnov, S.V.; Kazanskii, A.G.; Kashkarov, P.K. Femtosecond laser pulse modification of amorphous silicon films: Control of surface anisotropy. *Laser Phys. Lett.* **2018**, *15*, 056001. [CrossRef]
13. Martsinovskiǐ, G.A.; Shandybina, G.D.; Smirnov, D.S.; Zabotnov, S.V.; Timoshenko, V.Y.; Kashkarov, P.K. Ultrashort Excitations of Surface Polaritons and Waveguide Modes in Semiconductors. *Opt. Spectrosc.* **2008**, *105*, 67–72. [CrossRef]
14. Bencherif, H.; Dehimi, L.; Pezzimenti, F.; Della Corte, F.G. Improving the efficiency of a-Si:H/c-Si thin heterojunction solar cells by using both antireflection coating engineering and diffraction grating. *Optik* **2019**, *182*, 682–693. [CrossRef]

15. Amasev, D.V.; Khenkin, M.V.; Drevinskas, R.; Kazansky, P.; Kazanskii, A.G. Anisotropy of optical, electrical, and photoelectrical properties of amorphous hydrogenated silicon films modified by femtosecond laser irradiation. *Tech. Phys.* **2017**, *62*, 925–929. [CrossRef]
16. Drevinskas, R.; Beresna, M.; Gecevičius, M.; Khenkin, M.; Kazanskii, A.G.; Matulaitienė, I.; Niaura, G.; Konkov, O.I.; Terukov, E.I.; Svirko, Y.P.; et al. Giant birefringence and dichroism induced by ultrafast laser pulses in hydrogenated amorphous silicon. *Appl. Phys. Lett.* **2015**, *106*, 171106. [CrossRef]
17. Emelyanov, A.V.; Khenkin, M.V.; Kazanskii, A.G.; Forsh, P.A.; Kashkarov, P.K.; Lyubin, E.V.; Khomich, A.A.; Gecevicius, M.; Beresna, M.; Kazansky, P.G. Structural and electrophysical properties of femtosecond laser exposed hydrogenated amorphous silicon films. *Proc. SPIE* **2012**, *8438*, 84381I. [CrossRef]
18. Lipatiev, A.S.; Fedotov, S.S.; Okhrimchuk, A.G.; Lotarev, S.V.; Vasetsky, A.M.; Stepko, A.A.; Shakhgildyan GYu Piyanzina, K.I.; Glebov, I.S.; Sigaev, V.N. Multilevel data writing in nanoporous glass by a few femtosecond laser pulses. *Appl. Optics* **2018**, *57*, 978–982. [CrossRef]
19. Fedotov, S.S.; Okhrimchuk, A.G.; Lipatiev, A.S.; Stepko, A.A.; Piyanzina, K.I.; Shakhgildyan, G.Y.; Presniakov, M.Y.; Glebov, I.S.; Lotarev, S.V.; Sigaev, V.N. 3-bit writing of information in nanoporous glass by a single sub-microsecond burst of femtosecond pulses. *Opt. Lett.* **2018**, *43*, 851–854. [CrossRef]
20. Lotarev, S.V.; Fedotov, S.S.; Kurina, A.I.; Lipatiev, A.S.; Sigaev, V.N. Ultrafast laser-induced nanogratings in sodium germanate glasses. *Opt. Lett.* **2019**, *44*, 1564–1567. [CrossRef]
21. Shuleiko, D.V.; Kashaev, F.V.; Potemkin, F.V.; Zabotnov, S.V.; Zoteev, A.V.; Presnov, D.E.; Parkhomenko, I.N.; Romanov, I.A. Structural Anisotropy of Amorphous Silicon Films Modified by Femtosecond Laser Pulses. *Opt. Spectrosc.* **2018**, *124*, 801–807. [CrossRef]
22. Dyukin, R.V.; Martsinovskii, G.A.; Shandybina, G.D.; Yakovlev, E.B. Electrophysical phenomena accompanying femtosecond impacts of laser radiation on semiconductors. *J. Opt. Technol.* **2011**, *78*, 88–92. [CrossRef]
23. Ambrosone, G.; Coscia, U.; Lettieri, S.; Maddalena, P.; Minarini, C. Optical, structural and electrical properties of μc-Si:H films deposited by SiH4+H2. *Mater. Sci. Eng. B* **2003**, *101*, 236–241. [CrossRef]
24. Choi, T.Y.; Hwang, D.J.; Grigoropoulos, C.P. Ultrafast laser-induced crystallization of amorphous silicon films. *Opt. Eng.* **2003**, *42*, 3383–3388. [CrossRef]
25. Sipe, J.; Young, J.F.; Preston, J.; Van Driel, H. Laser-induced periodic surface structure. I. Theory. *Phys. Rev. B* **1983**, *27*, 1141–1154. [CrossRef]
26. Bonse, J.; Rosenfeld, A.; Krüger, J. On the role of surface plasmon polaritons in the formation of laser-induced periodic surface structures upon irradiation of silicon by femtosecond-laser pulses. *J. Appl. Phys.* **2009**, *106*, 104910. [CrossRef]
27. Bonse, J.; Munz, M.; Sturm, H. Structure formation on the surface of indium phosphide irradiated by femtosecond laser pulses. *J. Appl. Phys.* **2005**, *97*, 013538. [CrossRef]
28. Bonse, J.; Rosenfeld, A.; Krüger, J. Femtosecond laser-induced periodic surface structures: Eecent approaches to explain their subwavelength periodicities. *Proc. SPIE* **2011**, *7994*, 79940M. [CrossRef]
29. Volodin, V.A.; Kachko, A.S.; Cherkov, A.G.; Latyshev, A.V.; Koch, J.; Chichkov, B. Femtosecond pulse crystallization of thin amorphous hydrogenated films on glass substrates using near ultraviolet laser radiation. *JETP Lett.* **2011**, *93*, 603–606. [CrossRef]
30. Viera, G.; Huet, S.; Boufendi, L. Crystal size and temperature measurements in nanostructured silicon using Raman spectroscopy. *J. Appl. Phys.* **2001**, *90*, 4175–4183. [CrossRef]
31. Bergren, M.R.; Simonds, B.J.; Yan, B.; Yue, G.; Ahrenkiel, R.; Furtak, T.E.; Collins, R.T.; Taylor, P.C.; Beard, M.C. Electron transfer in hydrogenated nanocrystalline silicon observed by time-resolved terahertz spectroscopy. *Phys. Rev. B* **2013**, *87*, 081301. [CrossRef]
32. Lui, K.P.H.; Hegmann, F.A. Fluence- and temperature-dependent studies of carrier dynamics in radiation-damaged silicon-on-sapphire and amorphous silicon. *J. Appl. Phys.* **2003**, *93*, 9012–9018. [CrossRef]
33. Bezhanov, S.G.; Ionin, A.A.; Kanavin, A.P.; Kudryashov, S.I.; Makarov, S.V.; Seleznev, L.V.; Sinitsyn, D.V.; Uryupin, S.A. Reflection of a probe pulse and thermal emission of electrons produced by an aluminum film heated by a femtosecond laser pulse. *J. Exp. Theor. Phys.* **2015**, *120*, 937–945. [CrossRef]
34. Kudryashov, S.I.; Ionin, A.A. Multi-scale fluence-dependent dynamics of front-side femtosecond laser heating, melting and ablation of thin supported aluminum film. *Int. J. Heat Mass Transf.* **2016**, *99*, 383–390. [CrossRef]
35. Huang, M.; Zhao, F.; Cheng, Y.; Xu, N.; Xu, Z. Origin of Laser-Induced Near-Subwavelength Ripples: Interference between Surface Plasmons and Incident Laser. *ACS Nano* **2009**, *3*, 4062–4070. [CrossRef] [PubMed]
36. Reddy, N.P.; Gupta, R.; Agarwal, S.C. Electrical conduction and Meyer–Neldel Rule in nanocrystalline silicon thin films. *J. Non-Cryst. Solids* **2013**, *364*, 69–76. [CrossRef]
37. Born, M.; Wolf, E. *Principles of Optics*, 4rd ed.; Pergamon Press: Oxford, UK, 1968.
38. Ionin, A.A.; Kudryashov, S.I.; Levchenko, A.O.; Nguyen, L.V.; Saraeva, I.N.; Rudenko, A.A.; Ageev, E.I.; Potorochin, D.V.; Veiko, V.P.; Borisov, E.V.; et al. Correlated topographic and structural modification on Si surface during multi-shot femtosecond laser exposures: Si nanopolymorphs as potential local structural nanomarkers. *Appl. Surf. Sci.* **2017**, *416*, 988–995. [CrossRef]
39. Bustarret, E.; Ligeon, M.; Ortega, L. Visible light emission at room temperature from partially oxidized amorphous silicon. *Solid State Commun.* **1992**, *83*, 461–464. [CrossRef]

40. Emelyanov, A.V.; Kazanskii, A.G.; Khenkin, M.V.; Forsh, P.A.; Kashkarov, P.K.; Gecevicius, M.; Beresna, M.; Kazansky, P.G. Visible luminescence from hydrogenated amorphous silicon modified by femtosecond laser radiation. *Appl. Phys. Lett.* **2012**, *101*, 081902. [CrossRef]
41. Boyd, I.W.; Wilson, J.I.B. Oxidation of silicon surfaces by CO_2 lasers. *Appl. Phys. Lett.* **1982**, *41*, 162–164. [CrossRef]
42. Pierce, D.T.; Spicer, W.E. Electronic Structure of Amorphous Si from Photoemission and Optical Studies. *Phys. Rev. B* **1972**, *5*, 3017–3029. [CrossRef]
43. Green, M.A. Self-consistent optical parameters of intrinsic silicon at 300K including temperature coefficients. *Sol. Energ. Mat. Sol. C* **2008**, *92*, 1305–1310. [CrossRef]

MDPI
St. Alban-Anlage 66
4052 Basel
Switzerland
Tel. +41 61 683 77 34
Fax +41 61 302 89 18
www.mdpi.com

Nanomaterials Editorial Office
E-mail: nanomaterials@mdpi.com
www.mdpi.com/journal/nanomaterials

www.ingramcontent.com/pod-product-compliance
Lightning Source LLC
LaVergne TN
LVHW070357100526
838202LV00014B/1335